VERTEBRATES: Adaptation

Readings from
SCIENTIFIC AMERICAN

VERTEBRATES: Adaptation

With Introductions by
Norman K. Wessells
Stanford University

W. H. Freeman and Company
San Francisco

Most of the SCIENTIFIC AMERICAN articles in *Vertebrates: Adaptation* are available as separate Offprints. For a complete list of articles now available as Offprints, write to W. H. Freeman and Company, 660 Market Street, San Francisco, California, 94104.

Library of Congress Cataloging in Publication Data

Wessells, Norman K   comp.

Vertebrates: adaptation.

  Bibliography: p.
  Includes index.
  1. Vertebrates—Addresses, essays, lectures.
2. Adaptation (Biology)—Addresses, essays, lectures.
I. Scientific American.   II. Title.
QL605.W42   1980   596'.05   80–188
ISBN 0–7167–1167–2
ISBN 0–7167–1168–0 pbk.

Printed in the United States of America

9 8 7 6 5 4 3 2 1

PREFACE

The unprecedented triumphs of molecular biology since the 1950s tend to make us forget the broad and ancient base on which modern life sciences are built. As we succeed in constructing mechanistic interpretations of life processes, it is sobering to realize how perceptive and accurate were the observations of those phenomena made hundreds of years ago. Although the following quotation was written by Pliny the Elder 1900 years ago, nearly every topic in it is discussed in this book and is still under active investigation.

> The swiftest of all animals, not only those of the sea, is the dolphin: it is swifter than a bird and darts faster than a javelin, and were not its mouth much below its snout, almost in the middle of its belly, not a single fish would escape its speed. But nature's foresight contributes delay, because they cannot seize their prey except by turning over on their backs. . . . They have a habit of sallying out on to the land for an unascertained reason, and they do not die at once after touching earth—in fact they die more quickly if the gullet is closed up. . . . For a voice they have a moan like that of a human being. . . . The dolphin is an animal that is not only friendly to mankind but is also a lover of music, and it can be charmed by . . . the sound of the water-organ. (Pliny, *Natural History*, IX, vii, viii.)

Today we believe the dolphin's swiftness is due to laminar flow of water over its skin, so that eddies and turbulence do not generate drag. The sallies of whales or dolphins onto beaches may stem from a failure of the echo-locating navigational system to detect shallow, sloping sea bottoms, or may be due to ear parasites that interfere with the hearing of echoes. The dolphin's "voice" is of course an integral part of the echo-locating system and is also used for communication between these remarkably "brainy" cetaceans. And, what else but "nature's forethought" (natural selection in our terms) could explain so well the origin of these adaptations or the equally marvelous ones described throughout this collection of articles from *Scientific American?*

This book and its companion volume—*Vertebrates: Physiology*—(W. H. Freeman and Company, 1980) are about the structures, functions, and adaptations of vertebrates. We shall study molecules, cells, organs, and organisms in these two books, and our purpose will be to see how the parts are orchestrated to yield the functioning individual vertebrate, adapted for its particular niche. Any overview of the kind endeavored in these books must be based on certain biases. The prejudice observed in these books is that it is just as sterile to study only bones and morphology as to concentrate only upon molecules or cells. A student can best achieve real understanding by sampling at all levels of organization, and integrating the knowledge thus obtained to gain a fuller picture of a trout, a toad, or a titmouse.

This book concentrates upon adaptations of vertebrates to the physical and biological world in which they live. Swimming, running, flying, navigating, mating, and communicating are but a few of the activities treated in the various articles from *Scientific American*. Emphasis is placed upon an integrated view of selected vertebrates (crocodiles, seals, kangaroos, and so on), so that the student may gain a better appreciation of how anatomical, physiological. and behavioral features of a vertebrate fit together to yield a creature adapted for a given niche. The companion book, *Vertebrates: Physiology*, provides important background about the most important organ systems and physiological problems of vertebrates, and is meant to be read by the student prior to study of this book.

The two collections of articles are designed to supplement courses in introductory biology, vertebrate biology, comparative anatomy, and physiology. As such, the books are meant to fill some of the gaps and answer some of the questions that might arise from a general survey of vertebrate biology. Orientation to the field and articles is provided by Introductions to the Sections of the book. Although *Scientific American* has not of course published articles on all the subjects that might interest a student, the scope is nevertheless quite broad; therefore references to Offprints on related subjects are included. In addition, a special list of recent references for each section is found in the Bibliography at the end of each book.

I have assumed that teachers and students will be able to consult the *Annual Review of Physiology, Biological Reviews,* or *Physiological Reviews,* for many useful papers that are not cited specifically here. The important book by P. W. Hochachka and G. N. Somero, *Strategies of Biochemical Adaptation,* and *Animal Physiology* by R. Eckert and D. Randall, are to my mind the most useful means of amplifying the many intriguing aspects of vertebrate life treated in these *Scientific American* articles. For those needing background in ancient vertebrates and paleontology, the book by B. J. Stahl, *Vertebrate History: Problems in Evolution,* is by far the best balanced evaluation and summary.

Thanks go to the many students in my Vertebrate Biology course at Stanford, whose interest, enthusiasm, and questioning have added so much to the pleasures of teaching. I am also indebted to Sharon Willy for the excellent index. Finally, thanks also go to Lois Wessells for aid and comfort in busy times.

November 1979   *Norman K. Wessells*

CONTENTS

VI   VERTEBRATES IN THEIR HABITATS

Note on cross-references: References to articles included in this book are noted by the title of the article and the page on which it begins: references to articles that are available as Offprints, but are not included here, are noted by the article's title and Offprint number; references to articles published by SCIENTIFIC AMERICAN but which are not available as Offprints, are noted by the title of article and the month and year of its publication.

ADAPTATIONS: GENERAL FEATURES

Every species of animal is marvellously cunning for its own interests.

Pliny
NATURAL HISTORY, VIII, xii

ADAPTATIONS: GENERAL FEATURES

I

INTRODUCTION

This book is concerned with the many types of adaptations of vertebrates to the world in which they live. The article "Adaptation" by R. C. Lewontin provides a superb general and theoretical treatment of this subject.

Lewontin emphasizes that the "fit" of organisms to their physical and biological environments may be regarded as a primary result of the evolutionary process. For example, the properties of water, particularly its density, place certain restrictions upon the animal body if rapid locomotion is to be attempted. We see the result in the common shapes of dolphins, penguins, ichthyosaurs (an extinct type of aquatic dinosaur with a body shape nearly identical to that of a dolphin), and seals; all of these distinctive creatures evolved from terrestrial vertebrates that re-entered the sea and adopted it as their primary habitat at different times in vertebrate history. Yet, all have attained the same general body configuration that is compatible with laminar flow, minimal turbulence, and efficient, rapid swimming. It is argued that natural selection has operated to yield this convergence in body shape and the resultant "fit" with the environment.

As Lewontin points out, however, a given characteristic of an organism (its color, size of chin, and so on) may derive from a complex of factors, not solely genetic variation and natural selection acting on the character *per se*. The history of an organism and its lineage, particularly the modes of development it uses to build its organs and tissues, is a key factor that sets limits upon the ways that the organism and its offspring can respond to changing environments. Lewontin emphasizes the pointlessness of attempting to construct ad hoc explanations for every bodily feature in terms of selective pressure. Such a practice could divert attention from real causes. The reader should keep Lewontin's caveats and theoretical framework in mind while studying the various morphological, physiological, and behavioral adaptations that are described in the articles of this book.

Adaptation

by Richard C. Lewontin
September 1978

The manifest fit between organisms and their environment is a major outcome of evolution. Yet natural selection does not lead inevitably to adaptation; indeed, it is sometime hard to define an adaptation

The theory about the history of life that is now generally accepted, the Darwinian theory of evolution by natural selection, is meant to explain two different aspects of the appearance of the living world: diversity and fitness. There are on the order of two million species now living, and since at least 99.9 percent of the species that have ever lived are now extinct, the most conservative guess would be that two billion species have made their appearance on the earth since the beginning of the Cambrian period 600 million years ago. Where did they all come from? By the time Darwin published *On the Origin of Species* in 1859 it was widely (if not universally) held that species had evolved from one another, but no plausible mechanism for such evolution had been proposed. Darwin's solution to the problem was that small heritable variations among individuals within a species become the basis of large differences between species. Different forms survive and reproduce at different rates depending on their environment, and such differential reproduction results in the slow change of a population over a period of time and the eventual replacement of one common form by another. Different populations of the same species then diverge from one another if they occupy different habitats, and eventually they may become distinct species.

Life forms are more than simply multiple and diverse, however. Organisms fit remarkably well into the external world in which they live. They have morphologies, physiologies and behaviors that appear to have been carefully and artfully designed to enable each organism to appropriate the world around it for its own life.

It was the marvelous fit of organisms to the environment, much more than the great diversity of forms, that was the chief evidence of a Supreme Designer. Darwin realized that if a naturalistic theory of evolution was to be successful, it would have to explain the apparent perfection of organisms and not simply their variation. At the very beginning of the *Origin of Species* he wrote: "In considering the Origin of Species, it is quite conceivable that a naturalist...might come to the conclusion that each species...had descended, like varieties, from other species. Nevertheless, such a conclusion, even if well founded, would be unsatisfactory, until it could be shown how the innumerable species inhabiting this world have been modified, so as to acquire that perfection of structure and coadaptation which most justly excites our admiration." Moreover, Darwin knew that "organs of extreme perfection and complication" were a critical test case for his theory, and he took them up in a section of the chapter on "Difficulties of the Theory." He wrote: "To suppose that the eye, with all its inimitable contrivances for adjusting the focus to different distances, for admitting different amounts of light, and for the correction of spherical and chromatic aberration, could have been formed by natural selection, seems, I freely confess, absurd in the highest degree."

These "organs of extreme perfection" were only the most extreme case of a more general phenomenon: adaptation. Darwin's theory of evolution by natural selection was meant to solve both the problem of the origin of diversity and the problem of the origin of adaptation at one stroke. Perfect organs were a difficulty of the theory not in that natural selection could not account for them but rather in that they were its most rigorous test, since on the face of it they seemed the best intuitive demonstration that a divine artificer was at work.

The modern view of adaptation is that the external world sets certain "problems" that organisms need to "solve," and that evolution by means of natural selection is the mechanism for creating these solutions. Adaptation is the process of evolutionary change by which the organism provides a better and better "solution" to the "problem," and the end result is the state of being adapted. In the course of the evolution of birds from reptiles there was a successive alteration of the bones, the muscles and the skin of the forelimb to give rise to a wing; an increase in the size of the breastbone to provide an anchor for the wing muscles; a general restructuring of bones to make them very light but strong, and the development of feathers to provide both aerodynamic elements and lightweight insulation. This wholesale reconstruction of a reptile to make a bird is considered a process of major adaptation by which birds solved the problem of flight. Yet there is no end to adaptation. Having adapted to flight, some birds reversed the process: the penguins adapted to marine life by changing their wings into flippers and their feathers into a waterproof covering, thus solving the problem of aquatic existence.

The concept of adaptation implies a preexisting world that poses a problem to which an adaptation is the solution. A key is adapted to a lock by cutting and filing it; an electrical appliance is adapted to a different voltage by a transform-

ADAPTATION is exemplified by "industrial melanism" in the peppered moth (*Biston betularia*). Air pollution kills the lichens that would normally colonize the bark of tree trunks. On the dark, lichenless bark of an oak tree near Liverpool in England the melanic (*black*) form is better adapted: it is better camouflaged against predation by birds than the light, peppered wild type (*top photograph on opposite page*), which it largely replaced through natural selection in industrial areas of England in the late 19th century. Now air quality is improving. On a nearby beech tree colonized by algae and the lichen *Lecanora conizaeoides*, which is itself particularly well adapted to low levels of pollution, the two forms of the moth are equally conspicuous (*middle*). On the lichened bark of an oak tree in rural Wales the wild type is almost invisible (*bottom*), and in such areas it predominates. The photographs were made by J. A. Bishop of the University of Liverpool and Laurence M. Cook of the University of Manchester.

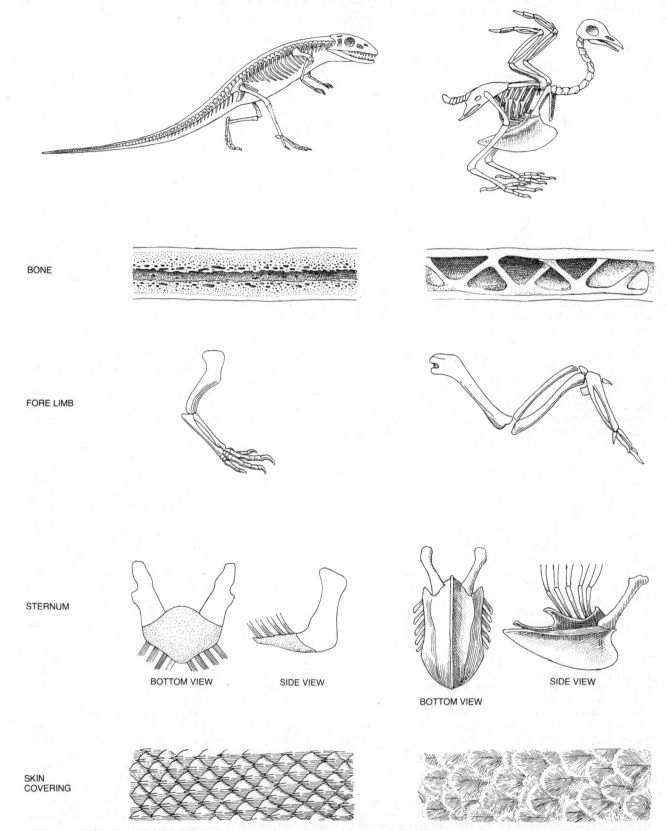

REPTILES   BIRDS

BONE

FORE LIMB

STERNUM

BOTTOM VIEW · SIDE VIEW   BOTTOM VIEW   SIDE VIEW

SKIN
COVERING

EVOLUTION OF BIRDS from reptiles can be considered a process of adaptation by which birds "solved" the "problem" of flight. At the top of the illustration the skeleton of a modern pigeon (*right*) is compared with that of an early reptile: a thecodont, a Triassic ancestor of dinosaurs and birds. Various reptile features were modified to become structures specialized for flight. Heavy, dense bone was restruc- tured to become lighter but strong; the forelimb was lengthened (and its muscles and skin covering were changed) to become a wing; the reptilian sternum, or breastbone, was enlarged and deepened to an- chor the wing muscles (even in *Archaeopteryx,* the Jurassic transi- tion form between reptiles and birds whose sternum is pictured here, the sternum was small and shallow); scales developed into feathers.

er. Although the physical world certainly predated the biological one, there are certain grave difficulties for evolutionary theory in defining that world for the process of adaptation. It is the difficulty of defining the "ecological niche." The ecological niche is a multidimensional description of the total environment and way of life of an organism. Its description includes physical factors, such as temperature and moisture; biological factors, such as the nature and quantity of food sources and of predators, and factors of the behavior of the organism itself, such as its social organization, its pattern of movement and its daily and seasonal activity cycles.

The first difficulty is that if evolution is described as the process of adaptation of organisms to niches, then the niches must exist before the species that are to fit them. That is, there must be empty niches waiting to be filled by the evolution of new species. In the absence of organisms in actual relation to the environment, however, there is an infinity of ways the world can be broken up into arbitrary niches. It is trivially easy to describe "niches" that are unoccupied. For example, no organism makes a living by laying eggs, crawling along the surface of the ground, eating grass and living for several years. That is, there are no grass-eating snakes, even though snakes live in the grass. Nor are there any warm-blooded, egg-laying animals that eat the mature leaves of trees, even though birds inhabit trees. Given any description of an ecological niche occupied by an actual organism, one can create an infinity of descriptions of unoccupied niches simply by adding another arbitrary specification. Unless there is some preferred or natural way to subdivide the world into niches the concept loses all predictive and explanatory value.

A second difficulty with the specification of empty niches to which organisms adapt is that it leaves out of account the role of the organism itself in creating the niche. Organisms do not experience environments passively; they create and define the environment in which they live. Trees remake the soil in which they grow by dropping leaves and putting down roots. Grazing animals change the species composition of herbs on which they feed by cropping, by dropping manure and by physically disturbing the ground. There is a constant interplay of the organism and the environment, so that although natural selection may be adapting the organism to a particular set of environmental circumstances, the evolution of the organism itself changes those circumstances. Finally, organisms themselves determine which external factors will be part of their niche by their own activities. By building a nest the phoebe makes the availability of dried grass an important part of its

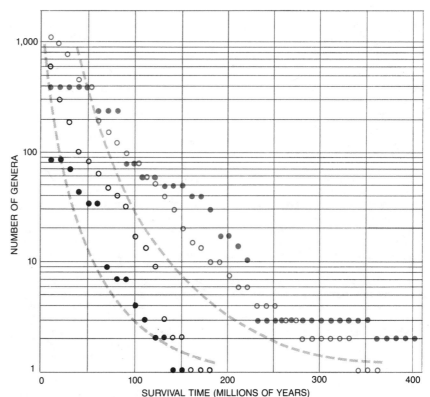

EXTINCTION RATES in many evolutionary lines suggest that natural selection does not necessarily improve adaptation. The data, from Leigh Van Valen of the University of Chicago, show the duration of survival of a number of living (*solid dots*) and extinct (*open circles*) genera of Echinoidea (*black*) and Pelecypoda (*color*), two classes of marine invertebrates. If natural selection truly fitted organisms to environments, the points should fall along concave curves (*broken-line curves*) indicating a lower probability of extinction for long-lived genera. Actually, points fall along rather straight lines, indicating constant rate of extinction for each group.

niche, at the same time making the nest itself a component of the niche.

If ecological niches can be specified only by the organisms that occupy them, evolution cannot be described as a process of adaptation because all organisms are already adapted. Then what is happening in evolution? One solution to this paradox is the Red Queen hypothesis, named by Leigh Van Valen of the University of Chicago for the character in *Through the Looking Glass* who had to keep running just to stay in the same place. Van Valen's theory is that the environment is constantly decaying with respect to existing organisms, so that natural selection operates essentially to enable the organisms to maintain their state of adaptation rather than to improve it. Evidence for the Red Queen hypothesis comes from an examination of extinction rates in a large number of evolutionary lines. If natural selection were actually improving the fit of organisms to their environments, then we might expect the probability that a species will become extinct in the next time period to be less for species that have already been in existence for a long time, since the long-lived species are presumably the ones that have been im-

proved by natural selection. The data show, however, that the probability of extinction of a species appears to be a constant, characteristic of the group to which it belongs but independent of whether the species has been in existence for a long time or a short one. In other words, natural selection over the long run does not seem to improve a species' chance of survival but simply enables it to "track," or keep up with, the constantly changing environment.

The Red Queen hypothesis also accounts for extinction (and for the occasional dramatic increases in the abundance and range of species). For a species to remain in existence in the face of a constantly changing environment it must have sufficient heritable variation of the right kind to change adaptively. For example, as a region becomes drier because of progressive changes in rainfall patterns, plants may respond by evolving a deeper root system or a thicker cuticle on the leaves, but only if their gene pool contains genetic variation for root length or cuticle thickness, and successfully only if there is enough genetic variation so that the species can change as fast as the environment. If the genetic variation is inadequate, the species will become extinct. The genetic resources of a species are finite, and eventually the environment will change so rapidly that the species is sure to become extinct.

The theory of environmental tracking seems at first to solve the problem of adaptation and the ecological niche. Whereas in a barren world there is no clear way to divide the environment into preexisting niches, in a world already occupied by many organisms the terms of the problem change. Niches are already defined by organisms. Small changes in the environment mean small changes in the conditions of life of those organisms, so that the new niches to which they must evolve are in a sense very close to the old ones in the multidimensional niche space. Moreover, the organisms that will occupy these slightly changed niches must themselves come from the previously existing niches, so that the kinds of species that can evolve are stringently limited to ones that are extremely similar to their immediate ancestors. This in turn guarantees that the changes induced in the environment by the changed organism will also be small and continuous in niche space. The picture of adaptation that emerges is the very slow movement of the niche through niche space, accompanied by a slowly changing species, always slightly behind, slightly ill-adapted, eventually becoming extinct as it fails to keep up with the changing environment because it runs out of genetic variation on which natural selection can operate. In this view species form when two populations of the same species track environments that diverge from each other over a period of time.

The problem with the theory of environmental tracking is that it does not predict or explain what is most dramatic in evolution: the immense diversification of organisms that has accompanied, for example, the occupation of the land from the water or of the air from the land. Why did warm-blooded animals arise at a time when cold-blooded animals were still plentiful and come to coexist with them? The appearance of entirely new life forms, of ways of making a living, is equivalent to the occupation of a previously barren world and brings us back to the preexistent empty niche waiting to be filled. Clearly there have been in the past ways of making a living that were unexploited and were then "discovered" or "created" by existing organisms. There is no way to explain and predict such evolutionary adaptations unless a priori niches can be described on the basis of some physical principles before organisms come to occupy them.

That is not easy to do, as is indicated by an experiment in just such a priori predictions that has been carried out by probes to Mars and Venus designed to detect life. The instruments are designed to detect life by detecting growth in nutrient solutions, and the solutions are prepared in accordance with knowledge of terrestrial microorganisms, so that the probes will detect only organisms whose ecological niches are like those on the earth. If Martian and Venusian life partition the environment in totally unexpected ways, they will remain unrecorded. What the designers of those instruments never dreamed of was that the reverse might happen: that the nature of the physical environment on Mars might be such that when it was provided with a terrestrial ecological niche, inorganic reactions might have a lifelike appearance. Yet that may be exactly what happened. When the Martian soil was dropped into the nutrient broth on the lander, there was a rapid production of carbon dioxide and then—nothing. Either an extraordinary kind of life began to grow much more rapidly than any terrestrial microorganism and then was poisoned by its own activity in a strange environment, or else the Martian soil is such that its contact with nutrient broths results in totally unexpected catalytic processes. In either case the Mars life-detection experiment has foundered on the problem of defining ecological niches without organisms.

Much of evolutionary biology is the working out of an adaptationist program. Evolutionary biologists assume that each aspect of an organism's morphology, physiology and behavior has been molded by natural selection as a solution to a problem posed by the

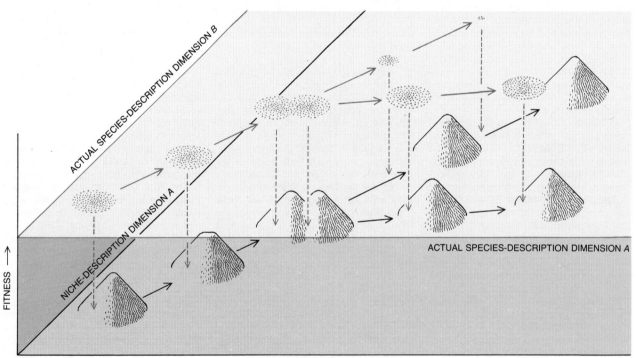

SPECIES TRACK ENVIRONMENT through niche space, according to one view of adaptation. The niche, visualized as an "adaptive peak," keeps changing (moving to the right); a slowly changing species population (*colored dots*) just manages to keep up with the niche, always a bit short of the peak. As the environment changes, the single peak becomes two distinct peaks, and two populations diverge to form distinct species. One species cannot keep up with its rapidly changing environment, becomes less fit (lags farther behind changing peak) and extinct. Here niche space and actual-species space have only two dimensions; both of them are actually multidimensional.

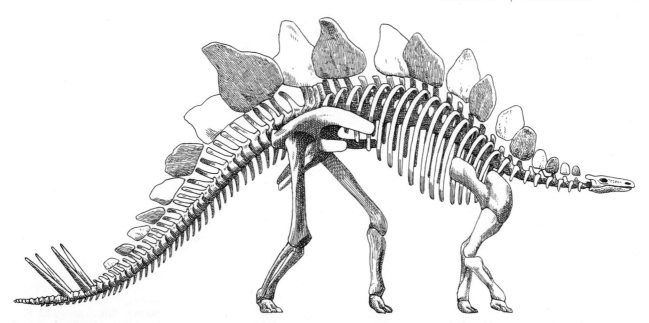

STEGOSAURUS, a large herbivorous dinosaur of the Jurassic period, had an array of bony plates along its back. Were they solutions to the problem of defense, courtship recognition or heat regulation? An engineering analysis reveals features characteristic of heat regulators: porous structure (suggesting a rich blood supply), particularly large plates over the massive part of the body, staggered arrangement along the midline, a constriction near the base and so on. This skeleton in the American Museum of Natural History is 18 feet long.

environment. The role of the evolutionary biologist is then to construct a plausible argument about how each part functions as an adaptive device. For example, functional anatomists study the structure of animal limbs and analyze their motions by time-lapse photography, comparing the action and the structure of the locomotor apparatus in different animals. Their interest is not, however, merely descriptive. Their work is informed by the adaptationist program, and their aim is to explain particular anatomical features by showing that they are well suited to the function they perform. Evolutionary ethologists and sociobiologists carry the adaptationist program into the realm of animal behavior, providing an adaptive explanation for differences among species in courting pattern, group size, aggressiveness, feeding behavior and so on. In each case they assume, like the functional anatomist, that the behavior is adaptive and that the goal of their analysis is to reveal the particular adaptation.

The dissection of an organism into parts, each of which is regarded as a specific adaptation, requires two sets of a priori decisions. First one must decide on the appropriate way to divide the organism and then one must describe what problem each part solves. This amounts to creating descriptions of the organism and of the environment and then relating the descriptions by functional statements; one can either start with the problems and try to infer which aspect of the organism is the solution or start with the organism and then ascribe adaptive functions to each part.

For example, for individuals of the same species to recognize each other at mating time is a problem, since mistakes about species mean time, energy and gametes wasted in courtship and mating without the production of viable offspring; species traits such as distinctive color markings, special courtship behavior, unique mating calls, odors and restricted time and place of activity can be considered specific adaptations for the proper recognition of potential mates. On the other hand, the large, leaf-shaped bony plates along the back of the dinosaur *Stegosaurus* constitute a specific characteristic for which an adaptive function needs to be inferred. They have been variously explained as solutions to the problem of defense (by making the animal appear to be larger or by interfering directly with the predator's attack), the problem of recognition in courtship and the problem of temperature regulation (by serving as cooling fins).

The same problems that arose in deciding on a proper description of the ecological niche without the organism arise when one tries to describe the organism itself. Is the leg a unit in evolution, so that the adaptive function of the leg can be inferred? If so, what about a part of the leg, say the foot, or a single toe, or one bone of a toe? The evolution of the human chin is an instructive example. Human morphological evolution can be generally described as a "neotenic" progression. That is, human infants and adults resemble the fetal and young forms of apes more than they resemble adult apes; it is as if human beings are born at an earlier stage of physical development than apes and do not

mature as far along the apes' development path. For example, the relative proportion of skull size to body size is about the same in newborn apes and human beings, whereas adult apes have much larger bodies in relation to their heads than we do; in effect their bodies "go further."

The exception to the rule of human neoteny is the chin, which grows relatively larger in human beings, whereas both infant and adult apes are chinless. Attempts to explain the human chin as a specific adaptation selected to grow larger failed to be convincing. Finally it was realized that in an evolutionary sense the chin does not exist! There are two growth fields in the lower jaw: the dentary field, which is the bony structure of the jaw, and the alveolar field, in which the teeth are set. Both the dentary and the alveolar fields do show neoteny. They have both become smaller in the human evolutionary line. The alveolar field has shrunk somewhat faster than the dentary one, however, with the result that a "chin" appears as a pure consequence of the relative regression rates of the two growth fields. With the recognition that the chin is a mental construct rather than a unit in evolution the problem of its adaptive explanation disappears. (Of course, we may go on to ask why the dentary and alveolar growth fields have regressed at different rates in evolution, and then provide an adaptive explanation for that phenomenon.)

Sometimes even the correct topology of description is unknown. The brain is divided into anatomical divisions corresponding to certain separable nervous functions that can be localized, but memory is not one of those functions. The memory of specific events seems to be stored diffusely over large regions of the cerebrum rather than being localized microscopically. As one moves from anatomy to behavior the problem of a correct description becomes more acute and the opportunities to introduce arbitrary constructs as if they were evolutionary traits multiply. Animal behavior is described in terms of aggression, division of labor, warfare, dominance, slave-making, cooperation—and yet each of these is a category that is taken directly from human social experience and is transferred to animals.

The decision as to which problem is solved by each trait of an organism is equally difficult. Every trait is involved in a variety of functions, and yet one would not want to say that the character is an adaptation for all of them. The green turtle *Chelonia mydas* is a large marine turtle of the tropical Pacific. Once a year the females drag themselves up the beach with their front flippers to the dry sand above the high-water mark. There they spend many hours laboriously digging a deep hole for their eggs, using their hind flippers as trowels. No one who has watched this painful process would describe the turtles' flippers as adaptations for land locomotion and digging; the animals move on land and dig with their flippers because nothing better is available. Conversely, even if a trait seems clearly adaptive, it cannot be assumed that the species would suffer in its absence. The fur of a polar bear is an adaptation for temperature regulation, and a hairless polar bear would certainly freeze to death. The color of a polar bear's fur is another matter. Although it may be an adaptation for camouflage, it is by no means certain that the polar bear would become extinct or even less numerous if it were brown. Adaptations are not necessary conditions of the existence of the species.

For extinct species the problem of judging the adaptive status of a trait is made more difficult because both the trait and its function must be reconstructed. In principle there is no way to be sure whether the dorsal plates of *Stegosaurus* were heat-regulation devices, a defense mechanism, a sexual recognition sign or all these things. Even in living species where experiments can be carried out a doubt remains. Some modern lizards have a brightly colored dewlap under the jaw. The dewlap may be a warning sign, a sexual attractant or a species-recognition signal. Experiments removing or altering the dewlap could decide, in principle, how it functions. That is a different question from its status as an adaptation, however, since the assertion of adaptation implies a historical argument about natural selection as the cause of its establishment. The large dorsal plates of *Stegosaurus* may have evolved because individuals with slightly larger plates were better able to gather food in the heat of the day than other individuals. If, when the plates reached a certain size, they incidentally frightened off predators, they would be a "preadaptation" for defense. The distinction between the primary adaptation for which a trait evolved and incidental functions it may have come to have cannot be made without the reconstruction of the forces of natural selection during the actual evolution of the species.

The current procedure for judging the adaptation of traits is an engineering analysis of the organism and its environment. The biologist is in the position of an archaeologist who uncovers a machine without any written record and attempts to reconstruct not only its operation but also its purpose. The hypothesis that the dorsal plates of *Stegosaurus* were a heat-regulation device is based on the fact that the plates were porous and probably had a large supply of blood vessels, on their alternate placement to the left and right of the midline (suggesting cooling fins), on their large size over the most massive part of the body and on the constriction near their base, where they are closest to the heat source and would be inefficient heat radiators.

Ideally the engineering analysis can be quantitative as well as qualitative and so provide a more rigorous test of the

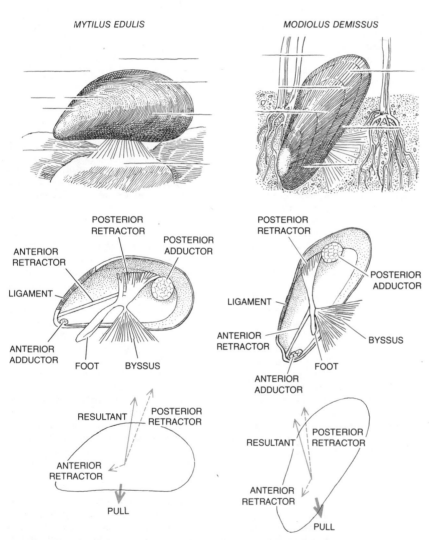

MYTILUS EDULIS

MODIOLUS DEMISSUS

FUNCTIONAL ANALYSIS indicates how the shape and musculature of two species of mussels are adapted to their particular environments. *Mytilus edulis* (*left*) attaches itself to rocks by means of its byssus, a beardlike group of threads (*top*). Its ventral, or lower, edge is flattened; the anterior and posterior retractor muscles are positioned (*middle*) so that their resultant force pulls the bottom of the shell squarely down to the substratum (*bottom*). Modiolus demissus (*right*) attaches itself to debris in marshes. Its ventral edge is sharply angled to facilitate penetration of the substratum; its retractor muscles are positioned to pull its anterior end down into the marsh. The analysis was done by Steven M. Stanley of Johns Hopkins University.

adaptive hypothesis. Egbert G. Leigh, Jr., of the Smithsonian Tropical Research Institute posed the question of the ideal shape of a sponge on the assumption that feeding efficiency is the problem to be solved. A sponge's food is suspended in water and the organism feeds by passing water along its cell surfaces. Once water is processed by the sponge it should be ejected as far as possible from the organism so that the new water taken in is rich in food particles. By an application of simple hydrodynamic principles Leigh was able to show that the actual shape of sponges is maximally efficient. Of course, sponges differ from one another in the details of their shape, so that a finer adjustment of the argument would be needed to explain the differences among species. Moreover, one cannot be sure that feeding efficiency is the only problem to be solved by shape. If the optimal shape for feeding had turned out to be one with many finely divided branches and protuberances rather than the compact shape observed, it might have been argued that the shape was a compromise between the optimal adaptation for feeding and the greatest resistance to predation by small browsing fishes.

Just such a compromise has been suggested for understanding the feeding behavior of some birds. Gordon H. Orians of the University of Washington studied the feeding behavior of birds that fly out from a nest, gather food and bring it back to the nest for consumption ("central-place foraging"). If the bird were to take food items indiscriminately as it came on them, the energy cost of the round trip from the nest and back might be greater than the energy gained from the food. On the other hand, if the bird chose only the largest food items, it might have to search so long that again the energy it consumed would be too great. For any actual distribution of food-particle sizes in nature there is some optimal foraging behavior for the bird that will maximize its net energy gain from feeding. Orians found that birds indeed do not take food particles at random but are biased in the direction of an optimal particle size. They do not, however, choose the optimal solution either. Orians' explanation was that the foraging behavior is a compromise between maximum energy efficiency and not staying away from the nest too long, because the young are exposed to predation when they are unattended.

The example of central-place foraging illustrates a basic assumption of all such engineering analyses, that of ceteris paribus, or all other things being equal. In order to make an argument that a trait is an optimal solution to a particular problem, it must be possible to view the trait and the problem in isolation, all other things being equal. If all

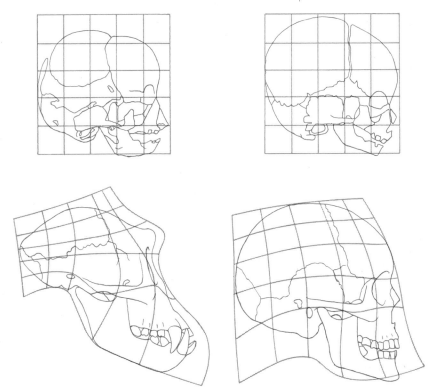

NEOTENY OF HUMAN SKULL is evident when the growth of the chimpanzee skull (*left*) and of the human skull (*right*) is plotted on transformed coordinates, which show the relative displacement of each part. The chimpanzee and the human skulls are much more similar at the fetal stage (*top*) than they are at the adult stage (*bottom*). The adult human skull also departs less from the fetal form than the adult chimpanzee skull departs from its fetal form, except in the case of the chin, which becomes relatively larger in human beings. The chin is a mental construct, however: the result of allometry, or differential growth, of different parts of human jaw.

other things are not equal, if a change in a trait as a solution to one problem changes the organism's relation to other problems of the environment, it becomes impossible to carry out the analysis part by part, and we are left in the hopeless position of seeing the whole organism as being adapted to the whole environment.

The mechanism by which organisms are said to adapt to the environment is that of natural selection. The theory of evolution by natural selection rests on three necessary principles: Different individuals within a species differ from one another in physiology, morphology and behavior (the principle of variation); the variation is in some way heritable, so that on the average offspring resemble their parents more than they resemble other individuals (the principle of heredity); different variants leave different numbers of offspring either immediately or in remote generations (the principle of natural selection).

These three principles are necessary and sufficient to account for evolutionary change by natural selection. There must be variation to select from; that variation must be heritable, or else there will be no progressive change from gen-

eration to generation, since there would be a random distribution of offspring even if some types leave more offspring than others. The three principles say nothing, however, about adaptation. In themselves they simply predict change caused by differential reproductive success without making any prediction about the fit of organisms to an ecological niche or the solution of ecological problems.

Adaptation was introduced by Darwin into evolutionary theory by a fourth principle: Variations that favor an individual's survival in competition with other organisms and in the face of environmental stress tend to increase reproductive success and so tend to be preserved (the principle of the struggle for existence). Darwin made it clear that the struggle for existence, which he derived from Thomas Malthus' *An Essay on the Principle of Population*, included more than the actual competition of two organisms for the same resource in short supply. He wrote: "I should premise that I use the term Struggle for Existence in a large and metaphorical sense.... Two canine animals in a time of dearth, may be truly said to struggle with each other which shall get food and live. But a plant on the edge of the desert is said to struggle for life against the drought."

The diversity that is generated by various mechanisms of reproduction and mutation is in principle random, but the diversity that is observed in the real world is nodal: organisms have a finite number of morphologies, physiologies and behaviors and occupy a finite number of niches. It is natural selection, operating under the pressures of the struggle for existence, that creates the nodes. The nodes are "adaptive peaks," and the species or other form occupying a peak is said to be adapted.

More specifically, the struggle for existence provides a device for predicting which of two organisms will leave more offspring. An engineering analysis can determine which of two forms of zebra can run faster and so can more easily escape predators; that form will leave more offspring. An analysis might predict the eventual evolution of zebra locomotion even in the absence of existing differences among individuals, since a careful engineer might think of small improvements in design that would give a zebra greater speed.

When adaptation is considered to be the result of natural selection under the pressure of the struggle for existence, it is seen to be a relative condition rather than an absolute one. Even though a species may be surviving and numerous, and therefore may be adapted in an absolute sense, a new form may arise that has a greater reproductive rate on the same resources, and it may cause the extinction of the older form. The concept of relative adaptation removes the apparent tautology in the theory of natural selection. Without it the theory of natural selection states that fitter individuals have more offspring and then defines the fitter as being those that leave more offspring; since some individuals will always have more offspring than others by sheer chance, nothing is explained. An analysis in which problems of design are posed and characters are understood as being design solutions breaks through this tautology by predicting in advance which individuals will be fitter.

The relation between adaptation and natural selection does not go both ways. Whereas greater relative adaptation leads to natural selection, natural selection does not necessarily lead to greater adaptation. Let us contrast two evolutionary scenarios. We begin with a resource-limited population of 100 insects of type *A* requiring one unit of food resource per individual. A muta-

tion to a new type *a* arises that doubles the fecundity of its bearers but does absolutely nothing to the efficiency of the utilization of resources. We can calculate what happens to the composition, size and growth rate of the population over a period of time [*see illustration below*]. In a second scenario we again begin with the population of 100 individuals of type *A*, but now there arises a different mutation *a*, which does nothing to the fecundity of its bearers but doubles their efficiency of resource utilization. Again we can calculate the population history.

In both cases the new type *a* replaces the old type *A*. In the case of the first mutation nothing changes but the fecundity; the adult population size and the growth rate are the same throughout the process and the only effect is that twice as many immature stages are being produced to die before adulthood. In the second case, on the other hand, the population eventually doubles its adult members as well as its immature members, but not its fecundity. In the course of its evolution the second population has a growth rate greater than 1 for a while but eventually attains a constant size and stops growing.

In which of these populations, if in either, would the individuals be better

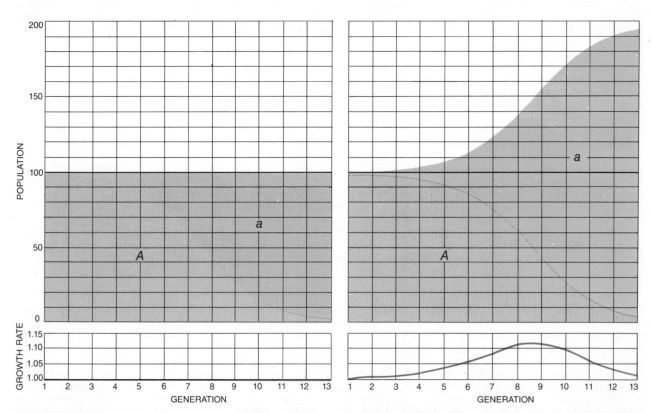

TWO DIFFERENT MUTATIONS have different demographic results for a resource-limited population of 100 insects. In one case (*left*) a mutation arises that doubles the fecundity of its bearers. The new type (*a*) replaces the old type (*A*), but the total population does not increase: the growth rate (*bottom*) remains 1.00. In the other case (*right*) a mutation arises that doubles the carrier's efficiency of resource utilization. Now the new population grows more rapidly, but only for a short time: eventually the growth rate falls back to 1.00 and the total population is stabilized at 200. The question is: Has either mutation given rise to a population that is better adapted?

adapted than those in the old population? Those with higher fecundity would be better buffered against accidents such as sudden changes in temperature since there would be a greater chance that some of their eggs would survive. On the other hand, their offspring would be more susceptible to the epidemic diseases of immature forms and to predators that concentrate on the more numerous immature forms. Individuals in the second population would be better adapted to temporary resource shortages, but also more susceptible to predators or epidemics that attack adults in a density-dependent manner. Hence there is no way we can predict whether a change due to natural selection will increase or decrease the adaptation in general. Nor can we argue that the population as a whole is better off in one case than in another. Neither population continues to grow or is necessarily less subject to extinction, since the larger number of immature or adult stages presents the same risks for the population as a whole as it does for individual families.

Unfortunately the concept of relative adaptation also requires the ceteris paribus assumption, so that in practice it is not easy to predict which of two forms will leave more offspring. A zebra having longer leg bones that enable it to run faster than other zebras will leave more offspring only if escape from predators is really the problem to be solved, if a slightly greater speed will really decrease the chance of being taken and if longer leg bones do not interfere with some other limiting physiological process. Lions may prey chiefly on old or injured zebras likely in any case to die soon, and it is not even clear that it is speed that limits the ability of lions to catch zebras. Greater speed may cost the zebra something in feeding efficiency, and if food rather than predation is limiting, a net selective disadvantage might result from solving the wrong problem. Finally, a longer bone might break more easily, or require greater developmental resources and metabolic energy to produce and maintain, or change the efficiency of the contraction of the attached muscles. In practice relative-adaptation analysis is a tricky game unless a great deal is known about the total life history of an organism.

Not all evolutionary change can be understood in terms of adaptation. First, some changes will occur directly by natural selection that are not adaptive, as for example the changes in fecundity and feeding efficiency in the hypothetical example I cited above.

Second, many changes occur indirectly as the result of allometry, or differential growth. The rates of growth of different parts of an organism are different,

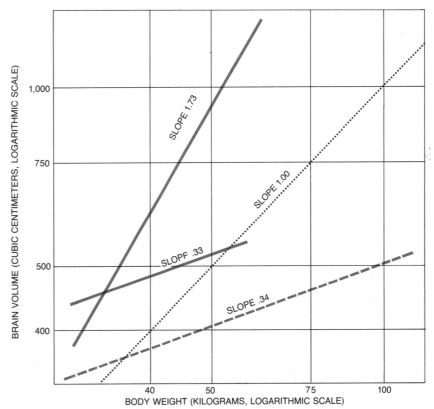

ALLOMETRY, or differential growth rates for different parts, is responsible for many evolutionary changes. Allometry is illustrated by this comparison of the ratio of brain size to body weight in a number of species of the pongids, or great apes (*broken black curve*), of *Australopithecus*, an extinct hominid line (*solid black*), and of hominids leading to modern man (*color*). A slope of less than 1.00 means the brain has grown more slowly than the body. The slope of more than 1.00 for the human lineage indicates a clear change in the evolution of brain size.

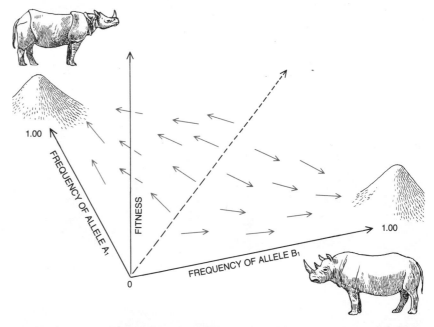

ALTERNATIVE EVOLUTIONARY PATHS may be taken by two species under similar selection pressures. The Indian rhinoceros has one horn and the African rhinoceros has two horns. The horns are adaptations for protection in both cases, but the number of horns does not necessarily constitute a specifically adaptive difference. There are simply two adaptive peaks in a field of gene frequencies, or two solutions to the same problem; some variation in the initial conditions led two rhinoceros populations to respond to similar pressures in different ways. For each of two hypothetical genes there are two alleles: A_1 and A_2, B_1 and B_2. A population of genotype A_1B_2 has one horn and a population of genotype A_2B_1 has two horns.

so that large organisms do not have all their parts in the same proportion. This allometry shows up both between individuals of the same species and between species. Among primate species the brain increases in size more slowly than the body; small apes have a proportionately larger brain than large apes. Since the differential growth is constant for all apes, it is useless to seek an adaptive reason for gorillas' having a relatively smaller brain than, say, chimpanzees.

Third, there is the phenomenon of pleiotropy. Changes in a gene have many different effects on the physiology and development of an organism. Natural selection may operate to increase the frequency of the gene because of one of the effects, with pleiotropic, or unrelated, effects being simply carried along. For example, an enzyme that helps to detoxify poisonous substances by converting them into an insoluble pigment will be selected for its detoxification properties. As a result the color of the organism will change, but no adaptive explanation of the color per se is either required or correct.

Fourth, many evolutionary changes may be adaptive and yet the resulting differences among species in the character may not be adaptive; they may simply be alternative solutions to the same problem. The theory of population genetics predicts that if more than one gene influences a character, there may often be several alternative stable equilibriums of genetic composition even when the force of natural selection remains the same. Which of these adaptive peaks in the space of genetic composition is eventually reached by a population depends entirely on chance events at the beginning of the selective process. (An exact analogy is a pinball game. Which hole the ball will fall into under the fixed force of gravitation depends on small variations in the initial conditions as the ball enters the game.) For example, the Indian rhinoceros has one horn and the African rhinoceros has two. Horns are an adaptation for protection against predators, but it is not true that one horn is specifically adaptive under Indian conditions as opposed to two horns on the African plains. Beginning with two somewhat different developmental systems, the two species responded to the same selective forces in slightly different ways.

Finally, many changes in evolution are likely to be purely random. At the present time population geneticists are sharply divided over how much of the evolution of enzymes and other molecules has been in response to natural selection and how much has resulted from the chance accumulation of mutations. It has proved remarkably difficult to get compelling evidence for changes in enzymes brought about by selection, not to speak of evidence for adaptive changes; the weight of evidence at present is that a good deal of amino acid substitution in evolution has been the result of the random fixation of mutations in small populations. Such random fixations may in fact be accelerated by natural selection if the unselected gene is genetically linked with a gene that is undergoing selection. The unselected gene will then be carried to high frequency in the population as a "hitchhiker."

If the adaptationist program is so fraught with difficulties and if there are so many alternative explanations of evolutionary change, why do biologists not abandon the program altogether?

WHALES

SEALS

PENGUINS

FISH

SEA SNAKES

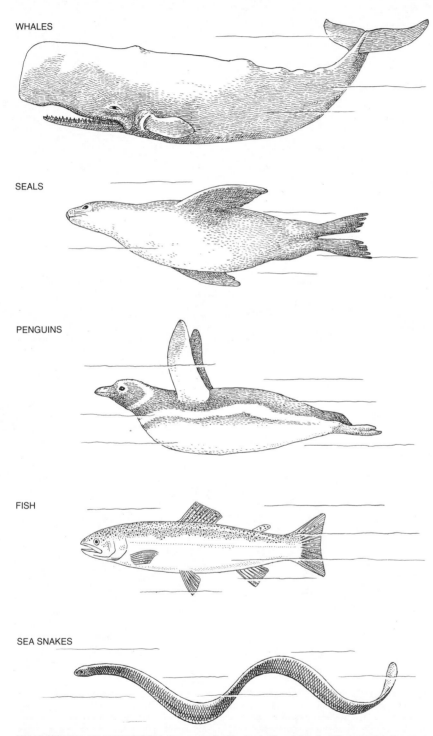

REALITY OF ADAPTATION is demonstrated by the indisputable fact that unrelated groups of animals do respond to similar selective pressures with similar adaptations. Locomotion in water calls for a particular kind of structure. And the fact is that whales and seals have flippers and flukes, penguins have paddles, fish have fins and sea snakes have a flat cross section.

There are two compelling reasons. On the one hand, even if the assertion of universal adaptation is difficult to test because simplifying assumptions and ingenious explanations can almost always result in an ad hoc adaptive explanation, at least in principle some of the assumptions can be tested in some cases. A weaker form of evolutionary explanation that explained some proportion of the cases by adaptation and left the rest to allometry, pleiotropy, random gene fixations, linkage and indirect selection would be utterly impervious to test. It would leave the biologist free to pursue the adaptationist program in the easy cases and leave the difficult ones on the scrap heap of chance. In a sense, then, biologists are forced to the extreme adaptationist program because the alternatives, although they are undoubtedly operative in many cases, are untestable in particular cases.

On the other hand, to abandon the notion of adaptation entirely, to simply observe historical change and describe its mechanisms wholly in terms of the different reproductive success of different types, with no functional explanation, would be to throw out the baby with the bathwater. Adaptation is a real phenomenon. It is no accident that fish have fins, that seals and whales have flippers and flukes, that penguins have paddles and that even sea snakes have become laterally flattened. The problem of locomotion in an aquatic environment is a real problem that has been solved by many totally unrelated evolutionary lines in much the same way. Therefore it must be feasible to make adaptive arguments about swimming appendages. And this in turn means that in nature the ceteris paribus assumption must be workable.

It can only be workable if both the selection between character states and reproductive fitness have two characteristics: continuity and quasi-independence. Continuity means that small changes in a characteristic must result in only small changes in ecological relations; a very slight change in fin shape cannot cause a dramatic change in sexual recognition or make the organism suddenly attractive to new predators. Quasi-independence means that there is a great variety of alternative paths by which a given characteristic may change, so that some of them will allow selection to act on the characteristic without altering other characteristics of the organism in a countervailing fashion; pleiotropic and allometric relations must be changeable. Continuity and quasi-independence are the most fundamental characteristics of the evolutionary process. Without them organisms as we know them could not exist because adaptive evolution would have been impossible.

ADAPTATIONS:
THE BODY AND
MOVEMENT

Almost all species except man and monkeys, both the viviparous and the oviparous, have tails corresponding to the requirements of their bodies, bare with the hairy species, like boars, small with the shaggy ones, like bears, very long with the bristly, like horses. With lizards and snakes when cut off they grow again. The tails of fishes steer their winding courses after the manner of a rudder, and even serve to propel them like a sort of oar by being moved to the right and left. Actual cases of two tails are found in lizards. Oxen's tails have a very long stem, with a tuft at the end, and in asses it is longer than in horses, but it is bristly in beasts of burden. A lion's tail is shaggy at the end, as with oxen and shrewmice, but not so with leopards; foxes and wolves have a hairy tail, as have sheep, with which it is longer. Pigs curl the tail, dogs of low breeds keep it between their legs.

Pliny
NATURAL HISTORY, XI, CXi

ADAPTATIONS: THE BODY AND MOVEMENT

II

INTRODUCTION

Life in water, life on land, life in the air—each places very different demands upon the vertebrate body. The muscle–bone complex used for support and locomotion and its neuronal control network have evolved into a bewildering variety of types. In this introduction, we shall discuss adaptations for each of the environmental habitats and see how the basic body components changed in time.

The elements of vertebrate swimming are illustrated diagrammatically in elongate fish, such as sharks or eels. When sharks are swimming (Figure 1), their long, thin bodies assume the shape of moving sine waves. This shape results from alternate series of contractions of the muscle masses (myotomes) that make up each side of the fish body. In swimming, a wave of muscle contraction starts in the most anterior muscle mass on one side of the body and then proceeds posteriorly, as sets of myotomes contract in turn. Then, before the first wave proceeds far, another wave starts at the anterior end on the opposite side of the body. The process continues, first on one side, then on the other, and the result is a bending motion like that of sine waves. This action causes each portion of the sides of the fish to be swept back and forth through the water as if it is a moving inclined plane. Forward propulsion results. In higher fishes, the sine-wave motions are much less obvious, and most of the propulsive thrust comes from the specialized tail region, where amplitude of the waves is greatest.

The pattern of the chains of contraction that occur in swimming is largely controlled by nerves in the spinal cord. If the spinal cord of a dogfish shark is severed from the brain, the organism can be made to swim. It will continue to do so for long periods. Some physiologists believe that the whole pattern of coordinated muscle activity is programmed in the network of motor nerves within the spinal cord, and that no sensory input (or exceedingly little) is necessary to coordinate the chains of contractions. Interestingly, the swimming speeds of such "spinal" animals are fairly constant: the spinal cord control system shows little capacity to alter the rate of muscle activity. If small portions of the cord's sensory input are intact, however, appropriate stimuli can evoke violent muscle contraction or faster swimming; such responses do not require presence of a brain. However, the brain (particularly the medulla) is necessary for normal initiation and control of swimming speed. Thus we may view the brain of a normal animal as the control circuit that can start, stop, or modify sets of activity in the motor nerve switchboard of the spinal cord. One might imagine, in fact, that the long neural tube (the spinal cord) originated in the prevertebrates as a motor control system for the segmental muscle masses.

Primitive vertebrate fishes probably propelled themselves as do today's sharks. Like sharks, they possessed asymmetric tails in which the dorsal half

Figure 1. **A diagrammatic representation of a shark swimming. The bends in the body are caused by waves of contraction of the muscle masses that start at the front and proceed toward the rear of the fish. The darkened muscle masses are those that are contracted; an instant later, these would be relaxed and more posterior muscle masses would be contracted. The next wave of contraction would start on the opposite side (right side in the drawing) of the animal.**

was longer and more massive than the ventral half. When such a tail sweeps to and fro through the water, it tends to drive water down and back; the tail, of course, reacts in the opposite direction: it is driven up and forward. The latter force provides propulsion; the former is a "lift" at the rear end of the body. As a result, the tail tends to rise and the head to sink during swimming. This was the motion of most agnathans, and it was perhaps a useful adaptation for easy movement along the bottom of the ocean or bodies of fresh water. For free swimming above the bottom, however, turning cartwheels in the water is hardly adaptive! Sharks overcome the difficulty with a broad head, which is flattened ventrally, and a set of wide, flat pectoral fins anterior to the center of gravity. Both flat surfaces tilt up at anterior edges, so they are like planes being pushed through the water; the result is a lift at the front that counteracts that at the rear. By controlling the angle of the anterior lift planes, the fish can direct its movement in the dorsoventral plane. Equivalent fin systems and flattened heads are seen in many types of early agnathan fishes. In addition, in those creatures and in almost all their descendants, there was at least one other pair of fins (the pelvics), located posterior to the center of gravity. The acquisition of the anterior and posterior pairs of fins was important not only for fishes but also for terrestrial vertebrates, because the paired fins later became the limbs of those creatures.

Most of the higher bony fish in the world differ from the sharks or ancestral fishes by having symmetrical tails (Figure 2). The reason is, no doubt, the existence of a swim bladder, which is a source of buoyancy. Unlike sharks, most of which sink if they stop moving forward in water, bony fishes can add or remove gas from their swim bladders to keep their bodies in neutral buoyancy with water at a given depth; they alter the total specific gravity of the body and so keep its mass exactly equal to the mass of the water it displaces. Clearly, any tendency to drive the body up or down, such as lift from an asymmetric tail, would hardly be advantageous to an organism that would otherwise be stable at one depth. So the tail fin became symmetrical, and, because the tail lift was eliminated as a result, the paired anterior fins were freed for functions other than counteracting tail lift. Hence, fins came to be used as brakes, tilting rudders, or even wings for gliding through the air! One wonders, in fact, what strange and wonderful differences might have taken place in vertebrate evolution if swim bladders and symmetric tails had been present in the earliest vertebrates so that the paired fins need never have developed.

Buoyancy considerations also have led to specialized changes in the shape of certain fishes. The reason is that protein itself is denser than water (1.33 versus 1.026 for sea water). In extreme deep-water fishes (bathypelagics), a common adaptation is a gross reduction in the size of the main body musculature (which is mostly protein in content), as well as in bone. A tiny body often seems to be associated with a relatively huge set of jaws in such weirdly shaped creatures (see Figure 3). Marine mammals have also been affected by buoyancy problems. Later, we shall see how sperm whales are reputed to adjust their buoyancy, in "The Head of the Sperm Whale," an article in Section III.

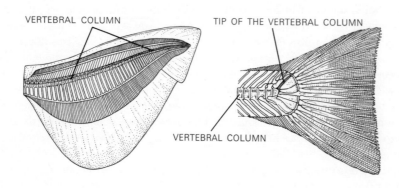

VERTEBRAL COLUMN

TIP OF THE VERTEBRAL COLUMN

VERTEBRAL COLUMN

Figure 2. The caudal "tail" fins of a shark and of a bony fish are seen in these sketches. The asymmetric shape of the shark fin results from an upward bend of the vertebral column and from a large flap of tissue that projects ventrally. The very tip of the bony fish's vertebral column also bends dorsally, perhaps reflecting its evolutionary derivation from an asymmetric tail. Nevertheless, the other bony fin supports are symmetrical so that the whole structure no longer generates a "lift," as does the shark fin.

It seems reasonable at first glance to assume that the speed of fish swimming, the power used in swimming, and the efficiency of swimming locomotion are merely functions of the fish's body itself. In fact, behavior may also play a role. Some fish and birds take advantage of properties of the fluid (water or air) in which they move to increase the efficiency of locomotion. For example, the biologist Evelyn Shaw has noted the great regularity in spacing of fish that swim in schools: analysis by D. Weihs of Cambridge University suggests that this spacing allows a considerable saving in swimming energy for properly placed trailing fish (see Figure 4). An analogous explanation has been offered to explain the V-shaped formation of migratory bird flocks. Rising vortices of air are created by the descending wing beats of birds in front; trailing birds in the V fly in the resultant updrafts and so conserve energy just as schooling fish do. Both these vertebrate types are exploiting a basic physical property of the fluid in which they move. That opportunity is not available to terrestrial creatures who must thrust against the solid earth for locomotion.

The swimming motion of fishes is still seen in the most primitive terrestrial vertebrates alive today, the salamanders (see Figure 5). A salamander uses its limbs for paddling or walking slowly but, if it is frightened, it tucks them in close to the body and swims or wriggles away by twisting its body in a series of sine-wavelike movements. Of course, the locomotory movement of snakes, as summarized by Carl Gans in "How Snakes Move," is also a utilization of the basic lateral undulations seen in fishes.

Life on a solid substrate led to major alterations in the vertebral column and the limbs of vertebrates. In the terrestrial vertebrate, the column shifted toward the dorsal surface, so that most of the body mass hangs below the chain of bones. Because this places great stresses on individual vertebrae, they tend to slip past each other in the dorsoventral plane. To counteract this tendency, individual vertebrae acquired various flanges of bone (zygapophyses and projections from centra) that brace the column and prevent slippage. Particularly in the ancient reptiles (the therapsids) that subsequently gave rise to mammals, the limbs—which had protruded outward somewhat perpendicular to the body axis—gradually moved in under the body: the hind limb turned forward and in, and the knee bent sharply so that the foot still pointed anteriorly; the forelimb initially bent back at the shoulder, so that a compensatory twist was necessary at the elbow; thus, the forearm rotated forward 180 degrees so that the forefoot also pointed anteriorly. These various bends produced the condition from which mammalian limbs radiated. (See Milton Hildebrand's "How Animals Run.")

Although walking on solid land has permitted a wide variety of limb shapes, sizes, stances, and gaits to evolve (as described by Hildebrand and by John Napier in "The Antiquity of Human Walking"), the neuronal control of basic limb movements is still a feature of the spinal cord, just as is the control of swimming in fishes. Keir Pearson ("The Control of Walking") gives evidence that, even in the most complex vertebrates (mammals such as cats), the basic rhythmical walking movements occur in "spinal" animals (ones in which the cerebrum is destroyed or in which the cord is severed). Since humans are conscious animals, we tend to jump to the conclusion that such complicated processes as walking or running must be controlled by our brains; in fact, the coordinated sequential use of limb muscles is dictated by networks in our spinal cords, just as swimming movements were in our ancestors.

Now let us turn to movement through the air, a medium that places special demands upon the flying vertebrate's body. Three types of vertebrates have been capable of flapping (power generated) flight: pterosaurs and birds, both derivatives of different groups of reptiles, and bats, derivatives of primitive insectivoran mammals. But, in addition, various other vertebrate groups have produced gliding flyers: the flying fishes and squirrels are two examples. Gliding may have been an early stage in the development of avian and chiropteran

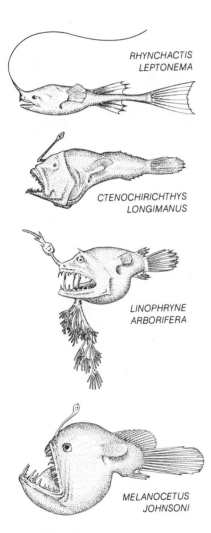

RHYNCHACTIS
LEPTONEMA

CTENOCHIRICHTHYS
LONGIMANUS

LINOPHRYNE
ARBORIFERA

MELANOCETUS
JOHNSONI

Figure 3. **Representative deep-sea fishes with accentuated jaw and head regions and reduced trunk and tail musculature. The antenna-like appendages extending from the heads of these so-called "angler" fish are apparently used to attract prey.**
[Courtesy of E. Bertelsen and *Dana Reports* 39–40, 1951–53. *Melanocetus* redrawn from Bertelsen, 1951; *Rhynchatis* and *Linophryne* redrawn from Regan, 1926; *Ctenochirichthys* redrawn from Regan and Trewavas, 1932.]

(bat) flight; for instance, it is hypothesized that descendents of small, bipedal, light-bodied dinosaurs such as those in the genus *Deinonychus*, may have run rapidly along the ground, spread their elongated forelimbs, and glided into the air for at least short distances. Presumably, elongate feathers must have been present on the forelimbs of those creatures in order to from the aerofoil that would be the source of lift. The earliest bird fossils of the genus *Archaeopteryx* have structures not unlike those hypothetical early gliders; and interestingly, in the *Archaeopteryx* fossils found to date a chest bone (sternum) with a substantial keel has not been found. Keels of modern birds serve as the site of origin of the major flight muscles that are so important in flapping flight.

Since flight places rigid demands on the construction of any animal body, it is not surprising that the body organization of pterosaurs, birds, and bats responded to those demands in quite similar ways. Light thin bones, elongate forelimbs and hands, reduced hind limbs, and increased wing surface area are features common to all three of these major flying groups. The wing of the pterosaur was a large, thin sheet of skin stretched behind an enormously elongate fourth digit. The wing of the bat, too, consists of sheets of skin stretched between digits two through five, all of which are greatly lengthened. The bird wing terminates in a set of bones—the second, third, and fourth digits—and the great bulk of the wing aerofoil is made up of the flight feathers. The feathered construction has obvious physiological advantages for temperature control since feathers are made of dead, keratinized cells; thus, feathers do not require a supply of the bird's warm blood with the attendant possibility of heat loss. In addition, feathers are more versatile aerodynamic surface than the sheets of skin comprising bat and pterosaur wings. As an example, wing "slots" for reducing drag can be created by twisting individual feathers. Such adjustments are impossible in bat or pterosaur wings. This is one of the reasons that bat flight appears so uneven and "fluttery" in comparison with the level flight of most birds.

The driving force of bird flight comes from the huge pectoralis muscles. Their action is a superb example of engineering design (see Figure 6). Since greater aerodynamic stability is gained if the center of gravity of a flying body is at or below the level of the wings, the location of a heavy mass of muscle above and between the wings of a bird would certainly present a problem. Thus the muscles that elevate the wings are actually located below the wings. Both the muscles that depress the wings (the pectoralis majors) and those that raise them (the pectoralis minors, or as some anatomists with a penchant for new names call them, the supracoracoideus muscles) are attached to a flange of bone (the keel) that projects downward from the central chest bone, the sternum. The depressor muscle, as might be expected, simply inserts on the ventral surface of the wing to lower it. The elevator muscle, however, ends at a tendon that proceeds dorsally over the shoulder joint to insert on the upper surface of the proximal wing bone. Thus, by a pulley action the muscles located well below the wings can raise them in flight! Bats and fossil pterosaurs, too, have keels from which the flight-depressor muscles originate—this is a good example of the evolutionary process producing the same functional condition at three different times in vertebrate history.

With this information as background, the reader can appreciate the specialized type of gliding flight described by C. J. Pennycuick in "The Soaring Flight of Vultures." Vultures and other soaring birds are adapted to utilize moving currents in the air in order to stay aloft with minimal expenditure of energy in flapping.

Let us now come back to earth, and consider one final type of vertebrate locomotion, the bipedal walking of humans and our near relatives. Our discussion will range widely since bipedal walking may have played an early pivotal role in freeing the forelimb to evolve into the arm, hand, and thumb;

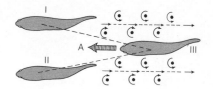

Figure 4. A view from above of three fish in a school. As the tails sweep back and forth in the water, columns of vortices of turbulent water are left in the wake of fishes I and II. Note that directly behind each fish the vortices induce a water velocity opposite to the fishes movement. But, between I and II the vortices create a "street" or channel of flow forward (A). If fish III swims in that street, it receives a boost and so conserves energy.
[After D. Weihs, *Nature*, 241, 1973.]

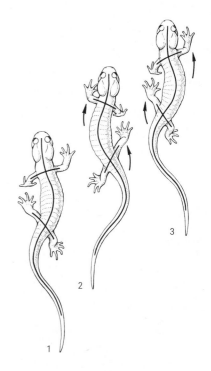

Figure 5. The body movement of a walking salamander. The vertebral column moves back and forth forming a "standing" sine wave first in one direction, then in the other. Though reminiscent of the moving sine waves of a shark (see Figure 1), the shapes are produced by different action of the musculature. In comparison with this salamander, when many lizards or mammals run, their vertebral column remains relatively straight, and the limbs move more independantly.
[After A. S. Romer, *The Vertebrate Body*, Third Edition, Saunders, 1962.]

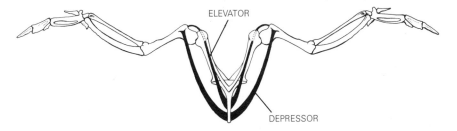

Figure 6. The major bones and muscle-tendon systems of a bird. The depressor muscles originate from the ventral keel. The elevator muscles start at the upper keel and extend over the shoulder joint to insert on the upper surface of the humerus, the proximal wing bone. [After Lois and Louis Darling, *Bird*, Houghton Mifflin, 1962.]

and that remarkable structure, with its great manipulative skills, in turn may have played an important part in the expansion of the primate brain.

Recent studies of so-called "bipedal" locomotion provide interesting perspective on Napier's discussions of the origins of human walking. When a chimp walks in an upright stance, it does so primarily by straightening up the pelvis relative to the femur and leg. The leg is held in the same position as during quadrupedel knuckle walking (see Figure 6). If one observes a chimpanzee in an upright stance from the front, it can be seen that the knee is held slightly lateral to the hip. Hence, the femur is said to be in an abducted position. In comparison, the femur of modern humans is directed inward at its lower end, so that our knees are closer together than are the hip joints (that is, the femur is adducted). Australopithecines, in contrast, held the femur in a directly vertical position or in a slightly adducted position, which reflects their closeness to modern humans in this important respect.

An interesting sidelight on the upright human stance comes from P. R. Zelazo and his colleagues who have studied "parental" behavior designed to accelerate motor development related to walking among African foraging peoples. Infants are commonly carried in a vertical position in a sling or held vertically in the mother's lap. A sample of such infants shows precocity in sitting, standing, and mature walking compared to non-African controls. Other workers have found that African children brought up in the European manner do not display this precocious development. This sort of phenomenon is another form of the ·social-biological interplay that may well have important

Figure 7. At the top are sketches of a chimpanzee's pelvic girdle and lower limbs during a single stride. Note that, in contrast to the case in humans, the legs are held well in advance of the girdle at all phases of the stride; thus, the body is not truly balanced above the limbs and feet. There is also a marked dorso-ventral movement of the pelvis (relative to the line of reference), instead of the much smoother forward motion in humans. At the bottom are views from the front of one-half of the pelvic girdle and the femur of (A) a chimpanzee, (B) an *Australopithecus robustus*, and (C) a modern human. Note how the lower end of the femur has pointed in, toward a "knock-kneed" condition (that is, an adducted one). This helps to bring the human foot in under the center of gravity both during a stride and when standing still.
[Courtesy of F. A. Jenkins Jr., and *Science*, vol. 178, pp. 877–879, Fig. 1, 2; 24 November 1972. Copyright 1972 by the American Association for the Advancement of Science.]

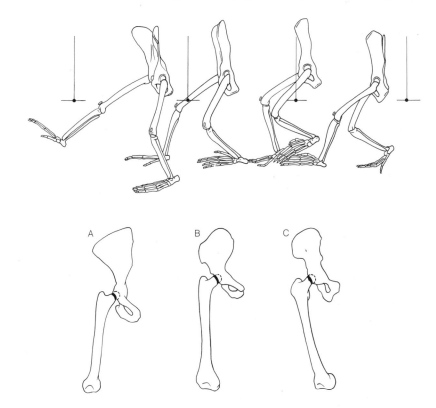

evolutionary consequences. We can only wonder whether such behavior contributed to the origin of human walking in the ancestors of the Australopithecines.

The series of primate fossils that has appeared since Raymond Dart, the pioneering anthropologist at the University of Witwatersrand, South Africa, discovered the first remains of the Australopithecines in 1924, has led to one surprising conclusion about the evolution of humans; appearance of an upright stance and "human" hand preceded development of the large brain. In the early Australopithecines who lived some three million years ago, the pelvic girdle and leg structure are indicative of a bipedal stance much like our own, and certainly not of the normal stooping stance of the apes. Coupled with this advanced posture and a prehensile hand was a remarkably small brain, which had a volume of about 510 cubic centimeters (about the same as an adult male gorilla). Until 1973, it was assumed that a series of humanoid fossils with progressively larger cranial capacities reflected the evolutionary line of man. For instance, *Homo habilis*, the next most advanced pre-human who was a contemporary of at least the later Australopithecines, had a cranial capacity of 680 cubic centimeters, and the next form, *Homo erectus* (*Pithecanthropus*) a cranial capacity of 975 cubic centimeters. Thereafter, as the various forms of *Homo sapiens* advanced, the brain volumes increased until the current average size of 1,400 cubic centimeters was reached.

The significance of this seeming sequence is thrown into confusion by Richard E. Leakey's discovery of a fossil skull that is clearly of the genus *Homo*, with a cranial capacity of about 800 cubic centimeters and an age of about 2.8 million years. This creature may have been a contemporary of Australopithecines (whose fossils date from about 3 to 1 million years ago). Leakey's specimen, identified simply as skull no. 1470, lacks the heavy brow ridges characteristic of our pongid relatives and other ancient ancestors. Leg bones from the same geological level as the new skull are even more like those of *Homo sapiens* than are ones of the Australopithecines. For example, the thickness of the shaft just beneath the head of the femur is thick in the new find and much thinner in Australopithecines; thus, it could probably support more sustained bipedal locomotion. What do these recent finds mean? Probably there were several ancient lines of terrestrial anthropoids living as contemporaries—the Australopithecines and *Homo* 1470 are two of these. Perhaps 1470 and its kin represent the ancestral relatives of *Homo erectus* and, so, indirectly of *sapiens*.

Although we tend to think of increasing brain size as being unique to primates and the human lineage, that is not correct. In his "Paleoneurology and the Evolution of the Mind," Harry J. Jerison assembles intriguing data demonstrating a positive correlation between brain weight and body weight in fishes, reptiles, and mammals. Even within a single group of mammals, say the ungulates, or the carnivores, the brain size becomes more variable and increases as one goes from archaic fossils to contemporary forms. But it is in the cetaceans (dolphins, whales) and primates that the trend toward relatively large brains is most marked. And that large size translates, of course, into more neurons and more neuronal connections—the features of the cerebral cortex believed to be responsible for the great capacities for learning and complex behavior that characterizes humans and our closest relatives.

At some point as the human brain expanded, language became possible. Sherwood L. Washburn (Offprint 1406, "The Evolution of Man") has stated: "Language, that marriage of speech and cognitive abilities, may well have been the critical new factor that provided a biological base for the acceleration of history." Language as we use it is not solely dependent upon the brain and central nervous system. Our capacity to produce certain vowel sounds— *a* or "ah" as in "not", *i* or "ee" as is "see," and *u* or "uu" as in "to"—results from the fact that the posterior part of our tongue projects backwards above the exit of the larynx and forms a moveable anterior wall of our pharynx. Those vowels, which are common to most if not all human languages, are shaped in that part of the vocal tract and not by the moving tip of the tongue or the lips as consonants are shaped. It has been noted that chimpanzees and newborn human infants cannot make such vowel sounds, and that their larynges open into the oral cavity almost at the level of the palate in the roof of the mouth. Some highly controversial interpretations of casts of Neanderthal fossils suggest to some paleontologists that those creatures had a tongue and pharynx more like that of apes than our own. If so, it is argued, language as we know it could not have been one of their capacities. Yet, some Neanderthal fossils equal or exceed modern *Homo sapiens* in brain volume. This is the kind of paradox that makes further scientific investigation so intriguing.

How Animals Run

by Milton Hildebrand
May 1960

*Many animals, both predators and prey, have evolved
the ability to run two or three times faster than a man
can. What are the adaptations that make these
impressive performances possible?*

A man (but not necessarily you or I!) can run 220 yards at the rate of 22.3 miles per hour, and a mile at 15.1 miles per hour. The cheetah, however, can sprint at an estimated 70 miles per hour. And the horse has been known to maintain a speed of 15 miles per hour not just for one mile but for 35 miles.

Other animals are capable of spectacular demonstrations of speed and endurance. Jack rabbits have been clocked at 40 miles per hour. The Mongolian ass is reported to have run 16 miles at the impressive rate of 30 miles per hour. Antelopes apparently enjoy running beside a moving vehicle; they have been reliably timed at 60 miles per hour. The camel has been known to travel 115 miles in 12 hours. Nearly all carnivorous mammals are good runners: the whippet can run 34 miles per hour; the coyote, 43 miles per hour; the red fox, 45 miles per hour. One red fox, running before hounds, covered 150 miles in a day and a half. A fox terrier rewarded with candy turned a treadmill at the rate of 5,000 feet per hour for 17 hours.

I have been attracted by such performances as these to undertake an investigation of how the living running-machine works. The subject has not been thoroughly explored. One study was undertaken by the American photographer Eadweard Muybridge in 1872. Working before the motion-picture camera was invented, Muybridge set up a battery of still cameras to make photographs in rapid sequence. His pictures are still standard references. A. Brazier Howell's work on speed in mammals and Sir James Gray's studies on posture and movement are well known to zoologists. Many investigators have added to our knowledge of the anatomy of running vertebrates, but the analysis of function has for the most part been limited to deductions

from skeletons and muscles. The movements of the running animal are so fast and so complex that they cannot be analyzed by the unaided eye.

In my study I have related comparative anatomy to the analysis of motion pictures of animals in action. The method is simple: Successive frames of the motion picture are projected onto tracing paper, where the movements of the parts of the body with respect to one another and to the ground can be analyzed. The main problem is to get pictures from the side of animals running at top speed over open ground. With an electric camera that exposes 200 frames per second I have succeeded in photographing the movements of a cheetah that had been trained by John Hamlet of Ocala, Fla., to chase a paper bag in an enclosure 65 yards long. However, the animal never demonstrated its top speed, but merely loped along at about 35 miles per hour. I have used the same

STRIDE OF A CANTERING HORSE is shown in these photographs from Eadweard Muybridge's *The Horse in Motion*, published in 1878. The sequence runs right to left across the

camera to make pictures of horses running on race tracks, and I am presently collecting motion-picture sequences of other running animals from commercial and private sources.

All cursorial animals (those that can run far, fast and easily) have evolved from good walkers, and in doing so have gained important selective advantages. They are able to forage over wide areas. A pack of African hunting dogs, for example, can range over 1,500 square miles; the American mountain lion works a circuit some 100 miles long; individual arctic foxes have on occasion wandered 800 miles. Cursorial animals can seek new sources of food and water when their usual supplies fail. The camel moves from oasis to oasis, and in years of drought the big-game animals of Africa travel impressive distances. The mobility of cursorial animals enables them to overcome seasonal variations in climate or in food supply. Some herds of caribou migrate 1,600 miles each year. According to their habit, the predators among the cursorial animals exploit superior speed, relay tactics, relentless endurance or surprise to overtake their prey. The prey species are commonly as swift as their pursuers, but sometimes they have superior endurance or agility.

Speed and endurance are the capacities that characterize all cursorial vertebrates. But one could not make a definitive list of the cursorial species without deciding quite arbitrarily how fast is fast and how far is far. Even then the list would be incomplete, because there are reliable data on speed for only a few animals; in most cases authors quote authors who cite the guesses of laymen. Many cursors are extinct. On the basis of fossils, however, we can surmise that many dinosaurs were excellent runners; that some extinct rhinoceroses, having had long and slender legs, were very fast; and that certain extinct South American grazing animals, having evolved a horselike form, probably had horselike speed.

In order to run, an animal must overcome the inertia of its body and set it into motion; it must overcome the inertia of its legs with every reversal in the direction of their travel; it must compensate for forces of deceleration, including the action of the ground against its descending feet. A full cycle of motion is called a stride. Speed is the product of length of stride times rate of stride. The giraffe achieves a moderate speed with a long stride and a slow rate of stride; the wart hog matches this speed with a short stride and a rapid rate. High speed requires that long strides be taken at a rapid rate, and endurance requires that speed be sustained with economy of effort.

Although longer legs take longer strides, speed is not increased simply by the enlargement of the animal. A larger animal is likely to have a lower rate of stride. Natural selection produced fast runners by making their legs long in relation to other parts of the body. In cursorial animals the effective length of the leg—the part that contributes to length of stride—is especially enhanced. The segments of the leg that are away from the body (the foot, shank and forearm) are elongated with respect to the segments close to the body (the thigh and upper arm). In this evolutionary lengthening process the bones equivalent to the human palm and instep have become the most elongated.

Man's foot does not contribute to the length of his leg, except when he rises on his toes. The bear, the opossum, the raccoon and most other vertebrates that walk but seldom run have similar plantigrade ("sole-walking") feet. Carnivo-

top row and continues across the bottom row. With these and similar photographs Muybridge settled the controversy of whether or not a horse "even at the height of his speed [has] all four of his feet . . . simultaneously free from contact with the ground."

rous mammals, birds, running dinosaurs and some extinct hoofed mammals, on the other hand, stand on what corresponds to the ball of the human foot; these animals have digitigrade ("finger-walking") feet. Other hoofed mammals owe an even further increase in the effective length of their legs to their unguligrade ("hoof-walking") posture, resembling that of a ballet dancer standing on the tips of her toes. Where foot posture and limb proportions have been modified for the cursorial habit, the increased length and slenderness of the leg is striking [see illustration on page 32].

The effective length of the front limb of many runners is also increased by the modification of the structure and function of the shoulder. The shoulder joint of amphibians, reptiles and birds is virtually immobilized by the collarbone, which runs from the breast bone to each shoulder blade, and by a second bone,

the coracoid bone. Because mammals do not have a coracoid bone their shoulder blade has some freedom of movement. In the carnivores this freedom is increased by the reduction of the collarbone to a vestige; in the ungulates the collarbone is eliminated. In both carnivores and ungulates the shoulder blade is oriented so that it lies against the side of a narrow but deep chest rather than against the back of a broad but shallow chest, as it does in man. Thus mounted, the shoulder blade pivots from a point about midway in its length, and the shoulder joint at its lower end is free to move forward and backward with the swing of the leg. The exact motion is exceedingly difficult to ascertain in a running animal, but I have found that it adds about 4.5 inches to the stride of the walking cheetah.

The supple spine of the cat and the dog increases the length of stride of these animals still further. The body of such an animal is several inches longer

when the back is extended than when it is flexed. By extending and flexing its back as its legs swing back and forth the animal adds the increase in its body length to its stride. Timing is important in this maneuver. If the animal were to extend its back while its body was in mid-air, its hindquarters would move backward as its forequarters moved forward, with no net addition to the forward motion of the center of mass of its body. In actuality the running animal extends its back only when its hind feet are pushing against the ground. The cheetah executes this maneuver so adeptly that it could run about six miles per hour without any legs.

With the extra rotation of its hip and shoulder girdles and the measuring-worm action of its back, the legs of the running cursor swing through longer arcs, reaching out farther forward and backward and striking and leaving the ground at a more acute angle than they would if the back were rigid. This clear-

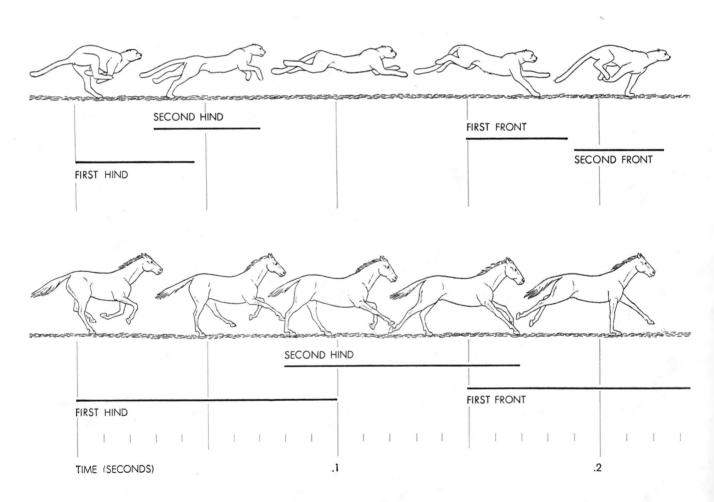

STRIDES OF THE CHEETAH AND THE HORSE in full gallop are contrasted in these illustrations. The sequence and duration of their footfalls, indicated by the horizontal lines under each animal, relate to the time-scale at bottom, which is calibrated in 10ths of a second. The cheetah has two unsupported periods, which account for about half its stride; the horse has one unsupported period,

ly increases stride length, but it also aggravates a problem. The body of the animal tends to rise when its shoulders and hips pass over its feet, and tends to fall when its feet extend to the front or rear. Carnivores offset this bobbing motion by flexing their ankles and wrists, thus shortening their legs. Ungulates do the same by sharply flexing the fetlock joint at the moment that the body passes over the vertical leg. The cheetah, a long-striding back-flexer, supplements its wrist-flexing by slipping its shoulder blade up its ribs about an inch, and thus achieves a smooth forward motion.

Since running is in actuality a series of jumps, the length of the jump must be reckoned as another important increment in the length of the stride. Hoofed runners have one major unsupported period, or jump, in each stride: when the legs are gathered beneath the body. The galloping carnivore has two major unsupported periods: when the back is flexed, and again when it is extended. In the horse all of these anatomical and functional adaptations combine to produce a 23-foot stride. The cheetah, although smaller, has a stride of the same length.

Fast runners must take their long strides rapidly. The race horse completes about 2.5 strides per second and the cheetah at least 3.5. It is plain that the higher the rate of stride, the faster the runner must contract its muscles. One might infer that cursorial animals as a group would have evolved the ability to contract their muscles faster than other animals. Within limits that is true, but there is a general principle limiting the rate at which a muscle can contract. Assuming a constant load on the muscle fibers, the rate of contraction varies inversely with any of the muscle's linear dimensions; the larger muscle therefore contracts more slowly. That is why an animal with a larger body has a slower rate of stride and so loses the advantage of its longer length of stride.

The familiar mechanical principle of gear ratio underlies the fast runner's more effective use of its trim musculature. In the linkage of muscle and bone the gear ratio is equal to the distance between the pivot of the motion (the shoulder joint, for example) and the point at which the motion is applied (the foot) divided by the perpendicular distance between the pivot and the point at which the muscle is attached to the bone. Cursorial animals not only have longer legs; their actuating muscles are also attached to the bone closer to the pivot of motion. Their high-gear muscles, in other words, have short lever-arms, and this increases the gear ratio still further. In comparison, the anatomy of walking animals gives them considerably lower gear-ratios; digging and swimming animals have still lower gear ratios.

But while high gears enable an automobile to reach higher speed, they do

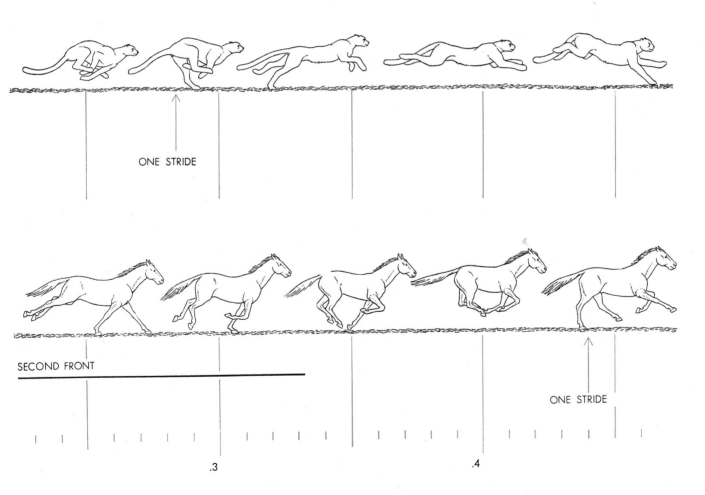

ONE STRIDE

SECOND FRONT

ONE STRIDE

.3   .4

which accounts for about a quarter of its stride. Although both the cheetah and the horse cover about 23 feet per stride, the cheetah attains speeds on the order of 70 miles per hour, to the horse's 43, because it takes about 3.5 strides to the horse's 2.5. The size of the horse has been reduced disproportionately in these drawings for the sake of uniformity in the stride-lines and time-scale.

SWIVELING SHOULDER BLADES of the horse and the cheetah add several inches to their stride length. The faster cheetah gains a further advantage from the flexibility of the spine, which in addition to adding the length of its extension to the animal's stride, adds the speed of its extension to the velocity if its travel. Horse's relatively longer leg partially compensates for its rigid spine.

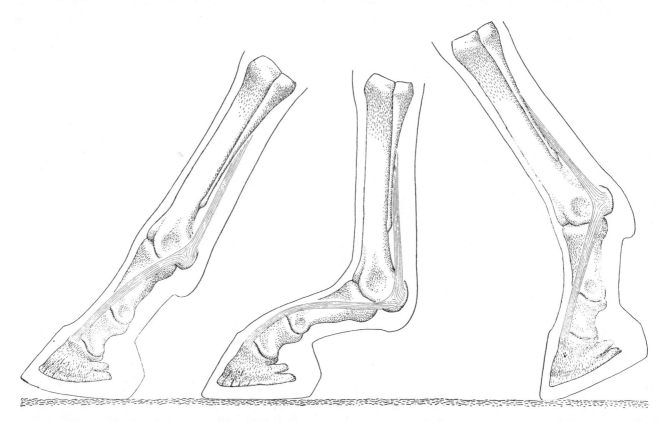

SPRING LIGAMENTS in the legs of horses, shown here, and other hoofed runners reduce the need for heavy muscles. Impact of the foot against the ground (*left*) bends the fetlock joint (*middle*) and stretches an elastic ligament (*shown in color*) that snaps back when the foot leaves the ground (*right*). The springing action at once straightens the foot and gives the leg an unward impetus.

so at the expense of power. The cursorial animal pays a similar price, but the exchange is a good one for several reasons. Running animals do not need great power: air does not offer much resistance even when they are moving at top speed. Moreover, as the English investigators J. M. Smith and R. J. G. Savage have noted, the animal retains some relatively low-gear muscles. Probably the runner uses its low-gear muscles for slow motions, and then shifts to its high-gear muscles to increase speed.

Since the speed at which a muscle can contract is limited, the velocity of the action it controls must be correspondingly limited, even though the muscle speed is amplified by an optimum gear-ratio. A larger muscle, or additional muscles, applied to action around the same joint can produce increased power but not greater speed. Several men together can lift a greater weight than one can lift alone, but several equally skilled sprinters cannot run faster together than one of them alone. The speed of a leg can be increased, however, if different muscles simultaneously move different joints of the leg in the same direction. The total motion they produce, which is represented by the motion of the foot, will then be greater than the motion produced by any one muscle working alone. Just as the total speed of a man walking up an escalator is the sum of his own speed plus that of the escalator, so the independent velocities of each segment of the leg combine additively to produce a higher total velocity.

The trick is to move as many joints as possible in the same direction at the same time. The evolution of the cursorial body has produced just this effect. By abandoning the flat-footed plantigrade posture in favor of a digitigrade or unguligrade one, the cursorial leg acquired an extra limb-joint. In effect it gained still another through the altered functioning of the shoulder blade. The flexible back of the cursorial carnivore adds yet another motion to the compound motion of its legs; the back flexes in such a way that the chest and pelvis are always rotating in the direction of the swinging limbs.

The supple spine of the carnivore contributes to stride rate by speeding up the motion of its body as well as of its legs. The spine is flexed when the runner's first hind foot strikes the ground, and by the time its second hind foot leaves the ground the animal has extended its spine and thus lengthened its body. In the brief interval when its hind

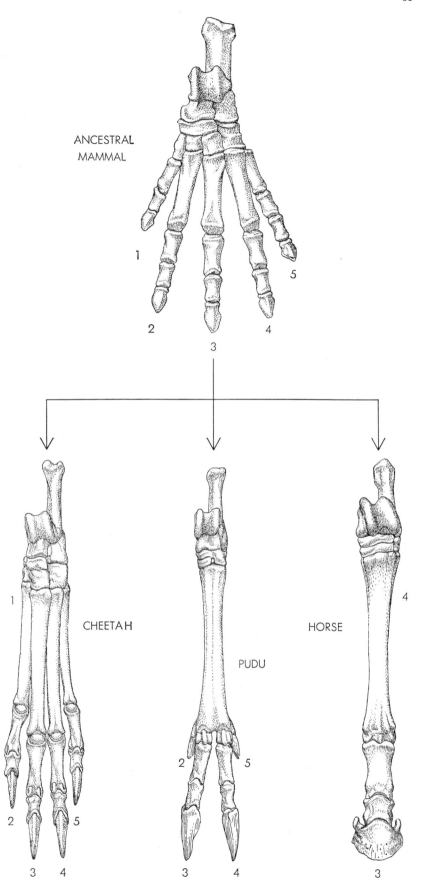

MODERN CURSORIAL FOOT EVOLVED from the broad, five-digited foot of an ancestral mammal (*top*). Lateral digits were lost and metatarsal bones, the longest in the foot, were further elongated. Resultant foot is lighter and longer. Pudu is a deer of the Andes.

feet are planted, the forequarters, riding on the extending spine, move farther and faster than the hindquarters. Similarly when the front feet are on the ground, the hindquarters move faster than the forequarters. So although the speed that the driving legs can impart to the forequarters or hindquarters is limited by their rate of oscillation, the body as a whole is able to exceed that limit. In a sense the animal moves faster than it runs. For the cheetah the advantage amounts to about two miles per hour—enough to add the margin of success in a close chase.

In addition to the obvious tasks of propelling the animal's body and supporting its weight, the locomotor muscles must raise the body to compensate for the falling that occurs during the unsupported phases of the stride. The load they must raise is proportional to the mass of the body, which is in turn proportional to the cube of any of its linear dimensions. A twofold increase in body length thus increases weight eightfold. The force that a muscle can exert, on the other hand, increases only as the square of its cross section. Thus against an eightfold increase of load, bigger muscles can bring only a fourfold increase of force. As body size increases, the capacity of the muscles to put the body in forward motion and to cause its legs to oscillate cannot quite keep up with the demands placed upon them. These factors in the nature of muscle explain why the largest animals can neither gallop nor jump, why small runners such as rabbits and foxes can travel as fast as race horses without having marked structural adaptations for speed and why the larger cursorial animals must be highly adapted in order to run at all.

If the bigger runners are to have endurance as well as speed, they must have not only those adaptations that increase the length and rate of their stride, but also adaptations that reduce the load on their locomotor structures and economize the effort of motion. In satisfying this requirement natural selection produced a number of large and fast runners that are able to travel for long distances at somewhat less than their maximum speed. In these animals the mass of the limbs is minimized. The muscles that in other animals draw the limbs toward or away from the midline of the body (the "hand-clapping" muscles in man) are smaller or adapted to moving the legs in the direction of travel, and the muscles that manipulate the digits or rotate the forearm have disappeared. The ulna in the forearm and the fibula in the shank —bones involved in these former motions—are reduced in size. The ulna is retained at the point where it completes the elbow joint, but elsewhere becomes a sliver fused to its neighbor; the fibula is sometimes represented only by a nubbin of bone at the ankle.

The shape of the cursorial limb embodies another load-reducing principle. Since the kinetic energy that must be alternately developed and overcome in oscillating the limb is equal to half the mass times the square of its velocity, the load on muscles causing such motions can be reduced not only by reducing the mass of the faster-moving parts of the limb but also by reducing the velocity of the more massive parts. Accordingly the fleshy parts of the limb are those close to the body, where they do not move so far, and hence not so fast, as the more distant segments. The lower segments, having lost the muscles and bones involved in rotation and in digit manipulation, are relatively light.

The rigor of design imposed by natural selection is especially evident in the feet of cursorial animals. The feet of other animals tend to be broad and pliable; the bones corresponding to those of the human palm and instep are rounded in cross section and well separated. In the foot of the cursorial carnivore, on the other hand, these bones are

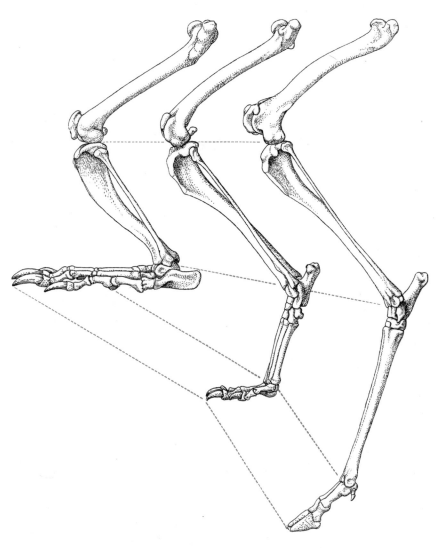

ADAPTATION OF THE LEG FOR SPEED is illustrated by the hind-leg bone of the slow badger (*left*), moderately fast dog (*middle*) and highly adapted deer (*right*). The lengthened metatarsus of the latter two has yielded a longer foot and an altered ankle posture that is better suited to running. The thigh bones of all three animals have been drawn to the same scale to show that the leg segments farthest from the body have elongated the most.

crowded into a compact unit, each bone having a somewhat square cross section. In the ungulates the ratio of strength to weight has been improved still further by reduction of the number of bones in the foot. The ungulates have tended to lose their lateral toes; sometimes the basal elements of the other toes are fused into a single bone. This process gave rise to the cannon bone: the shank of the hoofed mammals [*see illustration on page 31*]. In compensation for the bracing lost as the bones and muscles of their

lower limbs were reduced or eliminated, these animals evolved joints that are modified to function as hinges and allow motion only in the line of travel.

The burden on the muscles of hoofed animals is relieved by an especially elegant mechanism built into the foot. When the hoof of the running animal strikes the ground, the impact bends the fetlock joint and stretches certain long ligaments called the suspensory or springing ligaments [*see bottom illustration on page 30*]. Because the ligaments

are elastic, they snap back as the foot leaves the ground, thereby straightening the joint and giving the leg an upward push. Charles L. Camp of the University of California has found that these built-in pogo-sticks evolved from foot muscles at the time that the animals forsook river valleys for the open plains. The exchange was advantageous, for by means of this and the other adaptations, nature has reconciled the limitations of muscle mechanics with the exacting requirements of speed.

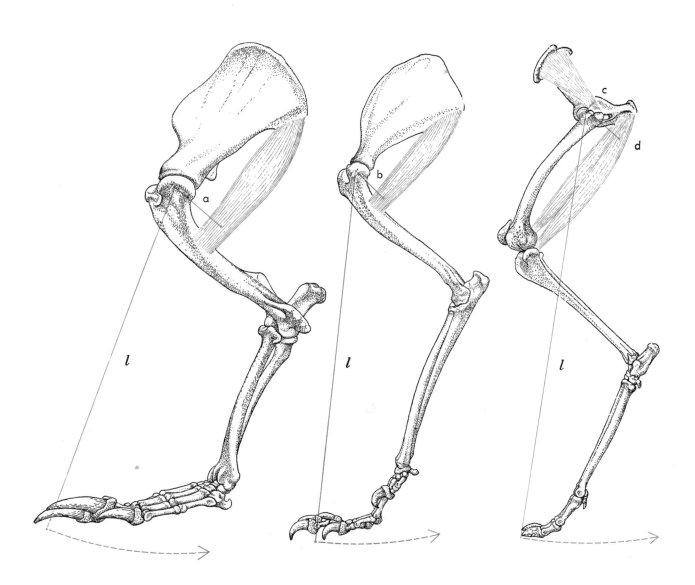

POWER AND SPEED are alternatively achieved in the badger (*left*) and the cheetah (*middle*) by placement of the teres major muscle. In the cheetah the small distance (b) between the muscle insertion and the joint it moves yields a higher rate of oscillation than in the badger, in which the distance (a) is greater. The higher oscillation rate, coupled with a longer leg (*l*), yields a faster stride. In the vicuña (*right*) the gluteus muscle (c) develops about five times the velocity but only a fifth the force of the larger semimembranosus muscle (d). The animal may use the latter to overcome inertia; the former, for high speed. Legs are not in same scale.

3

How Snakes Move

by Carl Gans
June 1970

They have four modes of progression, termed lateral, rectilinear, concertina and sidewinding. In lateral progression, the commonest mode, the snake uses its loops to push not downward but sideways

"The way of a serpent upon a rock" is one of four phenomena described in the Book of Proverbs as being beyond understanding. A snake undulates in a smooth and asymmetrical fashion, so that the size and curvature of the loops formed by its body change as the animal glides along. Because the parts of the snake's body that exert motive forces are never as clearly defined as those in species with limbs it is not obvious to a casual observer how such movement propels the animal forward. It would seem that the snake's undulations would be just as likely to move it backward or leave it thrashing in one place. Such considerations make the movement of snakes a challenging subject for investigation.

Slithering and related forms of motion employed by snakes are the consequence of the snakes' evolution. Snakes presumably evolved from animals that lost their appendages when they took up a burrowing existence. Since they lacked limbs, the entire body had to become an organ of locomotion. This function favored elongation of the trunk and necessitated a repackaging of the viscera. Both of these characteristics were further developed by most snakes when they emerged to radiate into a broad spectrum of terrestrial, arboreal and even aquatic niches. As the trunk lengthened it remained flexible because the number of bony segments in the vertebral column increased. Thus snakes have between 100 and 400 trunk vertebrae, a number that contrasts strikingly with the 16 or so found in a typical mammal. Ophidian trunk vertebrae are also joined by at least one extra set of articulations, and each vertebra bears a pair of ribs that reach down to form a flexible yet strong support for the sides of the elongated visceral cavity.

The increase in the number of the seg-

ments gives the snake its flexibility; it also complicates the snake's control problems. Some 30 years ago the late Walter Mosauer of the University of California at Los Angeles showed that a snake's vertebrae are joined in such a way that one vertebra can bend up and down in relation to another through approximately 28 degrees and swing from side to side through about 50 degrees. Although the bulk of tissue probably reduces a living snake's flexibility, the number of degrees through which a snake can bend is still remarkable. Even twisting movements can be approximated by a combination of up-and-down and side-to-side bending, in spite of the fact that the joints between any pair of vertebrae do not allow one segment actually to twist in relation to the other. The posture assumed by a snake at any one moment is established by the muscles that together with the skeleton form the axis of the body. The most important of these muscles are arrayed in 12 or more bundles that connect one vertebra to another and to the ribs on each side of the trunk, and in two to four bands that attach the skin to each side of the snake's main mass.

How does a snake utilize all these elements in order to move? An exact description of slithering requires that the position of each element at any given instant be specified. It is easily seen that investigators who seek such a detailed description must solve a truly staggering number of simultaneous equations. The task is complicated by the fact that as the snake travels its parts move and accelerate at different rates. The solution gains several magnitudes of complexity when one wants to know which of the snake's thousands of muscles are actually exerting the forces that induce and maintain motion.

One method of achieving a simpler solution is to work inward from the outside. The snake is treated as a special kind of "black box" capable of exerting forces on areas within an envelope of space considerably greater than its own volume. One can then deal with the snake's external responses without, for the moment, concerning oneself with the way they are generated internally. One first asks where and how these external forces are exerted; then it can be asked what internal forces and mechanisms are responsible for the externally observed efforts. Ultimately this approach allows investigation of the function of the specific muscle groups, and of the motor units and their control.

Since the immediate area of contact between the snake and the environment is the skin it is important to consider the texture of the skin's surface. The sides and back of most snakes are covered with longitudinal rows of lozenge-shaped and more or less rigid scales. Such scales are generally mounted, so that their trailing edges are free, on a flexible and accordion-pleated intermediate layer of skin that is often lined by a layer of muscle. The skin of a snake has a remarkable capacity for extension, as is evidenced when the animal swallows prey more than twice the diameter of its body. The snake's bottom surface is covered by a series of scutes: wide scales whose free trailing edges overlap like plates of armor. Each scute tends to be associated with a single row of scales on the snake's sides and back, and to a single vertebra and a single set of muscles; the scutes thus reflect the snake's internal organization. The side and back scales often have sensory pits of unknown function and may be ridged; the scutes do not have pits and are normally smooth.

It is generally accepted that the mis-

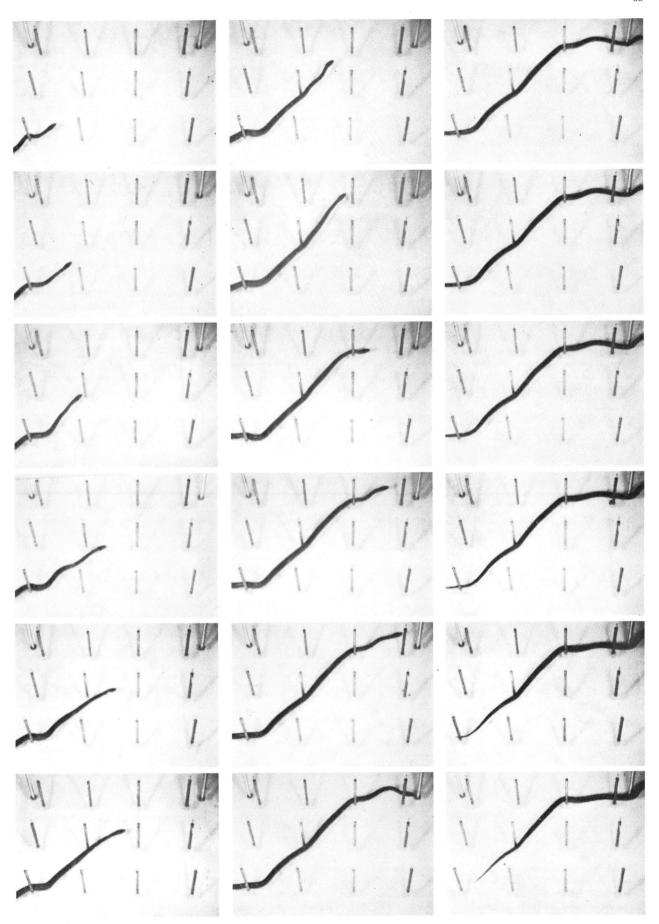

GARTER SNAKE slithering from left to right across an experimental surface demonstrates lateral undulation. These motion-picture frames show that the snake advances by forming curves with its body that it braces against the brass pegs projecting from surface.

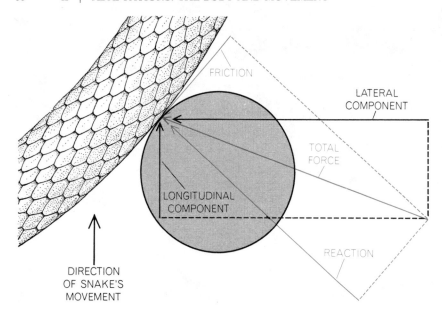

FORCES EXERTED by a snake's body against an object propel the animal. When the snake flexes its body, a reaction force is exerted that pushes the snake away from the peg, as indicated in diagram of forces (*color*). At the same time the snake slides past the peg because of pressure exerted elsewhere. This movement of its body against the peg produces a frictional force. Frictional force together with the reaction force equals the total force. This force consists of a propulsive longitudinal component and a lateral component.

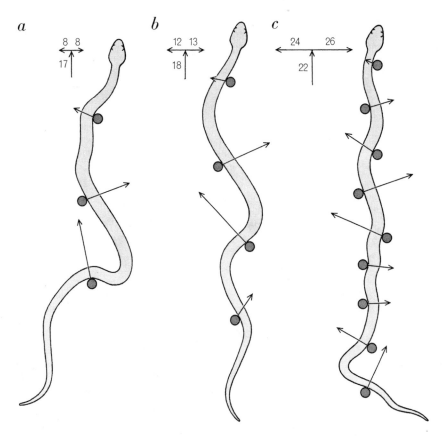

EFFICIENCY of lateral undulation, that is, the ratio of longitudinal forces propelling snake forward and lateral forces expended to the sides, declines as number of contact points increases. In *a* the snake propels itself forward by exerting force on two points at right and using one at left to balance the other two. Vector sum shows that forward-acting force is 17 grams and that 16 grams of force are expended to the sides. In *b* the snake uses four contact points and exerts 18 grams of force in a forward direction but expends about 25 grams of force to the sides. In *c* number of contact points has doubled and sum indicates that more force is expended to each side than is exerted in a forward direction.

cellaneous ways by which snakes propel themselves can be subdivided into four distinct patterns: lateral undulation, concertina progression, rectilinear locomotion and sidewinding. These categories are somewhat artificial. Most species of snakes can move in more than one of these patterns; indeed, a snake moves in more than one pattern at the same time. My colleagues and I at the State University of New York at Buffalo have investigated these patterns of movement. We are currently looking into their control mechanisms.

Lateral undulation is the kind of snake movement most frequently seen. It is the preferred movement not only of snakes but also of limbless lizards. Under certain conditions this kind of movement is even employed by limbed species, but their limbs are first folded against the body. The pattern can be shown to be the most primitive one in vertebrates. Lateral undulation involves the movement of the body along an S-shaped path. In an idealized situation the snake's track can be seen to be a single wavy line scarcely wider than the animal's body; each portion of the snake's trunk has traced this path. The snake begins to undulate laterally by bending the forward part of its body. This movement establishes a wavelike muscular contraction that travels down the snake's trunk.

How does this wave of muscular contraction cause the snake to move? Because human beings walk by exerting force directly on the surface under their feet, it is generally assumed that a snake too propels itself by somehow exerting a force directly on the ground. In actuality undulatory progression operates on a completely different principle. The wavelike contractions flowing down a snake's body do not propel it by acting on the ground under the snake's belly. Instead they cause the snake's body to exert force laterally on irregularities in the snake's path: small elevations and depressions, pebbles, tufts of grass and so on. The snake in effect pushes itself off from such points the way a man sitting in a chair with casters can push himself away from a desk. Since a snake can push only against a limited number of fixed objects, it sets its body in a corresponding pattern of loops, with the outside of each loop forming a contact point. The speed at which a loop progresses down the snake's body toward its tail equals the forward speed of the snake; the moving loops thus seem to be stationary with respect to the ground.

This analysis of lateral undulation has

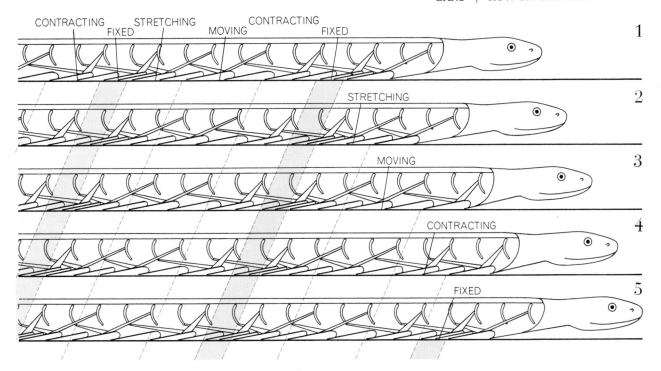

CONTRACTING STRETCHING CONTRACTING
FIXED MOVING FIXED

1

STRETCHING

2

MOVING

3

CONTRACTING

4

FIXED

5

RECTILINEAR LOCOMOTION enables a snake to advance in a straight line as it stalks prey or crosses a flat surface. In *1* the snake pulls together two sets of scutes, its large abdominal scales, and fixes them against the ground. By pulling against these fixed zones that have established frictional contact with the ground the snake propels itself forward. For this task it tenses the bands of muscle running from the scutes back toward the ribs. In *2* the fixed zones have moved toward the snake's tail (*colored bars*). In order to move the fixed zones the advancing snake picks up the forward scutes in each zone, using muscles that run from scutes forward to ribs, and pulls them clear of the ground (*stretching areas*). Simultaneously it pulls additional scutes forward from behind the fixed zones (*contracting areas*). In *3* snake has begun to contract scutes near its head. In *4* these scutes are being gathered into a new fixed zone, while the hindmost fixed zone has disappeared. In *5* two zones are again fixed at more forward sites and cycle is complete.

been confirmed by Hans W. Lissmann and Sir James Gray in a series of experiments they conducted to characterize the forces exerted against the ground by certain snakes. Their snakes crawled across a smooth board over which hung a regular array of heavy, pendulum-like pegs. Any lateral force exerted against a peg would cause it to swing, indicating both the magnitude of the applied force and its direction. Calculations of these forces showed them to be consistent with the force and speed of a snake's movement. When the pegs were taken away, the snakes were totally unable to move by lateral undulation; they slipped about on the smooth board because they had no objects to exert lateral force against.

Although a snake propels itself by exerting force laterally, it does apply other forces to the ground. One of them is a vertical force induced by the weight of the animal. Another is the horizontal force due to friction. Only the frictional force is exerted by muscle activity; it is the result of the movement produced when a snake curves its trunk in order to apply propelling forces against surrounding objects.

What effect do these forces have on a snake's movement? Friction is produced by lateral undulation in two ways: by the underside of the body sliding along the ground and by the flanks of the body sliding along objects to the side. The magnitude of such sliding friction equals the force pressing the two surfaces together multiplied by a coefficient determined by their nature. In the first instance this amounts to the weight of the snake times the coefficient of friction between the snake's belly skin and the ground. In the second instance it is a function of the force exerted against the object to the side times the coefficient between the flank skin and the object.

In both instances the frictional forces act in opposition to the force inducing motion; thus in lateral undulation friction has to be overcome by muscular effort. This being so, there is a selective evolutionary pressure to reduce the frictional coefficient of the snake's skin, and various kinds of snake scales have been shown to have the low frictional coefficient of about .35. That friction is unnecessary to the locomotion of snakes can be demonstrated by placing a snake in a virtually frictionless experimental environment: a smooth surface lubricat-

ed with powder and studded with vertical pegs that are fixed in position but able to rotate. The snake will traverse such a surface even more rapidly than it can cross one studded with nonrotating pegs.

Continuous forward progression by a snake requires a minimum of three contact sites at all times. Moreover, the contact sites must have a particular spatial arrangement. The snake pushes against two of the sites to generate force and uses the third to balance the forces produced at the other two so that its body can move in a particular direction. If the snake is to move, the contact points need to be on both sides of its body and the forces on the contact points must be exerted at an acute angle to the rear. Such considerations suggest that as the snake moves from one contact point to another it shifts its position slightly, and this is confirmed in films we have made. The transition from one set of three contact points to the next must be made quickly; otherwise the snake must utilize four or more contact points at a time and will lose efficiency.

The minimum number of three contact points represents a compromise between two opposing factors. One factor is the snake's flexibility. Since a snake is notably flexible it must expend a certain amount of energy to make its body rigid in order to transmit force from one contact point to another. The amount of energy applied for this purpose can be reduced if the number of contact points is increased; less energy is needed to maintain rigidity over several short distances than is needed over a few long ones. Yet the greater the number of contact points, the shorter the amplitude of the snake's curves, the wider the angle of force application and the greater the waste of force in the snake's direction of motion. The snake balances the requirement for internal force transmission against the requirement for efficient force application by establishing a number of contact points that is consistent with both. The number of points is also affected by the "desired" speed, the rela-

tive elongation (diameter divided by length) of the body and the spacing of objects to the side.

Experiments in which snakes of varying sizes are chased through a fixed array of pegs suggest that the speed of a snake is not proportional to its size. In fact, the limitations on the speed of snakes remain obscure. There may well be an optimum peg-spacing for a particular size of snake, where optimum means the facilitation of a maximum velocity rather than a minimum energy expenditure per traverse. No one has yet done experiments with different peg-spacings to see if such spacings would actually affect the snakes' velocities. Nor have there been studies relating speed to work output or speed to the time or distance a snake is capable of traveling.

Such questions about peg-spacings and efficient movement lead one to ask

further questions about control mechanisms. As we have seen, the body of a snake can twist around its long axis to some extent. Thus a snake is capable of lifting a loop over one contact point and making use of another, an ability that enables the snake to select advantageous contact points in its environment. What environmental factors affect such loop placement? Experiments indicate that the key factor may well be the relative resistance of contact points. A snake moving by lateral undulation through grass or sand will increase the size of its loops in order to displace loose surface material until it can exert the maximum force possible under the circumstances. Yet how does the snake monitor information about such resistance, and what is the feedback sequence that shifts the information from one segment of the snake's body to the next as the snake

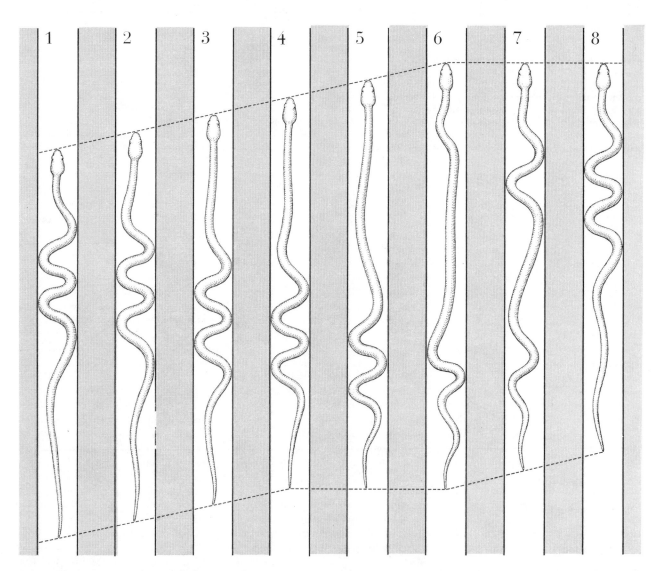

CONCERTINA MOVEMENT enables a snake to move in a narrow channel. In *1* the snake begins to move by bracing S-shaped loops against sides of channel. Snake then begins to extend its head and forebody (*2*) while pushing against passage walls until forward extension is complete (*6*). In order to bring tail section up snake forms a new S-curve behind its head (*7*) and braces against the sides of channel. Snake can now bring tail section along by pulling it forward and forming more loops against the passage walls.

moves past a particular contact point?

The lateral undulatory progression of a snake has the inherent advantage that the capacity for establishing staggered, backward-traveling waves of muscular contraction is phylogenetically much older than snakes. Snakes have only had to develop a feedback sequence for the control of such waves. This kind of progression is also effective because it permits the snake's body to travel at a relatively constant speed with only limited changes in its velocity. Yet lateral undulation does not work under all circumstances. It is useless for traveling down a tunnel with parallel sides: no backward-directed forces can be exerted against the sides of the tunnel. It is also ineffective on a flat surface that lacks elevations and depressions, and on a rounded surface such as a branch or a fallen tree trunk. Indeed, some snakes find it difficult to use lateral undulation at all. This mode of locomotion is well suited to relatively slender snakes, but it is less effective for short, stout ones. When a snake species has the option of invading an environmental niche that requires a relatively stout body, there is immediate selective pressure for a substitute method of locomotion.

An alternative is rectilinear movement. This mode of locomotion differs from lateral undulation in two respects: it involves the application of force somewhat downward instead of laterally, and it is effective only if friction is established between the snake's skin and the ground. Rectilinear locomotion is made possible by the very loose skin found in many snakes (perhaps as a result of such a specialization as prey constriction). In order to move in this mode the snake fixes several series of scutes and starts to move the skin between them. For example, it pulls together a series of scutes near its head, fixes them against the ground and then moves the rest of its body with respect to this fixed zone and several similar zones behind it. As the body of the snake moves forward the skin is stretched, pulling the forwardmost scutes of each series out of contact with the ground, while additional scutes are continuously pulled up to the rear edge of the series. In this way a constant length of belly surface remains fixed to the ground. Normally a snake fixes two or three of these zones to the ground at once, and they can be seen to move continuously to the rear as the snake progresses [see illustration on page 37].

In rectilinear locomotion the two sides of the snake must move symmetrically rather than in alternation. X-ray motion pictures have confirmed an earlier observation that in this mode of movement the ribs and the vertebrae do not move with respect to one another. The scutes are shifted forward by slender muscles that stretch forward from the sides of each scute to points high on the side of each rib. A second set of muscles runs to the rear at a shallower angle to attach the skin to the bottom ends of the ribs. Contraction of the muscles in the first group pulls the skin forward and up, out of sliding contact with the ground; contraction of the stouter and mechanically better-placed muscles of the second group accelerates the snake's mass and maintains its constant forward movement.

The fact that rectilinear locomotion relies on friction places a certain limitation on this form of movement. The particular form of friction involved is called static friction. Static friction is defined as the force that must be applied parallel to the contact surface between two objects in order to start them sliding past each other. Then static friction becomes sliding friction. It follows that a snake can only exert a force parallel to the ground that is less than the force of static friction; otherwise the animal will slip. A snake in rectilinear movement must accelerate smoothly, because even a temporary imbalance may shift a group of scutes from static to sliding contact and make the effort ineffective.

Even those snakes that have achieved the necessary body form to move by rectilinear progression find this mode of locomotion limiting because of the need to keep some parts in static contact. Rectilinear progression enables the snake to cross flat surfaces and to advance in a straight line when it is stalking prey. Such progression, however, is slow because each scale must establish stationary contact with the ground before it can transmit horizontal forces. It seems that motor control is again of critical importance in keeping the contact zones from sliding.

Most snakes that do not have the muscle and bone structures necessary for rectilinear progression can still use static friction in locomotion by employing concertina progression. In order to move in this way the snake draws itself into an S-shaped curve similar to the posture assumed in lateral undulation, and sets the curved portion of its body in static contact with the ground. Motion begins when the head, the neck and the forward part of the body are extended by forces transmitted to the ground in the zone that remains in stationary contact. These forces produce movement by acting against the force of friction generated by the weight of the snake's body. This reserve of friction is called the static-friction reservoir. It should be noted that the area of contact between the snake's body and the ground does not affect the "capacity" of the reservoir. Friction is a function solely of the force pressing two surfaces together multiplied by a coefficient that represents

SCUTES on the belly of this snake, a Japanese species called the *awo daishu* (blue general), correspond to rows of back and side scales. This pattern is found in all species of snakes.

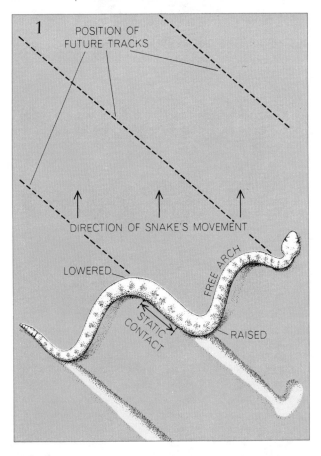

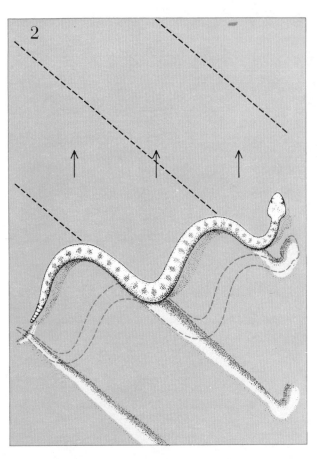

SIDEWINDING enables desert vipers to move rapidly. In *1* a snake with tail in first track (*lower left*) has lifted head and forebody from

hooked second track (*lower right*) and arched forward to begin a third track. In *2* head has risen from this new track at

their roughness. The force available for acceleration in the static-friction reservoir remains constant as long as the snake continues to support its weight on the stationary zone.

After the front end of such a snake has moved forward a short distance it stops. This establishes a new zone of stationary contact in which horizontal forces are exerted against the ground, and the rear end of the snake's body is pulled forward. The concertina progres-

sion of a snake is thus rather like the locomotion of an inchworm [*see illustration on page 38*].

As long as the moving portion of the snake's body can be lifted out of contact with the ground, the ratio of the moving portion to the stationary one depends only on the snake's stability. Yet if the moving parts slide forward or are dragged along they will induce sliding friction that must be overcome. Moreover, since each portion of the

snake's weight that induces sliding friction is unavailable for maintaining the static-friction reservoir, the snake will attempt to keep such areas to a minimum. It is therefore possible to predict the approximate fraction of a particular snake's body that can be kept in sliding motion over a particular kind of surface.

Snakes sometimes supplement weight-induced friction by muscular force, thereby enlarging the static-friction res-

SIDEWINDER (*Crotalus cerastes*) undulates at an angle of about 60 degrees to its direction of travel. This angling helps the snake, a western

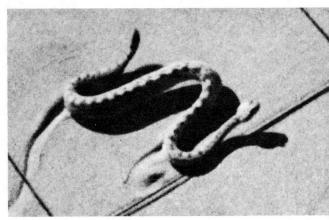

U.S. species, to avoid slipping. First the sidewinder sweeps out such a large area that it is likely to encounter objects to brace

Within image 1 labels: POSITION OF FUTURE TRACKS; DIRECTION OF SNAKE'S MOVEMENT; LOWERED; FREE ARCH; STATIC CONTACT; RAISED

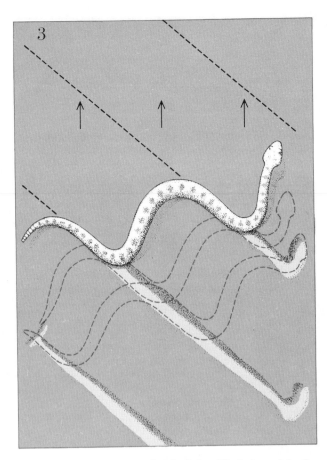

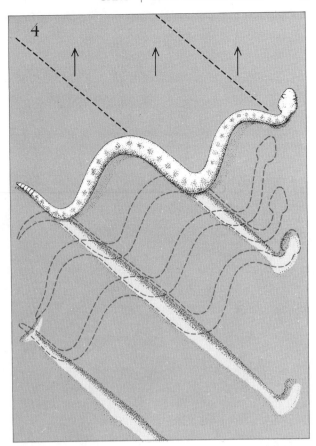

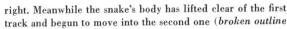

right. Meanwhile the snake's body has lifted clear of the first track and begun to move into the second one (*broken outline of snake's body indicates earlier position*). In *3* snake's body lies in both the second and the third track. In *4* a fourth track has been started.

ervoir. Such modified concertina progression is often used to traverse a straight-sided channel or to climb tree branches. In the first instance the snake widens the amplitude of its S-shaped loops, actively pushing them into contact with the channel's walls. In the second instance the snake forces its stationary surfaces into contact by constricting around a branch. Such enhancement of concertina progression probably arose among species that were already adapt-ed for it through the accident of having an overdeveloped axial musculature for the constriction of prey.

This observation pays an unexpected literary dividend. In "The Adventure of the Speckled Band" Sherlock Holmes solves a murder mystery by showing that the victim has been killed by a Russell's viper that has climbed up a bell rope. What Holmes did not realize was that Russell's viper is not a constrictor. The snake is therefore incapable of concer-tina movement and could not have climbed the rope. Either the snake reached its victim some other way or the case remains open.

Concertina progression is almost as common as lateral undulation, and one often sees a long snake combining the two kinds of movement. Various species of rat snake use this combination of movements to climb trees; they use the irregularities in the bark as channels. Many such snakes have a double keel on

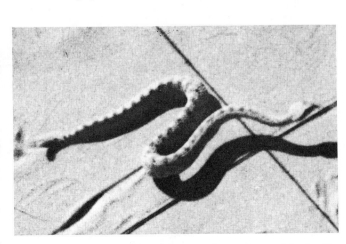

against. As the snake advances it also pushes sand back with its body until the sand piles up (*ridges in tracks*) and resists

further pressure. In this way the sidewinder is able to arrest the backward slippage of its loops and move its entire body forward.

SKELETON of *awo daishu* in this X-ray photograph has more than 300 vertebrae and 400 ribs. A large number of vertebrae enables a snake to bend and twist through many degrees.

their belly. The structure enables the snake to push laterally against irregularities in the surface it is climbing. Certain burrowing snakes increase their propulsive force by digging in their tail and then straightening their trunk.

Since concertina progression is effected by a bending of the body, it proceeds by asymmetrical contractions. Observation confirms that these intermittent contraction waves move along the body toward the tail. It thus appears that concertina progression is essentially a low-speed pattern of movement, because the animal must pause while it brings its rear end forward so that it can proceed again. Although concertina and rectilinear progression are suited to surfaces where lateral undulation would be ineffective, the price is lack of speed.

For pursuit—or escape—snakes need a mechanism that enables them to employ static friction without sacrificing speed. What is needed is a means of locomotion that keeps the zones that are in static contact from slipping, allows contraction

waves to pass continuously and regularly down the snake's body and prevents minor irregularities in the surface from affecting the pattern. Such a form of locomotion is sidewinding, which was first properly described by Walter Mosauer in 1932. A sidewinding snake achieves firm static contact by moving so that its body lies almost at right angles to the direction of its travel. In this orientation the snake is more likely to encounter rocks and other slight irregularities against which it can brace itself than it would if it were moving straight ahead, for the reason that it is sweeping out an area many times wider than the width of its loops.

The track of a sidewinding snake that has traversed a smooth flat surface appears as a series of straight parallel lines, each inclined some 60 degrees to the snake's direction of motion and each about as long as the snake itself. The rear end of each line is bent into a short, forward-pointing hook. The track of the scutes is well defined; slippage, if any,

can therefore only be sideways.

The series of actions producing these tracks can be visualized by assuming that the snake is lying so that its tail points about 60 degrees away from the direction in which the animal is going to move. The snake's head is raised and turned through an obtuse angle, so that it faces in the direction of travel. Only the bend of the snake's neck is in contact with the ground; this produces the hook at the rear end of each track.

As the snake starts to move forward it lifts its head and neck off the ground. In order to reach the next track the forward part of the body has to curve in a smooth loop. The extension of the snake's front end continues until approximately a quarter of the trunk is cantilevered out of contact with the ground. The head then arches downward as the cantilevered trunk remains off the ground; in making the first contact the neck bends at the next track of the sequence, so that another hook is produced. Successive sections of the body and then the tail follow along the new track, which parallels the preceding one. Considerably before the tail has been pulled into the second track the snake's front end starts into the third track. The body of a sidewinding snake thus lies on two or three separate tracks, with the body parts between tracks held off the ground.

Although the snake's body swings through loops that are reminiscent of lateral undulations, one can see that the force-transmission pattern is more like that of the concertina sequence. The force for the initial acceleration of the front part of the snake, and for maintaining the velocity of the moving parts, must be transmitted by static friction of the snake's belly. It is this pattern that gives sidewinding its peculiar advantages. When a nonsidewinding snake travels over loose sand, its lateral force shifts the sand and its track becomes a trough. Moreover, a snake that exerts a lateral force exceeding the resistance capacity of the force points tends to slip sideways instead of moving forward. Similarly, a snake in rectilinear locomotion will just dig in, slipping within its track, if the force exerted by the contact zones exceeds the limit of the static-friction reservoir. For the sidewinder such slippage proceeds sideways and causes the snake's body to dig in, piling up sand along its length and increasing the forces that produce forward movement. In other words, a slipping sidewinder quickly stabilizes itself by converting sliding friction into static friction.

H. Mendelssohn of the University of

Tel Aviv and I are currently analyzing the feedback mechanisms that control sidewinding. The energy required for sidewinding is apparently reflected in the height to which the traveling loops are cantilevered off the ground. The loop height is lowest for those surfaces that are smoothest and these therefore require the least amount of energy. Indeed, when the snakes are allowed to travel over a smooth, low-friction surface such as a polished terrazzo floor, they will slide rather than lift the moving parts of their body. If their tracks were visible, they would tend to be connected. When the snakes traverse a smooth but high-friction surface such as sandpaper, the tracks are quite separate. A rough surface such as a layer of crushed, sharp-edged aggregate causes a sidewinding snake to lift its loops quite high. Frame-by-frame analysis of films suggests that the loops continuously change their height.

Sidewinding has the further advantage that the contraction waves can start at the neck and pass down the snake's lateral musculature. Each motor sequence is thus continuous even though parts of the body stop and start. The snake is accordingly able to cross relatively flat areas, liberated from the need to search for lateral irregularities to propel itself.

Sidewinding seems to be a method of locomotion available to any kind of snake. Various conditions elicit it, and no special structural modification is required. Yet although many snakes sidewind under stress, only a few species do it effectively, and the beautiful control and minimal energy utilization described here were observed only in desert vipers. Such snakes seem to have found this method advantageous for traveling quickly over flat, sandy and rocky surfaces. Sidewinding would seem to be particularly useful when the crossing has to be made at midday, when speed and minimum contact may prevent overheating.

The various sidewinders differ in their track angles and in the number of tracks they occupy at one time. One of the most spectacular of the movement patterns that have been observed so far is seen in juvenile specimens of the Southwest African desert viper, a species that has evolved a special escape sequence when it is faced with heat stress. The muscle-contraction waves travel down the snake's body at very high speed, and some specimens jump completely off the ground from track to track [see illustration at right].

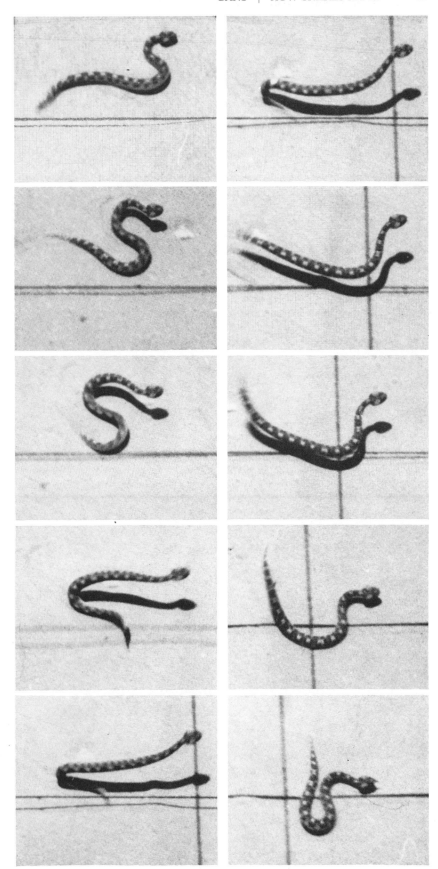

JUMPING SNAKE, the African desert viper *Bitis caudalis,* has developed this variation of the sidewinding pattern in order to escape intense heat. In the first of a series of frames from a motion-picture film this snake begins its leap by sending muscular contractions down its body toward its tail. In fourth frame down the impulses have lifted half of the snake from the ground. In the next two frames all of the snake except the tail is in the air. In the last frame the snake lands on a black line that is part of a measuring grid.

4

The Antiquity
of Human Walking

by John Napier
April 1967

*Man's unique striding gait may be the most significant
ability that sets him apart from his ancestors. A big-toe
bone found in Tanzania is evidence that his ability dates
back more than a million years*

Human walking is a unique activity during which the body, step by step, teeters on the edge of catastrophe. The fact that man has used this form of locomotion for more than a million years has only recently been demonstrated by fossil evidence. The antiquity of this human trait is particularly noteworthy because walking with a striding gait is probably the most significant of the many evolved capacities that separate men from more primitive hominids. The fossil evidence—the terminal bone of a right big toe discovered in 1961 in Olduvai Gorge in Tanzania—sets up a new signpost that not only clarifies the course of human evolution but also helps to guide those who speculate on the forces that converted predominantly quadrupedal animals into habitual bipeds.

Man's bipedal mode of walking seems potentially catastrophic because only the rhythmic forward movement of first one leg and then the other keeps him from falling flat on his face. Consider the sequence of events whenever a man sets out in pursuit of his center of gravity. A stride begins when the muscles of the calf relax and the walker's body sways forward (gravity supplying the energy needed to overcome the body's inertia). The sway places the center of body weight in front of the supporting pedestal normally formed by the two feet. As a result one or the other of the walker's legs must swing forward so that when his foot makes contact with the ground, the area of the supporting pedestal has been widened and the center of body weight once again rests safely within it. The pelvis plays an important role in this action: its degree of rotation determines the distance the swinging leg can move forward, and its muscles help to keep the body balanced while the leg is swinging.

At this point the "stance" leg—the leg still to the rear of the body's center of gravity—provides the propulsive force that drives the body forward. The walker applies this force by using muscular energy, pushing against the ground first with the ball of his foot and then with his big toe. The action constitutes the "push-off," which terminates the stance phase of the walking cycle. Once the stance foot leaves the ground, the walker's leg enters the starting, or "swing," phase of the cycle. As the leg swings forward it is able to clear the ground because it is bent at the hip, knee and ankle. This high-stepping action substantially reduces the leg's moment of inertia. Before making contact with the ground and ending the swing phase the leg straightens at the knee but remains bent at the ankle. As a result it is the

heel that strikes the ground first. The "heel strike" concludes the swing phase; as the body continues to move forward the leg once again enters the stance phase, during which the point of contact between foot and ground moves progressively nearer the toes. At the extreme end of the stance phase, as before, all the walker's propulsive thrust is delivered by the robust terminal bone of his big toe.

A complete walking cycle is considered to extend from the heel strike of one leg to the next heel strike of the same leg; it consists of the stance phase followed by the swing phase. The relative duration of the two phases depends on the cadence or speed of the walk. During normal walking the stance phase constitutes about 60 percent of the cycle and the swing phase 40 percent. Although

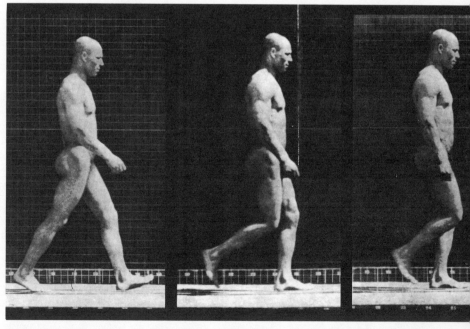

WALKING MAN, photographed by Eadweard Muybridge in 1884 during his studies of human and animal motion, exhibits the characteristic striding gait of the modern human.

the action of only one leg has been described in this account, the opposite leg obviously moves in a reciprocal fashion; when one leg is moving in the swing phase, the other leg is in its stance phase and keeps the body poised. Actually during normal walking the two phases overlap, so that both feet are on the ground at the same time for about 25 percent of the cycle. As walking speed increases, this period of double leg-support shortens.

Anyone who has watched other people walking and reflected a little on the process has noticed that the human stride demands both an up-and-down and a side-to-side displacement of the body. When two people walk side by side but out of step, the alternate bobbing of their heads makes it evident that the bodies undergo a vertical displacement with each stride. When two people walk in step but with opposite feet leading, they will sway first toward each other and then away in an equally graphic demonstration of the lateral displacement at each stride. When both displacements are plotted sequentially, a pair of low-amplitude sinusoidal curves appear, one in the vertical plane and the other in the horizontal [see *illustrations on next page*]. General observations of this kind were reduced to precise measurements during World War II when a group at the University of California at Berkeley led by H. D. Eberhart conducted a fundamental investigation of human walking in connection with requirements for the design of artificial legs. Eberhart and his colleagues found that a number of

functional determinants interacted to move the human body's center of gravity through space with a minimum expenditure of energy. In all they isolated six major elements related to hip, knee and foot movement that, working together, reduced both the amplitude of the two sine curves and the abruptness with which vertical and lateral changes in direction took place. If any one of these six elements was disturbed, an irregularity was injected into the normally smooth, undulating flow of walking, thereby producing a limp. What is more important, the irregularity brought about a measurable increase in the body's energy output during each step.

The Evidence of the Bones

What I have described in general and Eberhart's group studied in detail is the form of walking known as striding. It is characterized by the heel strike at the start of the stance phase and the push-off at its conclusion. Not all human walking is striding; when a man moves about slowly or walks on a slippery surface, he may take short steps in which both push-off and heel strike are absent. The foot is simply lifted from the ground at the end of the stance phase and set down flat at the end of the swing phase. The stride, however, is the essence of human bipedalism and the criterion by which the evolutionary status of a hominid walker must be judged. This being the case, it is illuminating to consider how the act of striding leaves its distinctive marks on the bones of the strider.

To take the pelvis first, there is a well-known clinical manifestation called Trendelenburg's sign that is regarded as evidence of hip disease in children. When a normal child stands on one leg, two muscles connecting that leg and the pelvis—the gluteus medius and the gluteus minimus—contract; this contraction, pulling on the pelvis, tilts it and holds it poised over the stance leg. When the hip is diseased, this mechanism fails to operate and the child shows a positive Trendelenburg's sign: the body tends to fall toward the unsupported side.

The same mechanism operates in walking, although not to the same degree. During the stance phase of the walking cycle, the same two gluteal muscles on the stance side brace the pelvis by cantilever action. Although actual tilting toward the stance side does not occur in normal walking, the action of the muscles in stabilizing the walker's hip is an essential component of the striding gait. Without this action the stride would become a slow, ungainly shuffle.

At the same time that the pelvis is stabilized in relation to the stance leg it also rotates to the unsupported side. This rotation, although small, has the effect of increasing the length of the stride. A familiar feature of the way women walk arises from this bit of anatomical mechanics. The difference in the proportions of the male and the female pelvis has the effect of slightly diminishing the range through which the female hip can move forward and back. Thus for a given length of stride women are obliged to rotate the pelvis through a greater

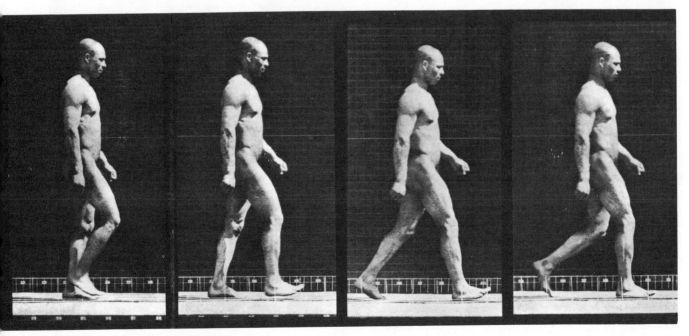

The free foot strikes the ground heel first and the body's weight is gradually transferred from heel to ball of foot as the opposite leg lifts and swings forward. Finally the heel of the stance foot rises and the leg's last contact with the ground is made with the big toe.

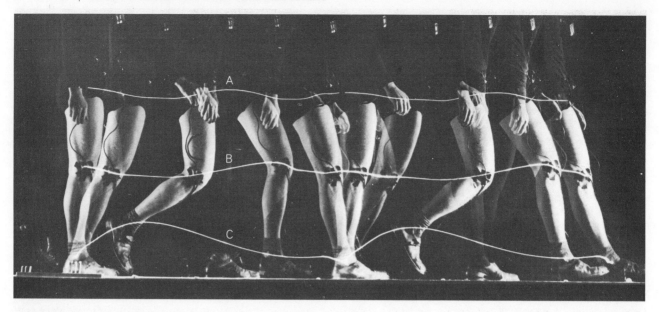

WALKING CYCLE extends from the heel strike of one leg to the next heel strike by the same leg. In the photograph, made by Gjon Mili in the course of a study aimed at improvement of artificial legs that he conducted for the U.S. Army, multiple exposures trace the progress of the right leg in the course of two strides. The ribbons of light allow analysis of the movement (see illustration below).

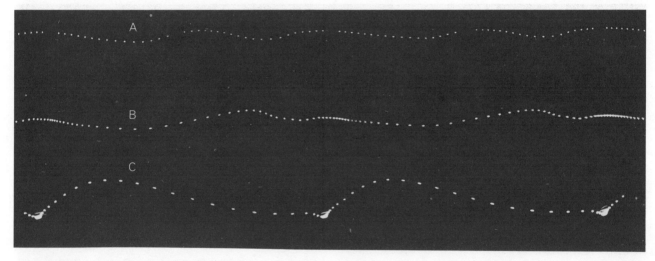

SINE CURVE described by the hip of a walking man was recorded on film by means of the experimental system illustrated above. An interrupter blade, passing in front of the camera lens at constant speed, broke the light from lamps attached to the walker into the three rows of dots. The speed of hip (a), knee (b) or ankle (c) during the stride is determined by measuring between the dots.

angle than men do. This secondary sexual characteristic has not lacked exploitation; at least in our culture female pelvic rotation has considerable erotogenic significance. What is more to the point in terms of human evolution is that both the rotation and the balancing of the pelvis leave unmistakable signs on the pelvic bone and on the femur: the leg bone that is joined to it. It is by a study of such signs that the walking capability of a fossil hominid can be judged.

Similar considerations apply to the foot. One way the role of the foot in walking can be studied is to record the vertical forces acting on each part of the foot while it is in contact with the ground during the stance phase of the walking cycle. Many devices have been built for this purpose; one of them is the plastic pedograph. When the subject walks across the surface of the pedograph, a motion-picture camera simultaneously records the exact position of the foot in profile and the pattern of pressures on the surface. Pedograph analyses show that the initial contact between the striding leg and the ground is the heel strike. Because the foot is normally turned out slightly at the end of the swing phase of the walking cycle, the outer side of the back of the heel takes the brunt of the initial contact [see illustration on opposite page]. The outer side of the foot contin-ues to support most of the pressure of the stance until a point about three-fifths of the way along the sole is reached. The weight of the body is then transferred to the ball of the foot and then to the big toe. In the penultimate stage of push-off the brunt of the pressure is under the toes, particularly the big toe. Finally, at the end of the stance phase, only the big toe is involved; it progressively loses contact with the ground and the final push-off is applied through its broad terminal bone.

The use of pedographs and similar apparatus provides precise evidence about the function of the foot in walking, but every physician knows that much the same information is recorded on the soles of everyone's shoes. Assuming that the shoes fit, their pattern of wear is a true record of the individual's habitual gait. The wear pattern will reveal a limp that one man is trying to hide, or unmask one that another man is trying to feign, perhaps to provide evidence for an insurance claim. In any case, just as the form of the pelvis and the femur can disclose the presence or absence of a striding gait, so can the form of the foot bones, particularly the form and proportions of the big-toe bones.

The Origins of Primate Bipedalism

Almost all primates can stand on their hind limbs, and many occasionally walk in this way. But our primate relatives are all, in a manner of speaking, amateurs; only man has taken up the business of bipedalism intensively. This raises two major questions. First, how did the basic postural adaptations that permit walking—occasional or habitual—arise among the primates? Second, what advantages did habitual bipedalism bestow on early man?

With regard to the first question, I have been concerned for some time with the anatomical proportions of all primates, not only man and the apes but also the monkeys and lower primate forms. Such consideration makes it possible to place the primates in natural groups according to their mode of locomotion. Not long ago I suggested a new group, and it is the only one that will concern us here. The group comprises primates with very long hind limbs and very short forelimbs. At about the same time my colleague Alan C. Walker, now at Makerere University College in Uganda, had begun a special study of the locomotion of living and fossil lemurs. Lemurs are among the most primitive offshoots of the basic primate stock. Early in Walker's studies he was struck by the frequency with which a posture best described as "vertical clinging" appeared in the day-to-day behavior of living lemurs. All the animals whose propensity for vertical clinging had been observed by Walker showed the same proportions—that is, long hind limbs and short forelimbs—I had proposed as forming a distinct locomotor group.

When Walker and I compared notes, we decided to define a hitherto unrecognized locomotor category among the primates that we named "vertical clinging and leaping," a term that includes both the animal's typical resting posture and the essential leaping component

in its locomotion. Since proposing this category a most interesting and important extension of the hypothesis has become apparent to us. Some of the earliest primate fossils known, preserved in sediments laid down during Eocene times and therefore as much as 50 million years old, are represented not only by skulls and jaws but also by a few limb bones. In their proportions and details most of these limb bones show the same characteristics that are displayed by the living members of our vertical-clinging-and-leaping group today. Not long ago Elwyn L. Simons of Yale University presented a reconstruction of the lemur-like North American Eocene primate *Smilodectes* walking along a tree branch in a quadrupedal position [see the article "The Early Relatives of Man," by Elwyn L. Simons; SCIENTIFIC AMERICAN Offprint 622]. Walker and I would prefer to see *Smilodectes* portrayed in the vertical clinging posture its anatomy unequivocally indicates. The fossil evidence, as far as it goes, suggests to us that vertical clinging and leaping was a major primate locomotor adaptation that took place some 50 million years ago. It may even have been the initial dynamic adaptation to tree life from which the subsequent locomotor patterns of all the living pri-

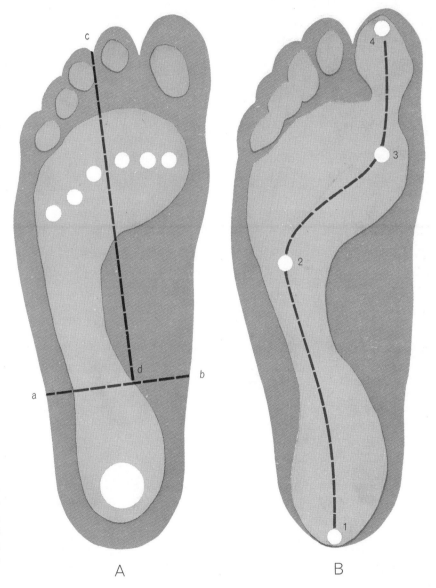

A   B

DISTRIBUTION OF WEIGHT in the human foot alters radically as action takes the place of rest. When motionless (*A*), the foot divides its static load (half of the body's total weight) between its heel and its ball along the axis *a–b*. The load on the ball of the foot is further divided equally on each side of the axis *c–d*. When striding (*B*), the load (all of the body's weight during part of each stride) is distributed dynamically from the first point of contact (*1, heel strike*) in a smooth flow via the first and fifth metatarsal bones (*2, 3*) that ends with a propulsive thrust (*4, push-off*) delivered by the terminal bone of the big toe.

mates, including man, have stemmed. Walker and I are not alone in this view. In 1962 W. L. Straus, Jr., of Johns Hopkins University declared: "It can safely be assumed that primates early developed the mechanisms permitting maintenance of the trunk in the upright position.... Indeed, this tendency toward truncal erectness can be regarded as an essentially basic primate character." The central adaptations for erectness of the body, which have been retained in the majority of living primates, seem to have provided the necessary anatomical basis for the occasional bipedal behavior exhibited by today's monkeys and apes.

What we are concerned with here is the transition from a distant, hypothetical vertical-clinging ancestor to modern, bipedal man. The transition was almost

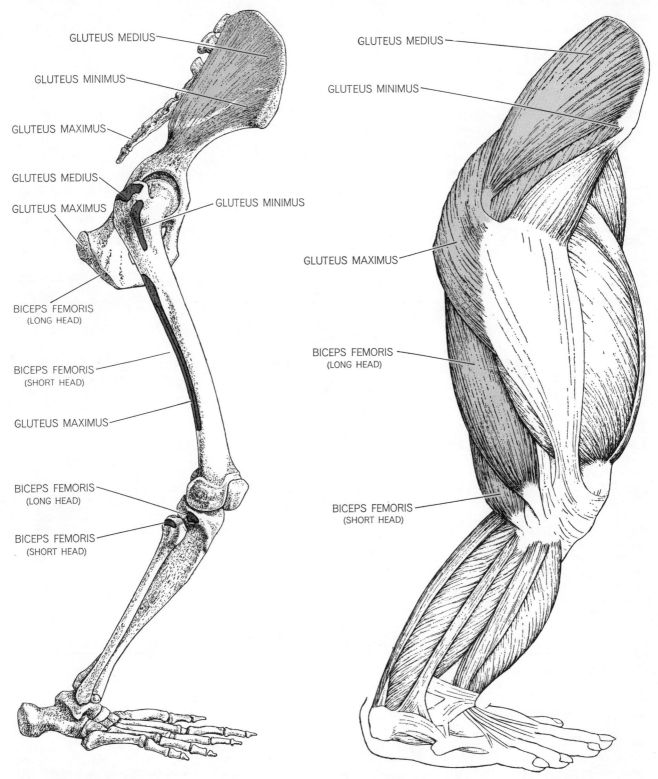

QUADRUPEDAL POSTURE needs two sets of muscles to act as the principal extensors of the hip. These are the gluteal group (the gluteus medius and minimus in particular), which connects the pelvis to the upper part of the femur, and the hamstring group, which connects the femur and the lower leg bones. Of these only the biceps femoris is shown in the gorilla musculature at right. The skeletal regions to which these muscles attach are shown in color at left. In most primates the gluteus maximus is quite small.

certainly marked by an intermediate quadrupedal stage. Possibly such Miocene fossil forms as *Proconsul*, a chimpanzee-like early primate from East Africa, represent such a stage. The structural adaptations necessary to convert a quadrupedal ape into a bipedal hominid are centered on the pelvis, the femur, the foot and the musculature associated with these bones. Among the nonhuman primates living today the pelvis and femur are adapted for four-footed walking; the functional relations between hipbones and thigh muscles are such that, when the animal attempts to assume a bipedal stance, the hip joint is subjected to a stress and the hip must be bent. To compensate for the resulting forward shift of the center of gravity, the knees must also be bent. In order to alter a bent-hip, bent-knee gait into

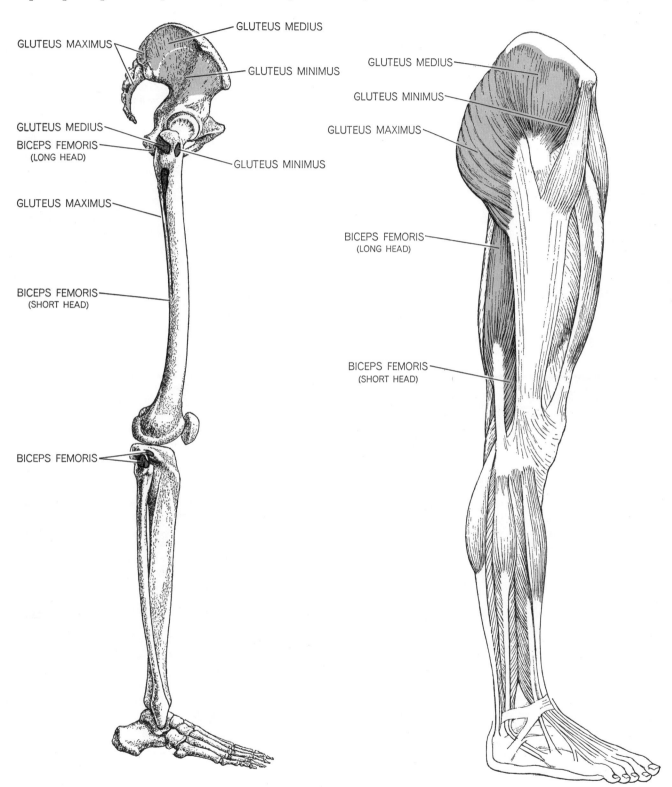

BIPEDAL POSTURE brings a reversal in the roles played by the same pelvic and femoral muscles. Gluteus medius and gluteus minimus have changed from extensors to abductors and the function of extending the trunk, required when a biped runs or climbs, has been assumed by the gluteus maximus. The hamstring muscles, in turn, now act mainly as stabilizers and extensors of the hip. At right are the muscles as they appear in man; the skeletal regions to which their upper and lower ends attach are shown in color at left.

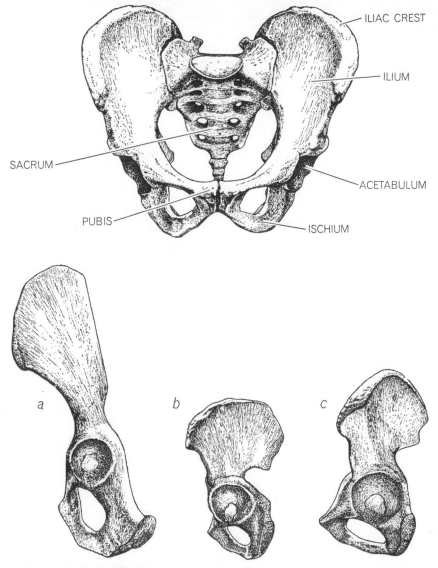

ILIAC CREST

ILIUM

SACRUM

ACETABULUM

PUBIS

ISCHIUM

a   *b*   *c*

COMPONENTS OF THE PELVIS are identified at top; the bones are those of the human pelvis. Below, ilium and ischium of a gorilla (*a*), of *Australopithecus* (*b*) and of modern man (*c*) are seen from the side (the front is to the left in each instance). The ischium of *Australopithecus* is longer than man's; this almost certainly kept the early hominid from striding in the manner of *Homo sapiens*. Instead the gait was probably a kind of jog trot.

man's erect, striding walk, a number of anatomical changes must occur. These include an elongation of the hind limbs with respect to the forelimbs, a shortening and broadening of the pelvis, adjustments of the musculature of the hip (in order to stabilize the trunk during the act of walking upright), a straightening of both hip and knee and considerable reshaping of the foot.

Which of these changes can be considered to be primary and which secondary is still a matter that needs elucidation. Sherwood L. Washburn of the University of California at Berkeley has expressed the view that the change from four-footed to two-footed posture was initiated by a modification in the form and function of the gluteus maximus, a thigh muscle that is powerfully

developed in man but weakly developed in monkeys and apes [*see illustrations on preceding two pages*]. In a quadrupedal primate the principal extensors of the trunk are the "hamstring" muscles and the two upper-leg muscles I have already mentioned: the gluteus medius and gluteus minimus. In man these two muscles bear a different relation to the pelvis, in terms of both position and function. In technical terms they have become abductor muscles of the trunk rather than extensor muscles of the leg. It is this that enables them to play a critical part in stabilizing the pelvis in the course of striding. In man the extensor function of these two gluteal muscles has been taken over by a third, the gluteus maximus. This muscle, insignificant in other primates, plays a sur-

prisingly unimportant role in man's ability to stand, or even to walk on a level surface. In standing, for example, the principal stabilizing and extending agents are the muscles of the hamstring group. In walking on the level the gluteus maximus is so little involved that even when it is paralyzed a man's stride is virtually unimpaired. The gluteus maximus comes into its own in man when power is needed to give the hip joint more play for such activities as running, walking up a steep slope or climbing stairs [*see illustration on page 52*]. Its chief function in these circumstances is to correct any tendency for the human trunk to jackknife on the legs.

Because the gluteus maximus has such a specialized role I believe, in contrast to Washburn's view, that it did not assume its present form until late in the evolution of the striding gait. Rather than being the initial adaptation, this muscle's enlargement and present function appear to me far more likely to have been one of the ultimate refinements of human walking. I am in agreement with Washburn, however, when he states that changes in the ilium, or upper pelvis, would have preceded changes in the ischium, or lower pelvis [see the article "Tools and Human Evolution," by Sherwood L. Washburn; SCIENTIFIC AMERICAN Offprint 601]. The primary adaptation would probably have involved a forward curvature of the vertebral column in the lumbar region. Accompanying this change would have been a broadening and a forward rotation of the iliac portions of the pelvis. Together these early adaptations provide the structural basis for improving the posture of the trunk.

Assuming that we have now given at least a tentative answer to the question of how man's bipedal posture evolved, there remains to be answered the question of why. What were the advantages of habitual bipedalism? Noting the comparative energy demands of various gaits, Washburn points out that human walking is primarily an adaptation for covering long distances economically. To go a long way with a minimum of effort is an asset to a hunter; it seems plausible that evolutionary selection for hunting behavior in man was responsible for the rapid development of striding anatomy. Gordon W. Hewes of the University of Colorado suggests a possible incentive that, acting as an agent of natural selection, could have prompted the quadrupedal ancestors of man to adopt a two-footed gait. In Hewes's view the principal advantage of bipedalism over quadrupedalism would be the free-

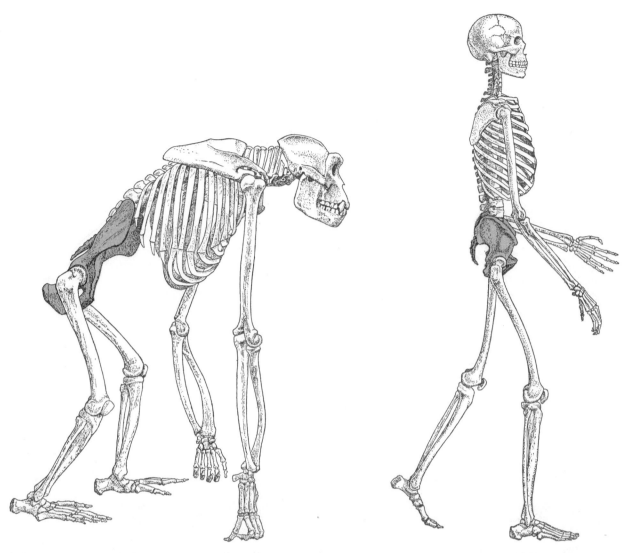

SHAPE AND ORIENTATION of the pelvis in the gorilla and in man reflect the postural differences between quadrupedal and bipedal locomotion. The ischium in the gorilla is long, the ilium extends to the side and the whole pelvis is tilted toward the horizontal (*see illustration on opposite page*). In man the ischium is much shorter, the broad ilium extends forward and the pelvis is vertical.

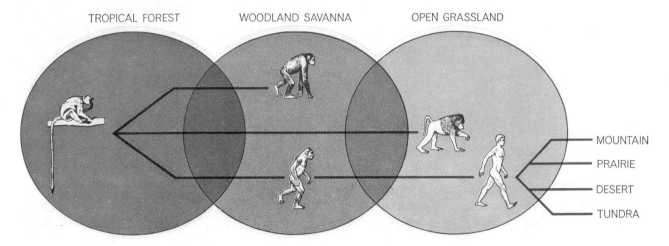

TROPICAL FOREST   WOODLAND SAVANNA   OPEN GRASSLAND

MOUNTAIN

PRAIRIE

DESERT

TUNDRA

ECOLOGICAL PATHWAY to man's eventual mastery of all environments begins (*left*) with a quadrupedal primate ancestor living in tropical forest more than 20 million years ago. During Miocene times mountain-building produced new environments. One, a transition zone between forest and grassland, has been exploited by three groups of primates. Some, for example the chimpanzees, have only recently entered this woodland savanna. Both the newly bipedal hominids and some ground-living quadrupedal monkeys, however, moved beyond the transition zone into open grassland. The quadrupeds, for example the baboons, remained there. On the other hand, the forces of natural selection in the new setting favored the bipedal hominid hunters' adaptation of the striding gait typical of man. Once this adaptation developed, man went on to conquer most of the earth's environments.

ing of the hands, so that food could be carried readily from one place to another for later consumption. To assess the significance of such factors as survival mechanisms it behooves us to review briefly the ecological situation in which our prehuman ancestors found themselves in Miocene times, between 15 and 25 million years ago.

The Miocene Environment

During the Miocene epoch the worldwide mountain-building activity of middle Tertiary times was in full swing. Many parts of the earth, including the region of East Africa where primates of the genus *Proconsul* were living, were being faulted and uplifted to form such mountain zones as the Alps, the Himalayas, the Andes and the Rockies. Massive faulting in Africa gave rise to one of the earth's major geological features: the Rift Valley, which extends 5,000 miles from Tanzania across East Africa to Israel and the Dead Sea. A string of lakes lies along the floor of the Rift Valley like giant stepping-stones. On their shores in Miocene times lived a fantastically rich fauna, inhabitants of the forest and of a new ecological niche—the grassy savanna.

These grasslands of the Miocene were the domain of new forms of vegetation that in many parts of the world had taken the place of rain forest, the dominant form of vegetation in the Eocene and the Oligocene. The savanna offered new evolutionary opportunities to a variety of mammals, including the expanding population of primates in the rapidly shrinking forest. A few primates— the ancestors of man and probably also the ancestors of the living baboons— evidently reacted to the challenge of the new environment.

The savanna, however, was no Eldorado. The problems facing the early hominids in the open grassland were immense. The forest foods to which they were accustomed were hard to come by; the danger of attack by predators was immeasurably increased. If, on top of everything else, the ancestral hominids of Miocene times were in the process of converting from quadrupedalism to bipedalism, it is difficult to conceive of any advantage in bipedalism that could have compensated for the added hazards of life in the open grassland. Consideration of the drawbacks of savanna living has led me to a conclusion contrary to the one generally accepted: I doubt that the advent of bipedalism took place in this environment. An environment neglected by scholars but one far better suited for the origin of man is the woodland-savanna, which is neither high forest nor open grassland. Today this halfway-house niche is occupied by many primates, for example the vervet monkey and some chimpanzees. It has enough trees to provide forest foods and ready escape from predators. At the same time its open grassy spaces are arenas in which new locomotor adaptations can be practiced and new foods can be sampled. In short, the woodland-savanna provides an ideal nursery for evolving hominids, combining the challenge and incentive of the open grassland with much of the security of the forest. It was probably in this transitional environment that man's ancestors learned to walk on two legs. In all likelihood, however, they only learned to stride when they later moved into the open savanna.

Moving forward many millions of years from Miocene to Pleistocene times, we come to man's most immediate hominid precursor: *Australopithecus*. A large consortium of authorities agrees that the shape of the pelvis in *Australopithecus* fossils indicates that these hominids were habitually bipedal, although not to the degree of perfection exhibited by modern man. A few anatomists, fighting a rearguard action, contend that on the contrary the pelvis of *Australopithecus*

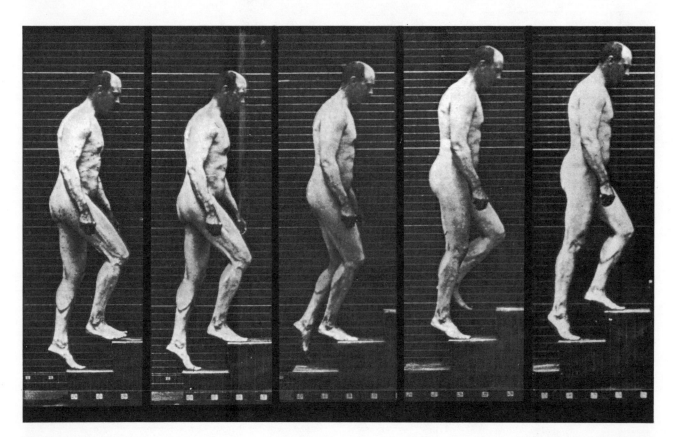

STAIR-CLIMBING, like running, is a movement that brings the human gluteus maximus into play. Acting as an extensor of the trunk, the muscle counteracts any tendency for the body to jackknife over the legs. Photographs are from Muybridge's collection.

shows that these hominids were predominantly quadrupedal. I belong to the first school but, as I have been at some pains to emphasize in the past, the kind of upright walking practiced by *Australopithecus* should not be equated with man's heel-and-toe, striding gait.

From Bipedalist to Strider

The stride, although it was not necessarily habitual among the earliest true men, is nevertheless the quintessence of the human locomotor achievement. Among other things, striding involves extension of the leg to a position behind the vertical axis of the spinal column. The degree of extension needed can only be achieved if the ischium of the pelvis is short. But the ischium of *Australopithecus* is long, almost as long as the ischium of an ape [*see illustration on page 51*]. Moreover, it has been shown that in man the gluteus medius and the gluteus minimus are prime movers in stabilizing the pelvis during each stride; in *Australopithecus* this stabilizing mechanism is imperfectly evolved. The combination of both deficiencies almost entirely precludes the possibility that these hominids possessed a striding gait. For *Australopithecus* walking was something of a jog trot. These hominids must have covered the ground with quick, rather short steps, with their knees and hips slightly bent; the prolonged stance phase of the fully human gait must surely have been absent.

Compared with man's stride, therefore, the gait of *Australopithecus* is physiologically inefficient. It calls for a disproportionately high output of energy; indeed, *Australopithecus* probably found long-distance bipedal travel impossible. A natural question arises in this connection. Could the greater energy requirement have led these early representatives of the human family to alter their diet in the direction of an increased reliance on high-energy foodstuffs, such as the flesh of other animals?

The pelvis of *Australopithecus* bears evidence that this hominid walker could scarcely have been a strider. Let us now turn to the foot of what many of us believe is a more advanced hominid. In 1960 L. S. B. Leakey and his wife Mary unearthed most of the bones of this foot in the lower strata at Olduvai Gorge known collectively as Bed I, which are about 1.75 million years old. The bones formed part of a fossil assemblage that has been designated by the Leakeys, by Philip Tobias of the University of the Witwatersrand and by me as possibly the earliest-known species of man: *Homo*

habilis. The foot was complete except for the back of the heel and the terminal bones of the toes; its surviving components were assembled and studied by me and Michael Day, one of my colleagues at the Unit of Primatology and Human Evolution of the Royal Free Hospital School of Medicine in London. On the basis of functional analysis the resem-

blance to the foot of modern man is close, although differing in a few minor particulars. Perhaps the most significant point of resemblance is that the stout basal bone of the big toe lies alongside the other toes [*see upper illustration on next page*]. This is an essentially human characteristic; in apes and monkeys the big toe is not exceptionally robust and

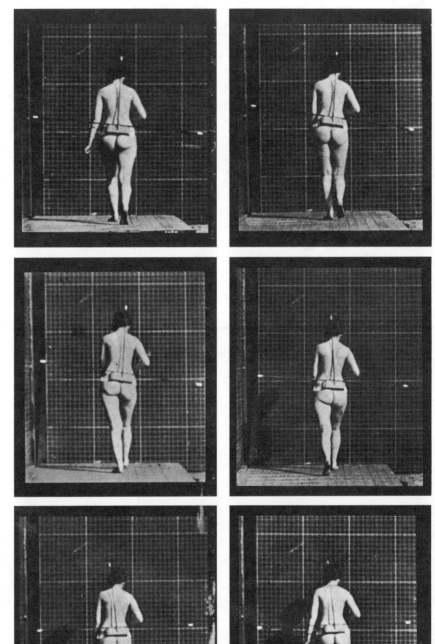

PELVIC ROTATION of the human female is exaggerated compared with that of a male taking a stride of equal length because the two sexes differ in pelvic anatomy. Muybridge noted the phenomenon, using a pole with whitened ends to record the pelvic oscillations.

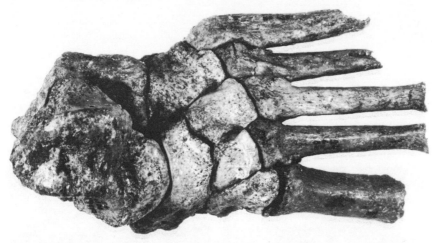

PRIMITIVE FOOT, complete except for the back of the heel and the tips of the toes, was unearthed from the lower level at Olduvai Gorge in Tanzania. Attributed to a very early hominid, *Homo habilis,* by its discoverer, L. S. B. Leakey, it is about 1.75 million years old. Its appearance suggests that the possessor was a habitual biped. Absence of the terminal bones of the toes, however, leaves open the question of whether the possessor walked with a stride.

BIG-TOE BONE, also discovered at Olduvai Gorge, is considerably younger than the foot bones in the top illustration but still probably more than a million years old. It is the toe's terminal bone (*bottom view at left, top view at right*) and bore the thrust of its possessor's push-off with each swing of the right leg. The tilting and twisting of the head of the bone in relation to the shaft is unequivocal evidence that its possessor walked with a modern stride.

diverges widely from the other toes. The foot bones, therefore, give evidence that this early hominid species was habitually bipedal. In the absence of the terminal bones of the toes, however, there was no certainty that *Homo habilis* walked with a striding gait.

Then in 1961, in a somewhat higher stratum at Olduvai Gorge (and thus in a slightly younger geological formation), a single bone came to light in an area otherwise barren of human bones. This fossil is the big-toe bone I mentioned at the beginning of this article [*see lower illustration at left*]. Its head is both tilted and twisted with respect to its shaft, characteristics that are found only in modern man and that can with assurance be correlated with a striding gait. Day has recently completed a dimensional analysis of the bone, using a multivariate statistical technique. He is able to show that the fossil is unquestionably human in form.

There is no evidence to link the big-toe bone specifically to either of the two recognized hominids whose fossil remains have been recovered from Bed I at Olduvai: *Homo habilis* and *Zinjanthropus boisei.* Thus the owner of the toe remains unknown, at least for the present. Nonetheless, one thing is made certain by the discovery. We now know that in East Africa more than a million years ago there existed a creature whose mode of locomotion was essentially human.

The Control of Walking

by Keir Pearson
December 1976

Recent experiments indicate that the mechanism by which the nervous system generates the rhythmic movements of the leg during walking is basically the same in animals as diverse as the cat and the cockroach

The scientific analysis of walking began a little more than 100 years ago as the result of a dispute, lingering from ancient times, about whether or not all four feet of a galloping horse are ever off the ground at the same time. In an effort to resolve the matter the pioneer action photographer Eadweard Muybridge in 1872 set up an electrically triggered series of cameras and succeeded in obtaining a fast sequence of photographs of a galloping horse at a racetrack in California. Reproductions of the photographs showing all four of the horse's feet off the ground appeared in many publications, including the October 19, 1878, issue of *Scientific American.* The clarity of these early pictures soon led Muybridge to realize that his photographic technique could be extended to analyze the walking movements of all kinds of animals in unprecedented detail. He subsequently photographed a host of walking and running animals, including cats, dogs, monkeys, camels, elephants, raccoons, pigs, kangaroos and birds. His remarkable collection of photographs was finally published in 1887 in a monumental 11-volume work titled *Animal Locomotion.*

How does an animal's nervous system generate the rhythmic walking movements of the leg that were so amply documented by Muybridge? The effort to answer the question began early in this century with the work of two British physiologists, C. S. Sherrington and T. Graham Brown. Sherrington first showed that rhythmic movements could be elicited from the hind legs of cats and dogs some weeks after their spinal cord had been severed. Since the operation had isolated from the rest of the nervous system the nervous centers that control the movement of the hind legs, it followed that the higher levels of the nervous system are not necessary for the organization of stepping movements.

Another notable discovery of this period, although it was not generally recognized as such until quite recently, was Graham Brown's demonstration in 1911 that rhythmic contractions of leg muscles, similar to those that occur during walking, could be induced immediately following transection of the spinal cord even in animals in which all input from sensory nerves in the legs had been eliminated. This discovery led Graham Brown to propose that mechanisms located entirely within the spinal cord are responsible for generating the basic rhythm for stepping in each leg. Earlier Sherrington had stressed the importance of sensory input from peripheral nerve receptors in initiating and coordinating such movements; he explained the generation of rhythmic leg movements by a series of "chain reflexes" (a reflex being a stereotyped movement elicited by the stimulation of a specific group of sensory receptors). Thus he conceived that the sensory input generated during any part of the step cycle elicits the next part of the cycle by a reflex action, producing in turn another sensory signal that elicits the next part of the cycle, and so on.

The two concepts of a spinal rhythm-generator for each leg and a triggering of leg movements by sensory signals are not incompatible. Indeed, Graham Brown considered that reflexes reinforced the centrally generated rhythm, a view that Sherrington later came to share. In 1924 Sherrington wrote: "From the observations of Professor Graham Brown, an intrinsic activity in spinal centres seems the essential nervous mechanism responsible for inconscient stepping, a central activity comparable with that of the respiratory centre in the bulb [the brain stem], and like the latter, highly regulable by reflex action." Recent experiments in a number of laboratories have yielded results that strongly support this dual view of the nervous mechanisms involved in walking.

For almost 40 years following the investigations of Sherrington and Graham Brown there were no significant advances in our understanding of how the nervous system controls walking. The major emphasis of research in this interlude was to determine the organization and the properties of the spinal reflexes. In general little effort was made to relate these findings to the normal behavior of the animal. Contemporary research on the nervous control of walking began in the 1960's with the recording of the patterns of muscle activity in a

EARLY STUDY of the walking and running movements of the horse is represented by the engraving on the following page, made from two rapid sequences of photographs obtained by the 19th-century photographer Eadweard Muybridge with the aid of an electrically triggered series of cameras; the illustration is reproduced from the opening page of the October 19, 1878, issue of SCIENTIFIC AMERICAN. The article accompanying the illustration, titled "A Horse's Motion Scientifically Determined," points out that the two series of "instantaneous photographs," forwarded by Muybridge following the appearance in the magazine of a short note about his experiments in California, show "the movement of the horse 'Abe Edington,' [the first] while walking . . . ; the second showing the same horse while trotting. . . . The exposure for each negative was about the two thousandth part of a second. The vertical lines on the background are twenty-eight inches apart; the heavy horizontal line represents the level of the track; the others mark elevations of four, eight, and twelve inches respectively. . . . It will be seen that the walking horse always has two feet on the ground, and, for a brief interval in each stride, three feet. The positions of the feet shown in Figs. A and E indicate a stride of four feet four inches. When trotting . . . the stride of the same horse is over 18 feet. Figs. 1 to 12 show the latter motion. In Figs. 4 and 5, and again in 9 and 10, the horse is entirely off the ground, literally flying through   a r. . . . The most careless observer of these figures will not fail to notice that the conventional figure of a trotting horse in motion does not appear in any of them, nor anything like it. Before these pictures were taken no artist would have dared to draw a horse as a horse really is when in motion, even if it had been possible for the unaided eye to detect his real attitude. . . Mr. Muybridge's ingenious and successful efforts to catch and fix the fleeting attitudes of moving animals thus not only make a notable addition to our stock of positive knowledge, but must also effect a radical change in the art of depicting horses in motion."

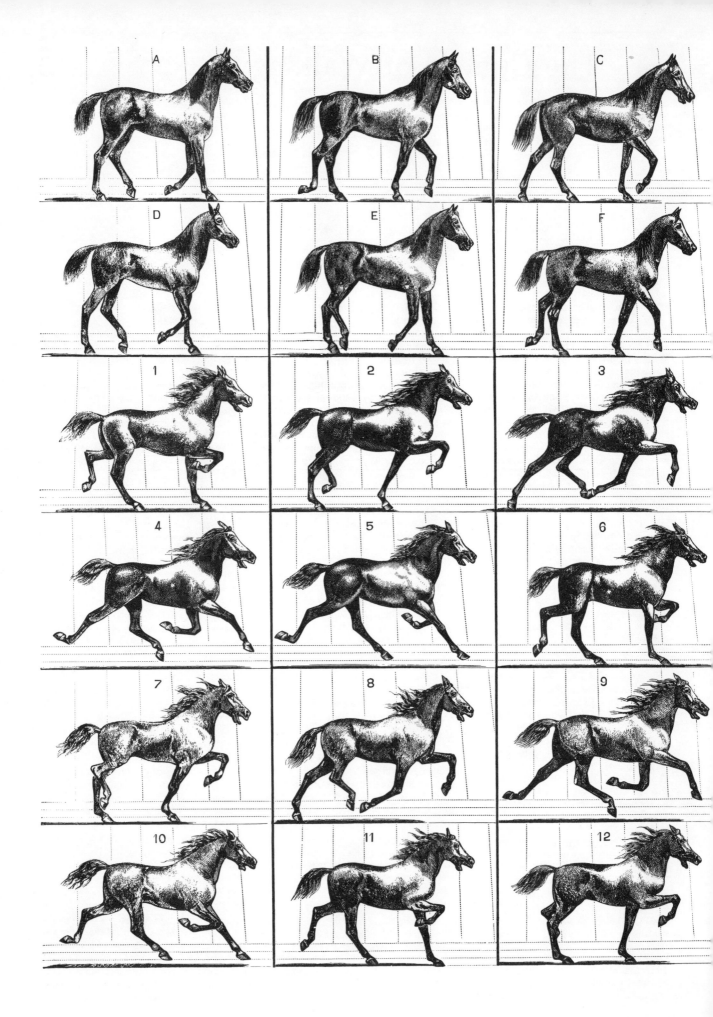

freely walking cat by Ingmar Engberg and Anders Lundberg in Sweden.

Then in 1965 the exciting discovery was made by the Russian workers M. L. Shik, F. V. Severin and G. N. Orlovskii that cats lacking the higher levels of the nervous system (the cerebral hemispheres and the upper brain stem) could be made to walk in a controlled manner on a treadmill. After the cat's upper brain stem is cut and the forebrain is removed electrical stimulation of the locomotor region in the remaining brain stem can cause the animal to walk when it is placed on a treadmill. The stepping movements on the treadmill are similar in all respects to those observed during normal walking; the gait the animal adopts depends on the speed of the treadmill and the strength of the stimulating current. With a low treadmill speed and a weak stimulus the animal will walk. Increasing the speed of the treadmill and the strength of the stimu-

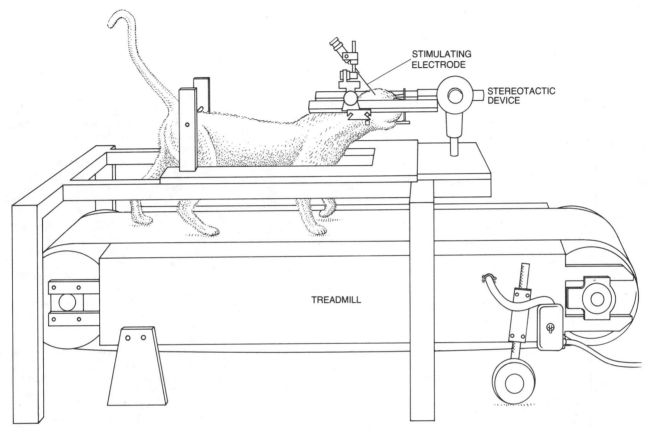

EXPERIMENTAL ARRANGEMENT employed in the author's laboratory at the University of Alberta to study the nervous control of walking in the cat was adapted from a method developed about a decade ago by a team of investigators in the U.S.S.R. The animal is first anesthetized and the higher regions of the nervous system are removed after severing the brain stem (*see illustration below*). Following this operation the animal is in a vegetative state, the only movements being those associated with respiration. By electrically stimulating the locomotor region of the remaining brain stem, however, the cat can be made to walk in a normal manner when placed on a treadmill. The rate of walking can be completely controlled by the experimenter by varying the speed of the treadmill and the strength of the stimulation. Because the animal is walking at a fixed position in space this arrangement, now used in a number of laboratories, makes it possible to record the electrical activity of the nerve cells within the spinal cord, the brain stem and the cerebellum.

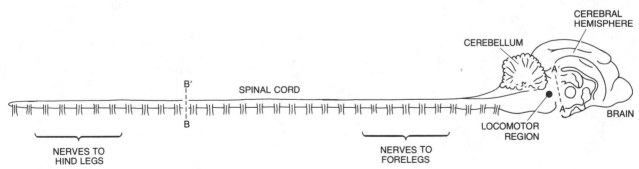

SPINAL CORD AND LOWER BRAIN STEM are isolated from the cerebral hemispheres and upper brain stem of the cat by cutting the brain stem at the line marked *A–A'* in this cross-sectional diagram of the cat's central nervous system. In an alternative experimental approach, developed early in this century by C. S. Sherring-ton, the hind-leg segments of the spinal cord are isolated by cutting the cord at the line marked *B–B'* in the diagram. In the latter case the hind legs will walk on a treadmill immediately after the operation if the animal is treated with certain drugs. Without drug treatment recovery period of a few weeks is required before hind legs will walk.

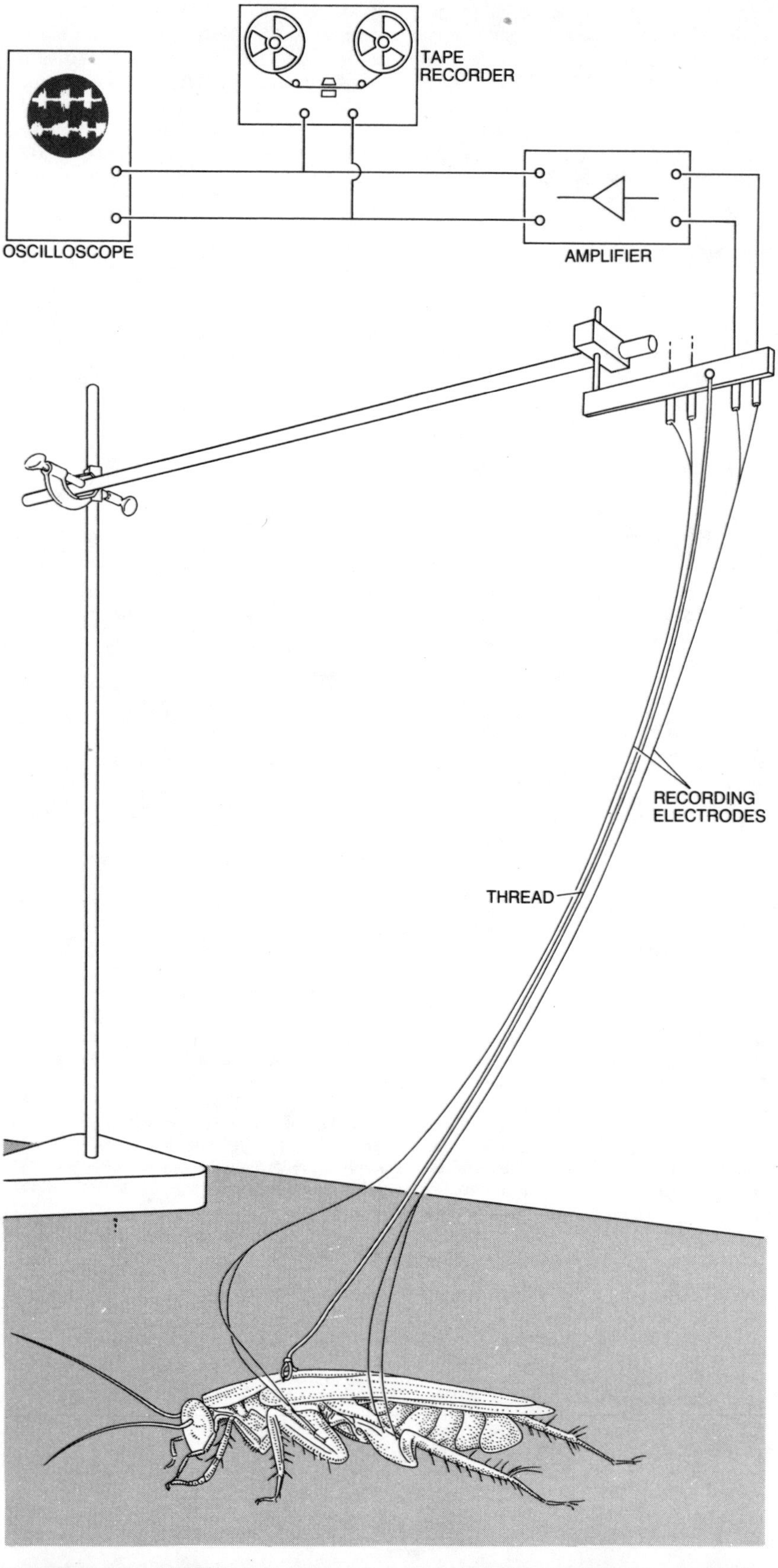

OSCILLOSCOPE

TAPE RECORDER

AMPLIFIER

RECORDING ELECTRODES

THREAD

ELECTRICAL ACTIVITY in the leg muscles of a freely walking cockroach is recorded by means of the apparatus shown here, devised by the author and his colleagues. Fine recording wires are implanted in the flexor and extensor muscles of the insect's leg and are anchored to the leg by tying them to the cuticle (the hard outer skeleton). The recording leads and a restraining thread are attached to a fixed terminal above the cockroach, and their length is adjusted so that the animal can walk about freely in a test area some two feet in diameter.

lus will cause the animal to trot, and with higher treadmill speeds and stronger stimuli the animal will gallop. Hence the walking behavior of the animal can be completely controlled by the experimenter. The great advantage of this experimental arrangement is that it enables one to make electrical recordings from nerve cells within the nervous system while the animal is walking, since the animal is fixed at one position in space.

Throughout this century there have also been numerous studies of the nervous control of walking in amphibians (such as toads and newts) and arthropods (insects, crabs and crayfish). Much of the early work on these animals was done at a time when the dominant concept in neurophysiology, originating with the work of Sherrington and I. P. Pavlov, was that reflexes were the basic units of behavior. Hence it is not surprising that the early hypotheses on the control of walking in amphibians and insects held that reflexes were essential for the generation of the stepping rhythm. Quite recently, however, substantial evidence has accumulated demonstrating that this notion is incorrect and that the basic rhythm is generated within the central nervous system and is modified by sensory input. One consequence of this demonstration is the realization that the basic stepping mechanism of the cockroach, say, is similar to that of the cat. Indeed, the relative simplicity of the walking system of the cockroach has provided much information on the mechanisms by which reflex activity modulates the centrally generated rhythm, which in turn has led to concepts that may make it possible to gain an understanding of the more complex systems.

Most current research on walking has been directed toward four goals: (1) gaining a precise description of leg movements and determining which muscles produce these movements, (2) establishing that mechanisms within the central nervous system generate the basic rhythm for stepping, (3) determining the function of the sensory input from leg receptors and (4) determining the nervous mechanisms for coordinating stepping in different legs. The two animals most intensively studied have been the cat and the cockroach. Here I shall summarize some of the main findings from experiments involving the two animals, emphasizing those similarities that suggest general principles about the control of walking.

In the cat there are four basic gaits: the walk, the trot, the pace and the gallop [see top illustration on page 61]. During walking, trotting and pacing the stepping movements of the two hind legs alternate with each other, as the stepping movements of the two forelegs do.

The differences between the three gaits lie in the timing of the stepping of the two legs on one side of the animal. During slow walking the left foreleg, for example, steps shortly after the left hind leg and before the right hind leg. Hence the stepping sequence is left hind leg, left foreleg, right hind leg, right foreleg, and so on.

The transition from walking to trotting is continuous, since with an increase in walking speed a foreleg begins to step before the opposite hind leg touches the ground. When the speed is such that diagonally opposite legs step at the same time, the animal is trotting. Pacing, which is done at the same speed as trotting or at a slightly higher speed, is characterized by the simultaneous stepping of the two legs on one side. During galloping opposite legs move almost synchronously and the forelegs alternate with the hind legs. The general feature that emerges from this analysis is that the coupling between adjacent legs (along and across the animal) is either in phase (stepping together) or out of phase (alternating).

The gaits commonly seen in amphibians and arthropods resemble the walking and trotting gaits of cats. In the cockroach, which of course has six legs, the stepping of adjacent legs alternates during moderate-speed and high-speed walking. Thus the stepping of the three legs right rear, left middle and right front alternates with the stepping of left rear, right middle and left front. The animal is therefore always supported by at least three legs, and for that reason the gait is referred to as the tripod gait. As the walking speed decreases, the gait is best described as a back-to-front sequence for the three legs on each side of the animal. At all times the two legs on any one segment of the insect strictly alternate.

For slow walking a feature common to the cat and the cockroach is the back-to-front sequence of stepping. The same is true of most other animals, a fact that is seen clearly in many-legged animals such as millipedes and centipedes, where "waves" of stepping appear to move from back to front.

The step cycle for an individual leg consists of two basic phases: the swing phase, when the foot is off the ground and moving forward, and the stance phase, when the foot is on the ground and the leg is moving backward with respect to the body. The propulsive force for progression is developed during the stance phase. A common feature of the step cycle in the cockroach and the cat (and in most other animals, including man) is that the duration of the swing phase remains comparatively constant as walking speed varies. Accordingly changes in the speed of progression are produced primarily by

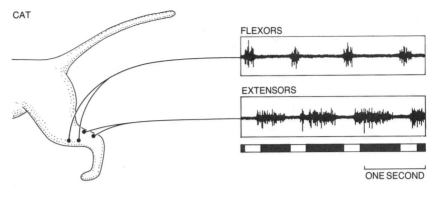

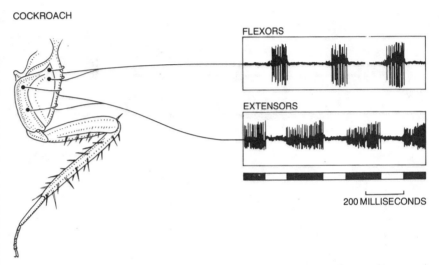

RECIPROCAL BURSTS of electrical activity are recorded in the flexor muscles and the extensor muscles during the swing phase (*white*) and the stance phase (*gray*) respectively, both for the cat walking on the treadmill (*top*) and for the cockroach walking about freely (*bottom*). The period of electrical activity corresponds to the time during which muscle is contracting.

changes in the time it takes for the legs to be retracted during the stance phase.

The simplest technique for determining what muscles are involved in giving rise to the swing and stance phases of the step cycle is to record the electrical activity in the leg muscles of a walking animal. Muscle contractions are initiated by a series of brief electric potentials propagated along the external membrane of each muscle fiber. The sum of the electrical events from different fibers can be recorded by means of fine wire electrodes implanted in the muscle. The records obtained by this method are called electromyograms.

To record electromyograms in the leg muscles of a freely walking cockroach the uninsulated ends of two very fine wires (two thousandths of an inch in diameter) are positioned in each of the leg muscles that produce movements around a joint. The other ends of the recording leads are connected to a terminal mounted above the animal, and the lengths of the leads are adjusted so

that the animal can walk about freely in an area two feet or so in diameter. The electrical signals from each muscle are then amplified and displayed on an oscilloscope, where they can be photographed. Storing the signals on magnetic tape enables one to do a computer analysis at a later time. To minimize any mechanical effect the recording wires might have on movement, recordings are usually made from only two muscles at a time. It is possible, however, to implant as many as 24 electrodes in a single animal. The technique for recording electromyograms from the muscles in other walking animals is similar in principle to the method developed for the cockroach.

Most of the muscles that account for the movements of the hind legs of the cat and the cockroach can be classified as either flexor muscles or extensor muscles according to whether the function of the muscle is to reduce the joint angle or to increase it. Electromyogram recordings from the muscles during walking have shown that the electrical activity of the flexors and extensors alternates. The alternating bursts of activity that are observed are well correlated with the movement of the leg. The flexors begin to discharge slightly before the initiation of the swing phase, and the resulting contractions in these muscles

rapidly lift the foot from the ground and move the leg forward with respect to the body. In the cat the extensors become active well before the end of the swing phase, causing the leg to straighten near the end of the swing, thereby producing an effective stride. The extensors remain active during most of the stance phase to

support the animal's weight and to develop the forward propulsive force. Similarly, in the hind leg of the cockroach activity in the extensor muscles during the stance phase straightens the leg and develops the forward thrust for walking.

In both animals the variations in the amplitude and the times of the onset of electrical activity in the many different flexor and extensor muscles in each leg are complex and depend on the speed of walking. So far little is known about how these complex patterns of electrical activity are created by the central nervous system. The apparently simpler problem of determining the mechanisms responsible for generating the basic alternating pattern of activity in flexors and extensors, however, has received much attention.

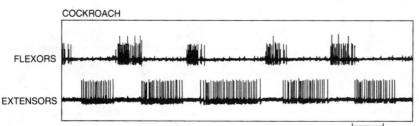

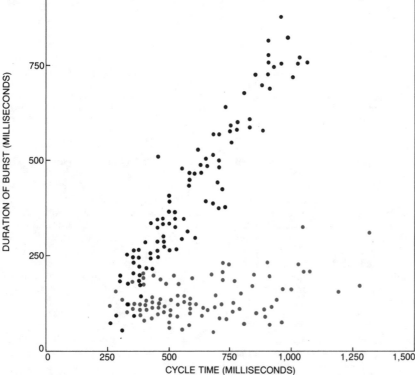

Within the past few years a great deal of evidence has accumulated to support Graham Brown's original proposal that the basic rhythm for stepping in each leg of the cat is generated by mechanisms within the spinal cord, that in effect there is a central rhythm-generator for each leg. Compelling evidence for the existence of these rhythm-generators comes from the extensive investigations of Sten Grillner and his colleagues in Sweden on the patterns of activity in the flexor and extensor muscles after the elimination of sensory input from the receptors in the legs of walking cats. (Sensory input can be easily eliminated by cutting the dorsal roots: distinct bundles of nerves that enter the spinal cord and contain almost all the sensory nerve fibers from the legs but none of the motor-nerve fibers supplying the muscles.) One of the most important results reported by the Swedish workers was that after the hind-leg segments have been isolated (by cutting either the spinal cord or the upper brain stem) the rhythmic reciprocal patterns of activity in flexors and extensors of the cat's hind leg could still be generated in the absence of sensory information from peripheral receptors in the leg. Moreover, they found that when such "spinal" or "decerebrate" cats were made to walk on a treadmill, the temporal sequence of activation of different hind-leg muscles was not altered by removing sensory input from the hind legs. The latter result showed that in addition to the basic rhythm being centrally generated, central mechanisms are responsible for determining the order in which different muscles are activated.

Similar results were obtained by John F. Iles and me in our investigations of the cockroach. After eliminating all sensory input from the legs of headless animals we found that the reciprocal patterning in hind-leg flexor and extensor motor neurons persists, the sequence of activation of different motor neurons re-

EXISTENCE OF A CENTRAL RHYTHM-GENERATOR for each leg of the cat and the cockroach is demonstrated by the fact that even after all sensory input from receptors in the hind legs had been disconnected, alternating rhythmic bursts of electrical activity were generated in the flexor muscles and the extensor muscles in the hind legs of both animals (*traces at top*). The graph at bottom shows that in the cockroach the duration of the centrally generated flexor bursts (*colored dots*) remained constant in contrast to the duration of the extensor bursts (*black dots*). The electrical data are consistent with the behavioral observation that during normal walking the duration of the swing phase of stepping remains comparatively constant.

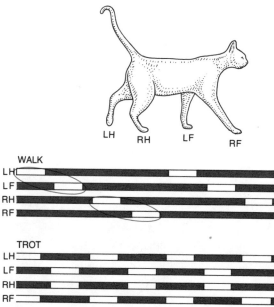

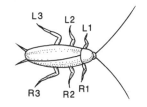

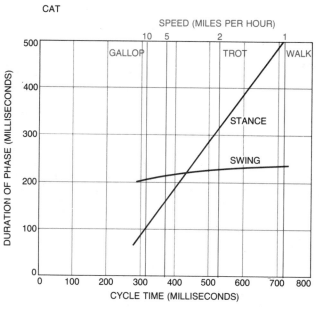

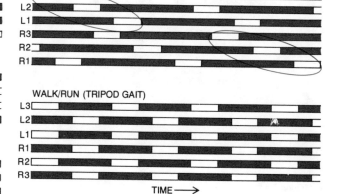

STEPPING PATTERNS of the cat and the cockroach are depicted schematically in these diagrams. Reading from left to right, each horizontal bar indicates for a single leg the time the foot is off the ground (*white parts of bar*) and on the ground (*gray parts of bar*). During slow walking there is a back-to-front sequence of stepping for both animals; the sequences are marked by the colored ellipses.

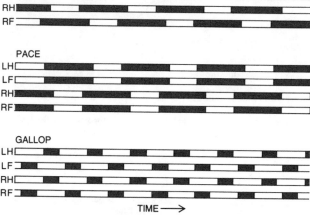

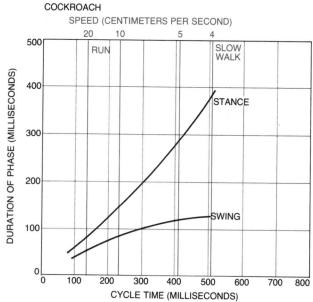

DURING NORMAL WALKING the duration of the swing phase of a step (when the foot is off the ground) remains fairly constant compared with changes in the duration of the stance phase (when the foot is on the ground) for most animals. These two graphs show the durations of the swing phase and the stance phase for the hind legs of the cat and the cockroach in relation to cycle time: the interval from the beginning of one swing phase to the beginning of the next. Colored scales indicate speed, measured in appropriate units for each animal.

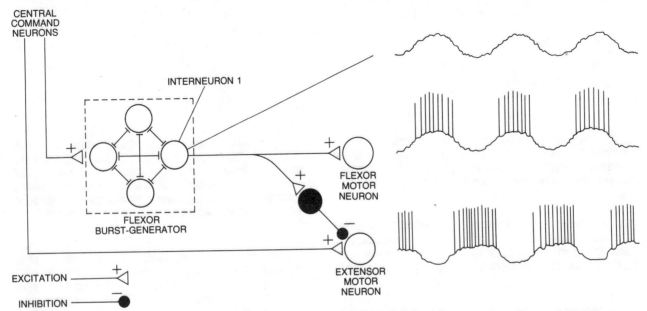

ORGANIZATION of the central rhythm-generating system for each hind leg of the cockroach is outlined here. The interactions of a system of interneurons, called collectively the flexor burst-generator, produce oscillations in the electric potential across the membrane of one of the members of this group of interneurons (*interneuron 1*). These oscillations in membrane potential in turn generate reciprocal bursts of electrical activity in the flexor motor neurons and at the same time inhibit activity in the extensor motor neurons. Diagram assumes that flexor burst-generator and extensor motor neurons are activated by central command interneurons from the brain.

mains the same as that in normal walking animals and the duration of the flexor activity remains fairly constant. Evidently there exist within the lower levels of the cockroach's nervous system networks of nerve cells that can generate the rhythmic sequence of electrical activity in flexors and extensors similar to the activity observed in these muscles during normal walking.

Centrally generated patterns of motor activity have now been found to underlie all kinds of stereotyped behavior in both vertebrate and invertebrate animals; behavior of this type includes respiration, eye movements, chewing, swallowing and flight. Neurophysiologists usually refer to such patterns of activity as motor programs.

The cellular mechanisms responsible for centrally generating rhythmic reciprocal bursts of activity in motor neurons are just beginning to be understood. Little progress has yet been made in determining such mechanisms in vertebrates, primarily because of the immense complexity of the nervous system of these animals. As a result most of our current ideas about the cellular basis of central rhythm-generation have come from studies of invertebrates. In these animals it has been found that the rhythmic activity in motor neurons can be generated in two qualitatively different ways: it can result from intrinsic oscillatory activity in individual nerve cells called pacemakers or it can emerge as a property of an interconnected group of nerve cells none of which have the property of intrinsic oscillation. There is no evi-

dence that in the walking systems of animals pacemaker nerve cells are responsible for generating the rhythmic motor patterns. There is evidence, however, that a group of interconnected nerve cells generates the reciprocal rhythm of activity in the hind leg of the cockroach.

The patterns of activity in cockroach flexor and extensor motor neurons, that is, the nerve cells that give rise to the contraction of the flexor and extensor muscles, led Iles and me to suggest in 1970 that the reciprocal burst activity is generated by a system of interneurons (which we call a flexor burst-generator) that periodically excites the flexor motor neurons and inhibits the extensor motor neurons. Subsequently Charles R. Fourtner and I succeeded in making intracellular recordings with microelectrodes from interneurons within the nervous system of the cockroach during rhythmic leg movements. Our results strongly supported the concept of a flexor burst-generating system. We identified a single interneuron in which the electric potential across the external membrane oscillated in phase with the rhythmic burst activity in flexor motor neurons. When similar oscillations in the membrane potential of this interneuron were artificially created by passing currents through the recording electrode, we were able to mimic the bursts of activity normally generated in the motor neurons during rhythmic leg movements. Moreover, the excitation of the flexor motor neurons was accompanied by inhibition of any spontaneous activity in the extensor motor neurons.

The interneuron exciting the flexors and inhibiting the extensors appears to belong to a network of nerve cells, and the interaction among the members of the network is responsible for the rhythmic activation of the flexors. The interesting feature of this rhythm-generating system in the cockroach is its asymmetry: there does not appear to be a system of interneurons responsible for generating the extensor bursts. The extensors are continuously excited, and their activity is periodically interrupted by inhibition from the flexor burst-generator. The flexor burst-generator gives rise to a burst of electrical activity that is fairly constant in duration, regardless of the rate at which bursts are produced; as a result the duration of the swing phase varies only slightly with walking speed.

Although less is known about how the walking rhythm is generated in the cat's spinal cord, we do know that the rhythm does not depend on direct coupling of motor neurons. As in the cockroach, interactions within a network of interneurons are responsible. A number of different models for how systems of interneurons can give rise to the walking rhythm, including one originally proposed by Graham Brown in 1914, are currently under consideration, but at present there is not enough information to assess the validity of any of them.

The existence of central rhythm-generation for each leg does not mean that sensory input is unimportant in the patterning of motor activity. In fact, sensory input is essential if the animal is to be able to adapt its stepping movements promptly to compensate for irregularities in the terrain on which it is walking. For example, the stepping of the hind legs of a cat that has undergone either a spinal or a cerebral transection adapts to match the speed of a treadmill for a wide range of treadmill speeds. Since there is little change in the duration of the swing phase, the adaptation of the step cycle to the treadmill speed results from a decrease in the duration of the stance phase. The explanation for this phenomenon is that a sensory signal indicates the completion of leg extension during the stance phase and switches the motor program from stance to swing. Hence following a stereotyped swing phase the foot is placed on the treadmill and moves backward with the speed of the treadmill until a sensory signal triggers the initiation of the next swing phase.

Two conditions seem to be necessary for the swing phase to be initiated. First, the hip joint must be extended; second, the extensor muscles must be unloaded. Both of these conditions are fulfilled near the end of the stance phase. The sensory switching is effected by means of circuits located within the hind-leg segments of the spinal cord. This important conclusion comes from the work of Grillner and his colleagues Serge Rossignol and Hans Forssberg, who showed that preventing extension of the hind leg of a spinal cat walking on a treadmill inhibits stepping in that leg. If the leg is allowed to extend slowly, the swing phase is eventually initiated when the leg is in a position close to the position where swing is initiated during normal walking.

The principle that a leg must be unloaded for swing to be initiated also ap-plies to the walking system of the cockroach. Activation of the receptors that detect forces in the cuticle (the hard external skeleton) of the leg during the stance phase inhibits the flexor burst-generating system of interneurons and thereby prevents the switching from stance to swing. As the leg is extended during the stance phase the stresses in the cuticle diminish, since the animal's weight is carried to a greater extent by other legs. This movement eliminates the inhibitory influence of the cuticle stress receptors from the flexor burst-generator, thus enabling the system to become active and so produce the swing phase.

In the cockroach the switching of the motor program from swing to stance is also triggered by a sensory input. A group of hair receptors is excited by flexion movements during the swing phase. The sensory input from these receptors inhibits the flexor motor neurons and the flexor burst-generating system. The latter effect terminates the flexor burst (and hence the swing phase) and enables the leg to initiate the stance phase. The initiation of stance is also facilitated by an excitatory connection from the hair receptors to the motor neurons of the extensor muscle. The inhibitory pathway from the hair receptors to the flexor burst-generator ensures that the leg position at the end of the swing phase remains constant regardless of the position from which the step was initiated.

Apart from the fact that the sensory input functions to switch the motor program from one phase to the other, it is clear that it also functions to modify the motor program during the execution of a single phase. Grillner and his colleagues described a functionally impor-tant reflex of this type in the hind leg of a spinal cat. They showed that electrical or mechanical stimulation of receptors in the skin on the top of one of the cat's hind feet during the swing phase causes a large increase in the electrical activity of the leg flexors and as a result the leg is lifted higher. The function of this reflex under natural conditions is easy to see, since the top of the foot will often hit a fixed object. The reflexive elevation of the foot will then enable the foot to avoid the object. A significant feature of the reflex is that it appears in the spinal animal. It follows that this automatic compensation in the step cycle for external perturbations does not necessarily require the participation of higher levels in the nervous system.

In the cockroach feedback from the stress receptors in the cuticle excites the extensor motor neurons during the stance phase. Since these motor neurons are active during the stance phase, the sensory input enhances the amplitude of the extensor activity. The effect can be regarded as a reinforcing reflex; its function is to compensate for variations in load when the resistance to extension unexpectedly changes. For example, if the load carried by the leg is suddenly increased because of irregular movement in one or more of the other legs, there is an increase in the activity of the cuticle stress receptors and thus an increase in extensor activity to resist any tendency of the additional load to decrease the rate of extension. In the cat's hind leg a reinforcing reflex may also function during the stance phase to help compensate for any unexpected increase in the load on the extensor muscles, since increasing the resistance to leg extension during the stance phase causes a marked increase in extensor ac-

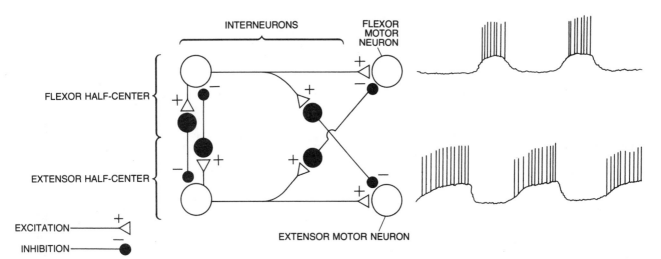

HALF-CENTER MODEL for explaining the central generation of rhythmic reciprocal activity in the flexor and extensor motor neurons of the cat's hind leg, originally proposed by the British neurophysiologist T. Graham Brown more than 60 years ago, is presented here in its modern version. According to this view, the flexor and extensor motor neurons, together with their associated driving interneurons, each constitute a "half-center." The two half-centers are **assumed to mutually inhibit each other, so that when one is active, the other is inactive. For rhythmic activity to be generated some mechanism is needed to switch the activity from one half-center to the other. One possible switching mechanism would be a comparatively slow intrinsic inactivation of the active half-center, which would have the effect of lessening the inhibition of the other half-center, enabling it to become active. Other models have also been proposed.**

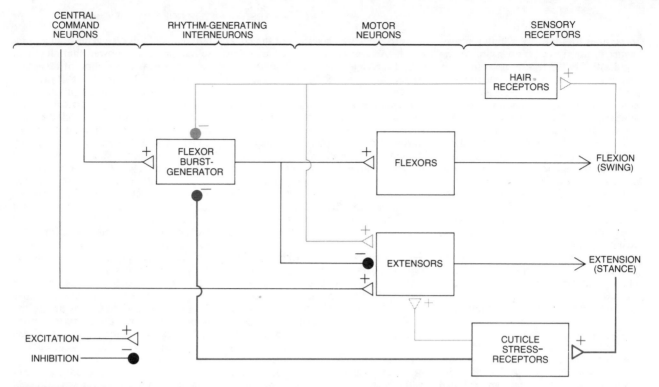

CENTRAL COMMAND NEURONS   RHYTHM-GENERATING INTERNEURONS   MOTOR NEURONS   SENSORY RECEPTORS

HAIR RECEPTORS

FLEXOR BURST-GENERATOR

FLEXORS

FLEXION (SWING)

EXTENSORS

EXTENSION (STANCE)

EXCITATION

INHIBITION

CUTICLE STRESS-RECEPTORS

REFLEX PATHWAYS in the hind leg of the cockroach are orga-nized to modify the centrally generated patterns of activity in the flexor and extensor motor neurons. During the stance phase of a step stress-receptors in the cuticle of the leg are excited and the input from these receptors inhibits the flexor burst-generator (*dark colored pathway*), thus preventing the initiation of the swing phase until the leg is unloaded at the end of stance. The reflex effect from the cuticle stress-receptors therefore controls the switching of the motor program from stance to swing. Activation of the hair recep-tors during swing inhibits the flexor burst-generator (*light colored pathway*) near the end of the swing phase, thereby causing the pro-gram to switch from swing to stance. This switching is facilitated by an excitatory effect of the hair receptors on the extensors. The cuti-cle stress-receptors also excite the extensor motor neurons during stance (*gray pathway*); this reflex reinforces central excitatory in-put to motor neurons. Central command neurons descend from brain.

tivity. This reflex compensation takes a significant amount of time and could function effectively only in a cat that is walking rather slowly.

In brief, it seems clear that reflexes have two quite different functions in controlling the step cycle of a single leg. The first is to switch the motor program from one phase to the other (that is, to initiate the swing phase or to initiate the stance phase) and the second is to modify the motor output within a single phase.

The mechanisms that coordinate step-ping in different legs are so far poor-ly understood. There are evidently three mechanisms that can function to coordi-nate stepping: (1) direct central coupling of rhythm-generators in each leg by pathways entirely within the nervous system; (2) indirect coupling of the rhythm-generators by a mechanical linkage such that movements of any one leg would alter the sensory influences on the rhythm-generators of the other legs; (3) direct sensory influences communi-cated from one leg to the rhythm-gener-ators of the other legs. Central inhibito-ry coupling between adjacent rhythm-generators has been demonstrated dur-ing walking in the cat and the cock-roach. Following the elimination of all sensory input from the hind legs of spi-nal cats, the motor activity in opposite

legs continues with a rhythmic alternat-ing pattern similar to the one observed during normal walking. In the cock-roach, on the other hand, flexor activity in adjacent legs rarely occurs synchro-nously, even after sensory input has been eliminated.

This observation in the cockroach, combined with an analysis of the electri-cal activity in the interneurons that con-nect adjacent segments of the insect, led Iles and me to conclude that adjacent flexor burst-generators mutually inhibit each other by means of central path-ways. Mutual inhibitory coupling be-tween adjacent flexor burst-generators explains the behavioral observation that adjacent legs never step at the same time. In addition to the central inhibito-ry coupling indirect coupling of rhythm-generators by means of mechanical link-ages is important in coordinating step-ping in the cockroach, particularly when the animal is walking on an uneven sur-face.

In a walking cockroach the legs that are loaded by the animal's weight are prevented from initiating the swing phase because the activity in the cuticle stress-receptors inhibits the flexor burst-generator associated with each of those legs. The swing phase can be initiated only in the legs that carry little or no load. When a stepping leg finds solid support, it will begin to carry some of

the animal's weight and will thus de-crease the load carried by one or more of the other legs. This effect may be suf-ficient to then allow the initiation of the swing phase in the other legs. Subject to the condition that no two adjacent legs can step at the same time (because of the mutual inhibition between adjacent flexor burst-generators), there may not necessarily be any precise and predict-able temporal relation between the step-ping movements of the different legs when the animal is walking on an un-even surface. The inhibition of the flexor burst-generators by sensory input from cuticle stress-receptors (signaling the load carried by the leg) is an efficient mechanism for enabling the animal to automatically adapt its gait to the ter-rain on which it is walking, so that the animal can walk stably at all times.

Each flexor burst-generator in the cockroach can be regarded as a simple decision-making element. The decision to be made is whether to initiate a step or not. The flexor burst-generator makes the decision on the basis of information from peripheral receptors signaling the state of that leg and information from other flexor burst-generators signaling the state of adjacent legs. The decision to step is made when the position of all the legs is such that a step will not lead to instability.

The outcome of the recent investigations into the nervous control of walking in many different animals has been to show that the basic nervous circuits for walking are located in the lower levels of the nervous system. The work on the cat, for example, has confirmed the conclusion of Sherrington and Graham Brown that stepping is an essentially autonomous act depending primarily on nervous circuits within the spinal cord. It appears that an important feature of walking systems in all animals is that the higher centers of the nervous system are not necessary for the basic patterning of motor activity.

This does not mean that the higher centers play no role in the control of walking. They are certainly involved in the initiation and maintenance of walking, but the pathways that convey the commands for walking from them have not yet been identified. Another important function of the higher centers, particularly in mammals and birds, is to modulate the basic walking motor program in response to sensory inputs from receptors in the head, such as the eyes and the vestibular apparatus (the receptor organ in the ear that senses the position and motion of the head). The modulation of the motor program by inputs from the receptors in the head presumably functions not only to control the direction of walking but also to help en-

sure that balance and stability are maintained at all times.

There are two further striking similarities between the mechanisms responsible for the control of walking in the cat and those responsible in the cockroach. The first is that the basic reciprocal rhythm of electrical activity in flexors and extensors can be generated in the absence of sensory input from receptors in the legs. Moreover, in both animals the duration of the centrally generated flexor bursts is fairly constant compared with the variation in the duration of the extensor bursts. This finding corresponds to the behavioral observation that the duration of the swing phase remains essentially constant for large changes in walking speed.

The second similarity is that reflexes function to switch the motor activity from extension to flexion. The initiation of the swing phase requires that the leg be extended and unloaded. Reflexes can also function to modulate the motor activity within either the swing phase or the stance phase. Both types of reflex function to modify the motor program in order to adjust the stepping movements to unpredictable variations in the external environment. It is also important to note that these reflexes can function in animals after the higher levels of the nervous system have been disconnected. Thus automatic adaptation of the stepping movement to the environment is partly achieved by reflex pathways confined to the lower levels of the nervous system.

At first the marked similarities between the basic mechanisms for the control of walking in the cat and those for the control of walking in the cockroach may seem surprising, particularly when it is considered that mammals and insects evolved separately from nonwalking wormlike ancestors. The similarities become more understandable, however, when it is remembered that in both groups of animals the walking system evolved for the same function, namely to move the animal rapidly and stably over uneven terrain. There are numerous examples in nature where evolutionarily unrelated animals rely on similar mechanisms for the same task: witness the similarity of the eye of cephalopod mollusks, such as the octopus, and the eye of vertebrates. The similarity of the walking systems in the cat and the cockroach suggests that the number of ways of optimally constructing a walking system is quite limited.

From the study of walking in diverse animals there are clearly beginning to emerge some common principles underlying the control of walking. A better understanding of these principles may in turn contribute to a better understanding of how the nervous system controls walking in man.

STEPPING IS COORDINATED in the six legs of a walking cockroach by means of the nervous pathways sketched here against the background of the insect's thoracic nerve cord (gray). For each leg there exists within the thoracic nerve cord a flexor burst-generator, which when active excites the flexors to produce the swing phase. Adjacent flexor burst-generators mutually inhibit each other; as a consequence no two adjacent legs can step at the same time. The existence of such pathways explains the basic alternating pattern of stepping in adjacent legs. An inhibitory input from the cuticle stress-receptors to the flexor burst-generators prevents any leg from stepping when it is loaded during stance (red pathways), thereby providing a mechanism for adapting the pattern of stepping when animal is walking on uneven surface.

The Soaring Flight of Vultures

by C. J. Pennycuick
December 1973

The six common vultures of East Africa can make a round trip of as much as 200 kilometers by skillfully riding updrafts. How they do so is examined with the aid of a powered glider

To watch a vulture soar effortlessly overhead for hours at a time is to become convinced that these birds are among the most skillful of fliers. Yet there are occasions when vultures cannot fly at all. Early in the morning on the East African plains one quite often meets little groups of vultures that had gathered at some small find the night before and then slept where they happened to be when night fell. If the birds are pursued, they take off, but they do not fly far before they land again. If they are forced to take off several times in rapid succession, they quickly become exhausted and can be caught by hand. Two of the commonest East African vultures, Rüppell's griffon (*Gyps rüppellii*) and the white-backed vulture (*Gyps africanus*), can easily be caught this way. Later in the morning, say after about 9:00 A.M., the technique sometimes still works on a vulture that is heavily gorged with food. More often the bird will fly straight ahead for a short distance, then turn sharply and at the same time start climbing. After turning in a few irregular narrow circles and intermittently flapping its wings, the vulture settles down to gliding in steady circles. It then continues to climb without flapping its wings and drifts downwind as it circles.

The reason these vultures have such difficulty flying under their own muscle power is that they are too big. There is a relation between the power required to fly, the power available from the muscles and the body weight that sets an upper limit to the weight of animals able to fly by muscle power. The larger vultures, storks and pelicans are quite near this limit and would barely be able to stay airborne were it not for their ability to extract energy from the atmosphere and use it for their locomotion.

The flight path of a bird that is gliding along at a steady speed (that is, not flap-ping its wings) is invariably inclined downward. The gliding bird therefore has a vertical downward component of velocity, called the "sinking speed." If the air through which the bird is flying happens to be rising at a speed greater than the sinking speed, the bird is carried up with it and acquires potential energy it can use later to glide through air that is not rising. Any system of maneuvers whose effect is to produce this result by taking advantage of some atmospheric process constitutes a method of soaring.

Soaring techniques can be classified according to the atmospheric process responsible for the rising air, or "lift" [see *illustration on page 69*]. The existence of lift in sufficient quantities, however, satisfies only part of the requirement for a successful soaring technique. Sufficient information must also be available to the bird to enable it to locate the lift and, having found it, to carry out the correct maneuvers to make use of it. If the bird executes the wrong maneuver, it will usually lose the lift.

A soaring bird can work with several different kinds of lift, for example slope lift, in which the air rises when a wind encounters a slope; thermal lift, in which columns or bubbles of air rise when heated from below by warm ground, and wave lift, in which the air rises in the course of undulatory motion downwind of an obstacle. The correct procedure for a bird using slope lift is for the bird to tack to and fro, remaining above the windward slope. If the lift happens to be due to a thermal, however, the bird must change over to circling, drifting downwind with the thermal instead of remaining over the slope. Wave lift, on the other hand, remains stationary with respect to the ground, and it must be worked in the same way as slope lift, except that there is no slope to indicate where the lift is. Inexperienced glider pilots often mistake wave lift for thermal; they circle in it and as a result drift downwind into the downgoing part of the wave. This error is apt to lead to a rapid and embarrassing descent *aux vaches*, as French pilots say. Present indications are that many, if not all, birds make the same mistake when they encounter wave lift.

Soaring birds are not the easiest birds to study, partly because the bird watcher, traditionally equipped with binoculars and rubber boots, cannot match their mobility and tends to lose track of them, and partly because one can never tell from the ground what kind of air movements a bird is soaring in. There is a limit to what one can deduce from even the most meticulously detailed ground observation, and in fact not a great deal of progress has been made with this approach since E. H. Hankin's classic book *Animal Flight* appeared in 1913.

Entirely new opportunities have been opened up with the advent in recent years of practical powered gliders. I have been extremely fortunate in having the use of one of the best of these, the Schleicher ASK-14, which was generously put at my disposal by Anglia Television and Okapia Film. The ASK-14 was developed from the well-known K-6E sailplane, and it is powered by a 26-horsepower engine. Because of its highly efficient aerodynamic design, the aircraft takes off and climbs in a lively manner with the aid of this tiny power plant. The engine is normally run for five or 10 minutes after takeoff until a thermal has been found, after which the engine is switched off and the propeller is feathered. The aircraft then becomes an excellent soaring glider, and the flight can be extended, usually for several hours, without further recourse to the engine. If the thermals should fail, or if the pilot gets too low over some unsuitable place, the engine can be restarted to extract him from the difficulty.

I have flown the ASK-14 mainly over the Serengeti National Park in northern Tanzania and neighboring areas of Tanzania and Kenya. This area enjoys good soaring weather throughout most of the year, and it is frequented by a remarkable variety of species of soaring birds. Many of the birds are common enough to be encountered during nearly every flight.

The common soaring birds of East Africa soar mainly in slope lift and thermals. Slope lift is, of course, useful only in the hillier areas and at comparatively low altitudes, and so in general thermals are the most important source of lift. When vultures are taking off from the ground, one can see at once that they are using some definite structure in the atmosphere rather than the randomly distributed kind of energy characteristic of turbulent air. As soon as one vulture is climbing successfully it is quickly joined by others (and often by other kinds of birds as well), until soon a group of birds forms, all circling around a common axis. This axis marks the "core" of a thermal, which at low altitudes is generally a vortex of the "dust devil," or columnar, type. "Dust devil" refers to the fact that over dry ground vigorous thermals of this type are often visible as whirling columns of dust. Higher up vortex-ring thermals may form, and the tops of many thermals of either type are marked by cumulus clouds. As far as the soaring bird or pilot is concerned, either type of thermal can be considered a circular patch of lift that drifts along with the wind. The appropriate soaring maneuver is to fly in steady circles of as small a radius as possible, in order to stay as close as possible to the middle of the core, where the lift is strongest.

A bird's ability to use thermals, either for staying airborne or for traveling cross-country, depends on its gliding performance. This is most often expressed in terms of its "glide polar," which is a graph of sinking speed plotted against forward speed. The glide polar for a glider can be measured directly. The same type of graph for a gliding bird can be produced by estimating the differences in horizontal and vertical speed between a glider with a known polar and the bird. In this way the glide polars of the ASK-14 and the white-backed vulture have been compared.

In a straight glide the glider can travel much faster than the vulture at a given gliding angle. This is owing partly to the glider's superior aerodynamic efficiency and partly to its higher wing loading (the ratio of weight to wing area). On the other hand, when the glide polar is translated into circling flight, the effect of the vulture's lower wing loading is that it can turn in much smaller circles at a similar rate of sink. This means that in a narrow thermal the bird can center its circle in the strongest part of the core, whereas the glider is obliged to fly in the weaker lift around the outside. Thus the vulture can often outclimb the glider, particularly at low altitude and early in the day, when the thermals tend to be weak and narrow, even though its sinking speed in straight flight is much the same. On one occasion, on a day of exceptionally narrow thermals, I was outclimbed by a tawny eagle (which has a still lower wing loading than the white-backed vulture), even though I had my engine running while the eagle was only gliding.

Many birds of prey are adapted primarily to using thermals as a means of remaining airborne in order to look out for food below. The African martial eagle (*Polemaetus bellicosus*), which preys on other birds, uses thermals in much the same way that smaller eagles and hawks use a rocky crag or a telephone pole. When foraging, it climbs by circling in a thermal to some modest height, usually from 300 to 600 meters above the ground, and then glides slowly along with its head pointing down, looking out for prey below.

Vultures use much the same technique in searching for carrion, but their tactics are based on the need to arrive promptly on the scene whenever and wherever a

POWERED GLIDER, its engine off and its propeller feathered, soars in rising air over the Serengeti National Park in northern Tanzania, accompanied by a pair of vultures that utilize the same lift to gain altitude effortlessly. The glider is a Schleicher ASK-14.

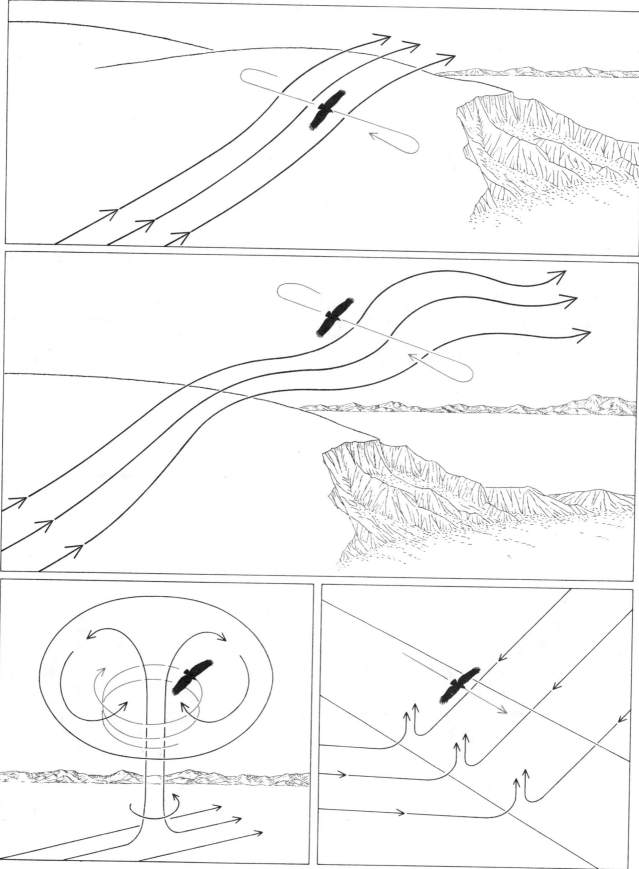

FOUR KINDS OF LIFT are exploited by different patterns of flight. Where air movement uphill provides "slope lift" (*top*), bird or sailplane should tack back and forth, heading sufficiently into the wind to stay within the same zone of rising air. The same pattern of flight is used to exploit "wave lift" on the leeward side of a slope (*middle*). To exploit a "thermal" (*bottom left*) the soarer travels downwind with the rising air column, circling within the thermal. When two air masses converge (*bottom right*), producing a line of "frontal lift," the soarer can either tack back and forth or set off in one direction and travel the entire length of the line.

dead animal happens to turn up. By alternately climbing in thermals and gliding straight in different directions, they patrol over likely areas, usually between 200 and 500 meters above the ground. Starting from a height of, say, 300 meters, a vulture can reach any point on the ground within a radius of about 4.5 kilometers within six minutes. Points nearer at hand can be reached more quickly in a fast, steep dive; the steeper the angle is, the faster the vulture can glide. The mammalian scavengers, mainly the spotted hyena (*Crocuta crocuta*), react like the vultures to signs of activity in the distance (including descending vultures), but they have to work much harder to get to the site and have lower maximum speeds. Thus the vultures, although they cannot drive off the hyenas in a direct confrontation, can still compete with them effectively through their advantage in arriving quickly at an unpredictable source of food.

The African vultures have evolved two different approaches to the strategy of searching for food. One is to learn the geography of a home range and search it very thoroughly. In areas where game is scarce no other approach is possible. One of the special features of the East African fauna, however, is the existence of large populations of migratory ungulates, and this opens up another niche for scavengers adapted to take advantage of it. In the Serengeti, for instance, hundreds of thousands of wildebeest make an annual migration around a circuit some 500 kilometers in length, and comparable populations of zebras and Thomson's gazelles make similar migrations [see "A Grazing Ecosystem in the Serengeti," by Richard H. V. Bell; SCIENTIFIC AMERICAN, Offprint 1228].

To a sedentary scavenger the arrival of the migratory herds represents a temporary superabundance of potential food, but the herds move somewhat erratically and are apt to move out of reach as suddenly as they came. A scavenger that is adapted primarily to feeding on migratory game must dispense with a fixed home range and follow the migrating herds wherever they go. Thus it cannot learn the ground as thoroughly as a sedentary scavenger.

It seems that among the four large species of East African vultures two are basically sedentary and two are itinerant. This was postulated several years ago by Hans Kruuk, on the basis of differences in the behavior of the birds at the carcasses on which they were feeding. The two sedentary species, the lappet-faced vulture (*Torgos tracheliotus*) and the white-headed vulture (*Trigonoceps occipitalis*), never gather in large numbers; it is rare to see more than eight of the former and two of the latter at the same gathering. The two species manage to arrive first at carcasses much more often than their relatively small numbers would lead one to expect, indicating that these sedentary birds pursue a "thorough search" strategy. In contrast, the itinerant Rüppell's griffon and its fellow member of the genus *Gyps*, the white-backed vulture, often gather in the hundreds at a large carcass, but in spite of their numbers they do not usually arrive first. It would be an exaggeration to say that they never find their own food, but they seem to rely more heavily on following carnivorous mammals or other vultures and less on searching directly by themselves. Unlike the other two species, they never fight in the air. They seem to have no territorial behavior whatever, although they do squabble a good deal when they are actually feeding.

In the air the difference in behavior is evident even from a powered aircraft. Once the thermals are under way lappet-faced vultures and white-headed vultures may be met almost anywhere, usually in pairs, but the griffons concentrate over the migratory herds (particularly the wildebeest), often in spectacular numbers. When griffons are encountered away from concentrations of game, they are usually traveling steadily cross-country, commuting between their feeding area, which is changeable, and their nesting area, which is necessarily fixed. Because the two areas may be widely separated, the griffons have to be able to use thermals for cross-country travel as well as for staying aloft while foraging.

The bird that travels the farthest is Rüppell's griffon, which nests in colonies on cliffs. This habit, in which it differs from the white-backed vulture (a tree-nester), is most probably a reflection of

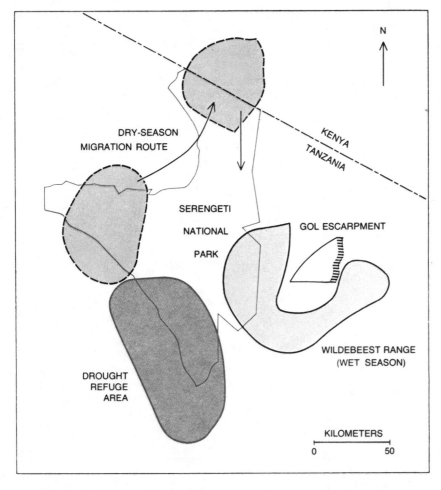

NESTING GROUND of the Rüppell's griffons that frequent the Serengeti National Park is the Gol Escarpment (*right*), a zone convenient to the grazing area preferred by wildebeest in the wet season when the griffons are raising their nestlings (*light color*). An unseasonable drought, however, will move the wildebeest to the west (*dark color*), forcing the griffons to travel 100 kilometers in each direction to get food for the nestlings. In the dry season, during the annual wildebeest migration (*gray zones*), the griffons follow them.

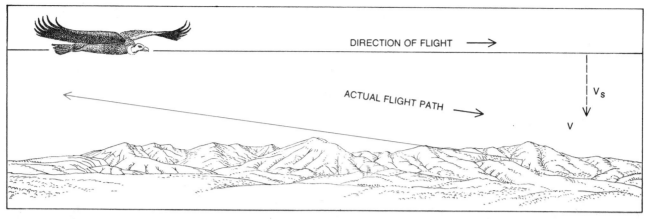

GLIDING BIRD in still air loses altitude at a "sinking speed" (V_s) while traveling forward at a "flight speed" (V); the ratio V/V_s is the bird's "glide ratio" and is equal to the distance traveled forward per unit of altitude lost. If the air, instead of being still, rises faster than the bird sinks, the bird will not lose but gain altitude and acquire potential energy proportional to the altitude gained.

the fact that, being mainly a bird of the more arid country to the north, it is near the edge of its range in the Serengeti. The only suitable cliffs in the Serengeti area are outside the national park along the eastern escarpment of the Gol Mountains, and here some 500 or so pairs of Rüppell's griffons nest. Their breeding season is so timed that they normally raise their young during the period from February to May. This is the rainy season, during which the main migratory ungulate populations are usually on the Serengeti and Salei plains, within easy reach of the nesting cliffs. It often hap-

pens, however, that the rains are interrupted by dry spells, and wildebeest and zebras are then forced to move away to the south and the west. That may oblige the griffons to travel 100 kilometers or more each way to get food for their young, which have to be fed daily by one parent or the other.

By following vultures on cross-country flights I have found that in good soaring weather, which usually prevails when the plains are dry, they can keep up average cross-country speeds of some 45 kilometers (28 miles) per hour. Thus they must travel two or three hours each

way between the nesting cliffs and the dry-weather areas. In windy weather the vultures take off and start slope soaring along the cliffs at the first light. The prevailing winds are easterly, and by moving from slope to slope the vultures can make their way westward across the hills to the edge of the Serengeti Plain. There they have to wait for convection to begin (usually between 8:00 and 9:00 A.M.) before they can drift farther downwind across the plains by circling in the first weak thermals.

The best soaring hours of the day are usually between 11:00 A.M. and 4:00

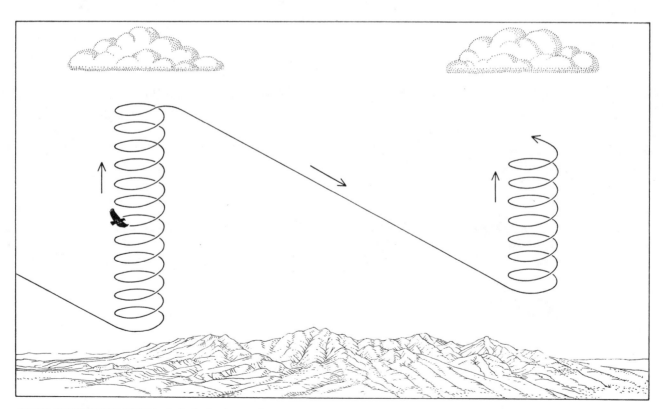

CROSS-COUNTRY FLIGHT is the behavior imposed on itinerant vultures such as Rüppell's griffon by the movements of prey. It re-

quires climbing in successive thermals and gliding in the desired direction. In nesting season 200-kilometer flights are common.

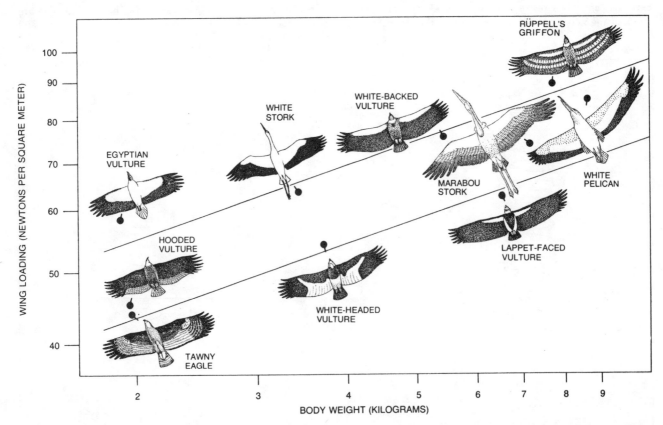

VULTURES' ADVANTAGE over scavenging quadrupeds when prey is sighted lies in being able to reach the carrion first. In this example the descent of a vulture in the distance (*right*) attracts the attention of a hyena and a lappet-faced vulture (*left*) to a carcass some 3.5 kilometers away. The vulture, targeting in at 70 kilometers per hour, reaches the carrion in three minutes, whereas the hyena, running at 40 kilometers per hour, needs 4.25 minutes to cover the same distance. A vulture flying 300 meters above the ground, as in this example, can reach any point 4.5 kilometers away in only six minutes and can get to closer points proportionately faster.

TEN SOARING BIRDS observed in East Africa fall mainly into two groups: five birds with a relatively heavy wing loading (*upper diagonal*) and four with a relatively light wing loading (*lower diagonal*). The wing loading of the 10th bird, the marabou stork, is intermediate on this double-logarithmic plot. The more lightly loaded birds, able to soar in thermals too small and weak for use by the others, hunt intensively over relatively small territories. The heavily loaded birds instead often fly cross-country for food.

P.M., when in dry weather frequent thermals provide rates of climb typically between two and four meters per second over an altitude range from very near the ground (say 1,600 meters above sea level) up to cloudbase (which is normally about 3,500 meters above sea level). A climb from the bottom to the top of this range usually takes 10 minutes or so, after which the bird can glide off straight in the direction it wants to go.

The speed on the straight glides is typically between 70 and 85 kilometers (45 and 55 miles) per hour. At these speeds the glide ratio is about 10 : 1, that is, the bird loses one meter of height for every 10 it travels forward. If it encountered no vertical motion in the air at all, it could glide some 18 kilometers in, say, 15 minutes before the need for another thermal became urgent. The total time needed to travel this distance, including the time for the climb, would be about 25 minutes, equivalent to an average cross-country speed of 43 kilometers per hour. In practice the vultures do not generally use the full altitude range available to them; they usually leave a thermal at between 2,500 and 3,000 meters above sea level. Moreover, they flatten their gliding angle considerably by slowing down as they fly through thermals and speeding up in between. On one occasion a Rüppell's griffon I was accompanying in the glider flew for 32 kilometers without circling by using this tactic and managed to lose only 520 meters of altitude—an achieved glide ratio of better than 60 : 1!

The griffon vultures' ability to forage at a distance of 100 kilometers from the nest gives them another competitive advantage over their formidable competitor the spotted hyena. Hyena pups are confined to a den for the first few months of their life, so that when the game concentrations are not in the immediate vicinity, the mother may have to travel many kilometers to get food, periodically coming back to suckle the young. A hyena's foraging radius cannot be reliably estimated, but it is certainly much less than that of a griffon vulture, because of the hyena's lower speed and the greater effort that travel on foot requires.

The biggest advantage low wing loading gives a bird is the ability to soar early in the day, when the first thermals are usually feeble dust devils, both narrow and weak. This advantage is important to foraging vultures: the earlier they become airborne in the morning, the better are their chances of getting pickings from the remains of animals that have

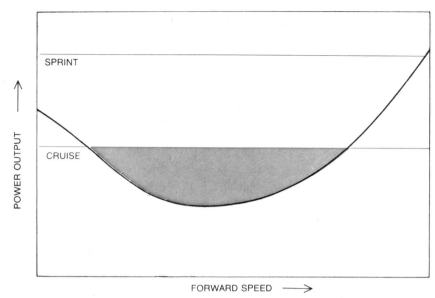

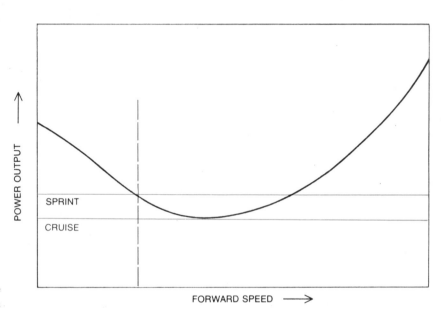

POWER REQUIRED FOR FLIGHT is less at moderate speeds (*trough of "u"*) and greater at hovering (*left*) and high (*right*) speeds. For a bird of medium size, such as a pigeon (*top*), continuously available muscle power (*lower line*) is enough for protracted flight over a wide range of speeds (*shaded area*) and sprint power (*upper line*) allows jump take-offs. For a big bird such as a vulture (*bottom*) continuously available muscle power is not enough to allow protracted flight at any speed. The big bird must run, or dive from a high perch, to reach minimum flying speed (*broken vertical line*) and then use sprint power to remain airborne long enough to reach lift and begin relatively exertionless soaring flight.

died or been killed by predators during the night. In the heat of the day the thermals are broader and stronger, and the advantage in rate of climb to be had from a very low wing loading becomes insignificant; the beneficial effect on cross-country speed is more than offset by the loss of speed on the straight glides. To some extent birds achieve the best of both worlds, because they can reduce their wing area for fast flight. Even so the cross-country speeds they can achieve are modest compared with those of man-made gliders.

When the wing loadings of various birds are plotted against their mass on a double-logarithmic scale, the different vulture species segregate into two groups [*see bottom illustration, opposite page*]. The sedentary lappet-faced vultures and white-headed vultures have low wing loadings; they are specialized for being able to stay airborne in the weakest possible thermals. The griffons, being cross-country fliers, have compromised by having somewhat higher wing loadings.

Not much is known about the ecology

of the two species of small vultures: the Egyptian vulture and the hooded vulture. Both frequent human habitations and garbage dumps, which the larger species do not. They also gather around dead animals and predator kills, and they tend to remain, picking up small scraps, after the bigger vultures have left. The Egyptian vulture's wing loading agrees with that of the high-loading group. The bird also resembles Rüppell's griffon in being a cliff-nester, although it does not form colonies. The hooded vulture, which falls in the low-loading group, sometimes gathers at carcasses in large numbers, but it is a solitary tree-nester.

Even in strong convection currents neither a bird nor a man-made glider can afford simply to glide along in one direction, relying on chance to bring it to the next thermal. All too often the ground is reached before a thermal is found. Thermals are mostly invisible, except for vigorous dust devils, but it is possible to increase the chance of encountering one by flying under a growing cumulus cloud or over a ground feature that looks likely to warm up differentially in the heat of the sun. A glider pilot's skill largely reflects his ability to notice such things, which in turn is based on his knowledge of atmospheric processes. Vultures and eagles are very good at finding the best lift, and they appear to make use of the visible signs in the same way the glider pilot does. For example, a common phenomenon in East Africa is the alignment of thermals into "streets" that are marked by lines of cumulus clouds. Here the thermals can be so close together that a bird or a glider pilot can fly from one to another, without circling and without losing height, for as much as 80 kilometers. Vultures regularly fly along thermal streets, and they will go out of their way to do so.

Vultures will also join other birds (or gliders) that are already climbing in a thermal, and often quite a large group will gather in this way. There is no "flock," however, in the sense of a continuing group. When the individual birds have gained enough height for their immediate needs, they leave the thermal separately in different directions, and the group disperses.

An entirely different kind of behavior is seen in the European white stork (*Ciconia ciconia*). This species is the longest-distance cross-country flier of them all, and it relies almost entirely on thermal soaring to make its annual migration between northern Europe and the southern half of Africa. On the wing

loading diagram it falls in the high-loading group. In some years many hundreds of these birds spend the northern winter in East Africa rather than pressing on farther south. There they can often be seen traveling about in search of the best feeding areas.

The white storks rely on coordinated social behavior to increase their chance of finding thermals. It is rare to see one of these birds flying by itself; usually there are at least 20 of them together, and big flocks numbering several hundred individuals are common. When the storks are between thermals, they press on in the direction they want to go regardless of the appearance of the sky ahead. They will make detours to avoid rain showers but not to follow thermal streets or to fly under active-looking cumulus clouds. The members of the flock spread out laterally into a loose formation, and they fly steadily along on parallel headings. As soon as one part of the flock happens to fly into a thermal, the birds in that part start rising with respect to the rest. The others then alter their headings to converge on those birds that are rising fastest. Soon all the storks are concentrated in a spiraling column in the strongest part of the thermal, each of them constantly adjusting the position of its circle by reference to the relative rates of climb of its neighbors. At the top of the thermal all the birds leave together and once again spread out in their lift-searching formation. The net effect of this behavior is to increase the probability of finding thermals by searching a path 200 or 300 meters wide.

White storks tend to spend more time at the relatively high altitudes near cloudbase than vultures. The thinner air at high altitudes has much the same effect on performance as an increase of wing loading, and it increases cross-country speed provided that the thermals are large and strong. Glider pilots often continue their climb up into a cumulus cloud, but successful use of this tactic calls for a compass and at least one gyroscopic instrument. It is not quite clear whether storks can make an extended climb in cloud. They do enter cumulus clouds from the bottom, but on some occasions (and perhaps always) they stop circling just after they have entered the cloud and fly straight, skimming half in and half out of cloudbase, until they come out at the edge.

One of the most interesting species of East African soaring birds is the marabou stork (*Leptoptilos crumeniferus*). Marabous spend much of their time dabbling about in marshes and shallow water like other storks, but in addition they

are partial to carrion, and they search for it like vultures. Their flight resembles that of white storks in some ways; for example, they sometimes travel in flocks. There is the curious difference, however, that although the marabous' formation on the straight glides is held just as steadily as the white storks', the marabou flock tends to spread out along the direction of travel rather than laterally. This behavior suggests that the marabous depend less than white storks on flock behavior for finding thermals. Supporting this notion is the fact that marabous often soar individually, like vultures, and will follow thermal streets.

The use of the flock as a thermal-searching unit is most highly developed in the white pelican (*Pelecanus onocrotalus*). On the interthermal glides the members of a pelican flock, which again may number several hundred individuals, spread out in an extended echelon, or multiple-V, formation, making a continuous line with no gaps. When the birds are circling, the entire flock turns in formation, so that from a distance one sees a periodic flash of white as they all catch the sun together. Pelican flocks travel in this way between the different lakes of the East African Rift Valley system.

The primary source of energy in soaring flight is of course that extracted from the motion of the atmosphere. For soaring birds, however, there is always some metabolic expense as well. The wings of the bird are not mechanically locked in the horizontal position but have to be held there by the pectoral muscles. All soaring birds, whatever their evolutionary origins, have a subdivided pectoralis muscle, and it appears that one division is a tonic muscle adapted for the function of sustaining tension. The metabolic cost of running this muscle is somewhat conjectural, but it is certainly much less than that needed for flapping flight. In small birds the actual saving is not very great, because the basal metabolism is large compared with the power needed for flight. In large birds the metabolic rate is relatively much less, and so a greater proportion of the total power can be saved by soaring. Some small birds, notably swifts, do soar, but soaring as the primary means of locomotion is chiefly characteristic of large species, most of which would have limited powers of flight without it. Like man, soaring birds extend their powers of locomotion by using a source of energy external to their own bodies; they are perhaps the only other group of animals that do so.

The Evolution of the Hand

by John Napier
December 1962

*In 1960 tools were found together with the hand bones
of a prehuman primate that lived more than a million
years ago. This indicates that the hand of modern
man has much later orgins than had been thought*

At Olduvai Gorge in Tanganyika two years ago L. S. B. Leakey and his wife Mary unearthed 15 bones from the hand of an early hominid. They found the bones on a well-defined living floor a few feet below the site at which in the summer of 1959 they had excavated the skull of a million-year-old man-ape to which they gave the name *Zinjanthropus*. The discovery of *Zinjanthropus* has necessitated a complete revision of previous views about the cultural and biological evolution of man. The skull was found in association with stone tools and waste flakes indicating that at this ancient horizon toolmakers were already in existence. The floor on which the hand bones were discovered has also yielded stone tools and a genuine bone "lissoir," or leather working

tool. Hence this even older living site carries the origins of toolmaking still further back, both in time and evolution, and it is now possible for the first time to reconstruct the hand of the earliest toolmakers.

Research and speculation on the course of human evolution have hitherto paid scant attention to the part played by the hand. Only last year I wrote: "It is a matter of considerable surprise to many to learn that the human hand, which can achieve so much in the field of creative art, communicate such subtle shades of meaning, and upon which the pre-eminence of *Homo sapiens* in the world of animals so largely depends, should constitute, in a structural sense, one of the most primitive and generalized parts of the human body." The im-

plication of this statement, which expresses an almost traditional view, is that the primate forebears of man were equipped with a hand of essentially human form long before the cerebral capacity necessary to exploit its potential had appeared. The corollary to this view is that the difference between the human hand and the monkey hand, as the late Frederic Wood Jones of the Royal College of Surgeons used to insist, is largely one of function rather than structure. Although broadly speaking it is true that the human hand has an extraordinarily generalized structure, the discovery of the Olduvai hand indicates that in a number of minor but nevertheless highly significant features the hand is more specialized than we had supposed.

Tool-using—in the sense of improvisa-

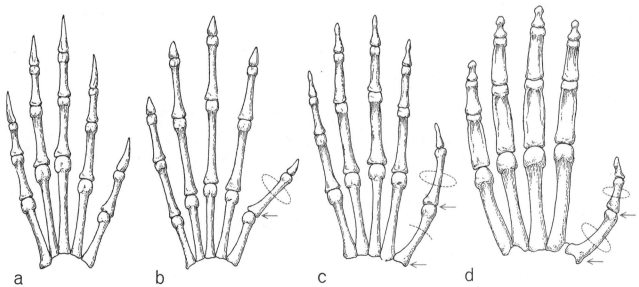

| a | b | c | d |

HANDS OF LIVING PRIMATES, all drawn same size, show evolutionary changes in structure related to increasing manual dexterity. Tree shrew (*a*) shows beginnings of unique primate possession, specialized thumb (*digit at right*). In tarsier (*b*) thumb is distinct and can rotate around joint between digit and palm. In capuchin monkey (*c*), a typical New World species, angle between thumb and finger is wider and movement can be initiated at joint at base of palm. Gorilla (*d*), like other Old World species, has saddle joint at base of palm. This allows full rotation of thumb, which is set at a wide angle. Only palm and hand bones are shown here.

tion with naturally occurring objects such as sticks and stones—by the higher apes has often been observed both in the laboratory and in the wild and has even been reported in monkeys. The making of tools, on the other hand, has been regarded as the major breakthrough in human evolution, a sort of status symbol that could be employed to distinguish the genus *Homo* from the rest of the primates. Prior to the discovery of *Zinjanthropus*, the South African man-apes (Australopithecines) had been associated at least indirectly with fabri-

cated tools. Observers were reluctant to credit the man-apes with being toolmakers, however, on the ground that they lacked an adequate cranial capacity. Now that hands as well as skulls have been found at the same site with undoubted tools, one can begin to correlate the evolution of the hand with the stage of culture and the size of the brain. By the same token one must also consider whether the transition from tool-using to toolmaking and the subsequent improvement in toolmaking techniques can be explained purely in

terms of cerebral expansion and the refinement of peripheral neuromuscular mechanisms, or whether a peripheral factor—the changing form of the hand—has played an equally important part in the evolution of the human species. And to understand the significance of the specializations of the human hand, it must be compared in action—as well as in dissection—with the hands of lower primates.

In the hand at rest—with the fingers slightly curled, the thumb lying in the plane of the index finger, the poise of the

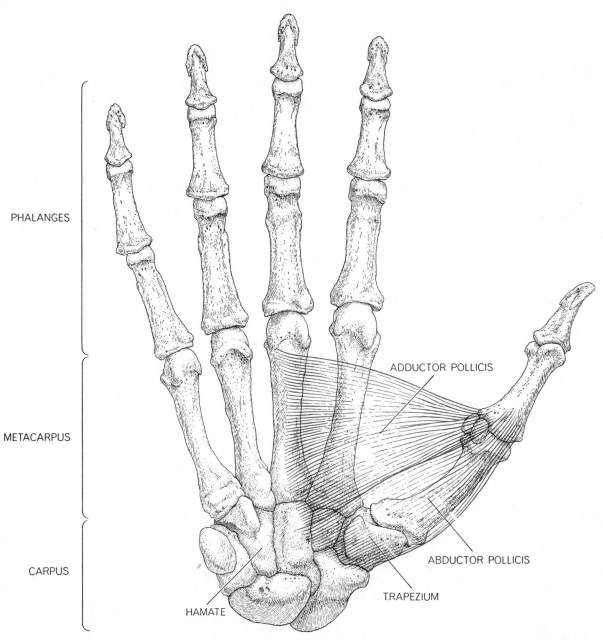

HAND OF MODERN MAN, drawn here actual size, is capable of precise movements available to no other species. Breadth of terminal phalanges (end bones of digits) guarantees secure thumb-to-finger grip. Thumb is long in proportion to index finger and is set at very wide angle. Strong muscles (*adductor pollicis* and

abductor pollicis) implement movement of thumb toward and away from palm. Saddle joint at articulation of thumb metacarpal (a bone of the palm) and trapezium (a bone of the carpus, or wrist) enables thumb to rotate through 45 degrees around its own longitudinal axis and so be placed in opposition to all the other digits.

whole reflecting the balanced tension of opposing groups of muscles—one can see something of its potential capacity. From the position of rest, with a minimum of physical effort, the hand can assume either of its two prehensile working postures. The two postures are demonstrated in sequence by the employment of a screw driver to remove a screw solidly embedded in a block of wood [*see illustration below*]. The hand first grips the tool between the flexed fingers and the palm with the thumb reinforcing the pressure of the fingers; this is the "power grip." As the screw comes loose, the hand grasps the tool between one or more fingers and the thumb, with the pulps, or inner surfaces, of the finger and thumb tips fully opposed to one another; this is the "precision grip." Invariably it is the nature of the task to be performed, and not the shape of the tool or object grasped, that dictates which posture is employed. The power grip is the grip of choice when the full strength of the hand must be applied and the need for precision is subordinate; the precision grip comes into play when the need for power is secondary to the demand for fine control.

The significance of this analysis becomes apparent when the two activities are correlated with anatomical structure. The presence or absence of these structural features in the hands of a lower primate or early hominid can then be

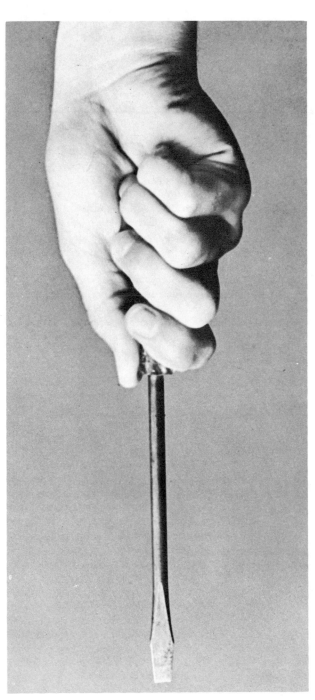

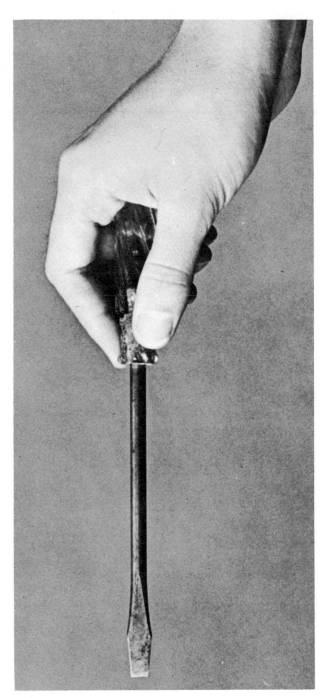

POWER GRIP is one of two basic working postures of human hand. Used when strength is needed, it involves holding object between flexed fingers and palm while the thumb applies counterpressure.

PRECISION GRIP is second basic working posture and is used when accuracy and delicacy of touch are required. Object is held between tips of one or more fingers and the fully opposed thumb.

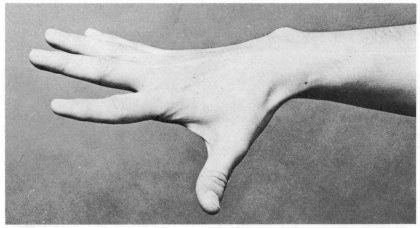

DIVERGENCE, generally associated with weight-bearing function of hand, is achieved by extension at the metacarpophalangeal joints. All mammalian paws are capable of this action.

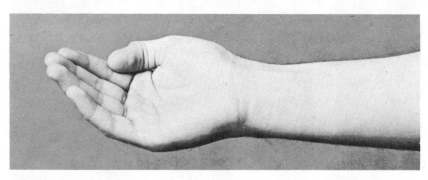

CONVERGENCE is achieved by flexion at metacarpophalangeal joints. Two convergent paws equal one prehensile hand; many mammals hold food in two convergent paws to eat.

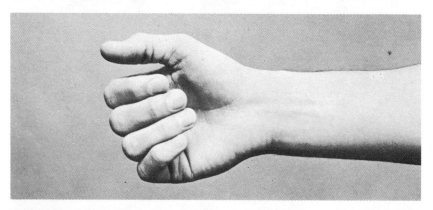

PREHENSILITY, the ability to wrap the fingers around an object, is a special primate characteristic, related to the emergence of the specialized thumb during evolutionary process.

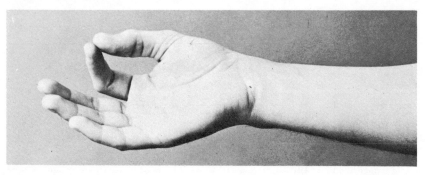

OPPOSABILITY is ability to sweep thumb across palm while rotating it around its longitudinal axis. Many primates can do this, but underlying structures are best developed in man.

taken to indicate, within limits, the capabilities of those hands in the cultural realm of tool-using and toolmaking. In the case of the hand, at least, evolution has been incremental. Although the precision grip represents the ultimate refinement in prehensility, this does not mean that more primitive capacities have been lost. The human hand remains capable of the postures and movements of the primate foot-hand and even of the paw of the fully quadrupedal mammal, and it retains many of the anatomical structures that go with them. From one stage in evolution to the next the later capability is added to the earlier.

The study of primate evolution is facilitated by the fact that the primates now living constitute a graded series representative of some of its principal chapters. It is possible, at least, to accept a study series composed of tree shrews, tarsiers, New World monkeys, Old World monkeys and man as conforming to the evolutionary sequence. In comparing the hands of these animals with one another and with man's, considerable care must be taken to recognize specializations of structure that do not form part of the sequence. Thus the extremely specialized form of the hand in the anthropoid apes can in no way be regarded as a stage in the sequence from tree shrew to man. The same objection does not apply, however, to certain fossil apes. The hand of the Miocene ancestral ape *Proconsul africanus* does not, for example, show the hand specializations of living apes and can legitimately be brought into the morphological sequence that branches off on the man-ape line toward man.

In the lowliest of the living primates —the tree shrew that inhabits the rain forests of the East Indies and the Malay Archipelago—the hand is little more than a paw. It exhibits in a primate sense only the most rudimentary manual capability. This is the movement of convergence that brings the tips of the digits together by a flexion of the paw at the metacarpophalangeal joints, which correspond in man to the knuckles at the juncture of the fingers and the rest of the hand. The opposite movement— divergence—fans the digits outward and is related to the pedal, or weight-bearing, function of the paw. With its paws thus limited the tree shrew is compelled to grasp objects, for example its insect prey, in two-handed fashion, two convergent paws being the functional equivalent of a prehensile hand. For purposes of locomotion in its arboreal

STONE TOOLS to left of center are similar to those found at Olduvai Gorge, Tanganyika, in conjunction with the hand bones of an early hominid. Such crude tools can be made by using the power grip, of which the Olduvai hand was capable. Finely flaked Old Stone Age tools at right can be made only by using the precision grip, which may not have been well developed in Olduvai hand.

habitat, this animal does not require prehensility because, like the squirrel, it is small, it has claws on the tips of its digits and is a tree runner rather than a climber. Even in the tree shrew, however, the specialized thumb of the primate family has begun to take form in the specialized anatomy of this digit and its musculature. Occasionally tree shrews have been observed feeding with one hand.

The hand of the tarsier, another denizen of the rain forests of the East Indies, exhibits a more advanced degree of prehensility in being able to grasp objects by bending the digits toward the palm. The thumb digit also exhibits a degree of opposability to the other digits. This is a pseudo opposability in that the movement is restricted entirely to the metacarpophalangeal joint and is therefore distinct from the true opposability of man's thumb. The movement is facilitated by the well-developed abductor and adductor muscles that persist in the hands of the higher primates. With this equipment the tarsier is able to support its body weight on vertical stems and to grasp small objects with one hand.

The tropical rain forests in which these animals live today are probably not very different from the closed-canopy forests of the Paleocene epoch of some 70 million years ago, during which the first primates appeared. In the wide variety of habitats that these forests provide, ecologists distinguish five major strata, superimposed like a block of apartments. From the top down these are the upper, middle and lower stories (the last being the main closed canopy), the shrub layer and the herb layer on the ground. To these can be added a sixth deck: the subterrain. In the emergence of prehensility in the primate line the three-dimensional arrangement of this system of habitats played a profound role. Prehensility is an adaptation to arboreal life and is related to climbing. In animals that are of small size with respect to the branches on which they live and travel, such as the tree shrew, mobility is not hampered by lack of prehensility. They can live at any level in the forest, from the forest floor to the tops of the tallest trees, their stability assured by the grip of sharp claws and the elaboration of visual and cerebellar mechanisms.

The tree-climbing as opposed to the tree-running phase of primate evolution may not have begun until the middle of the Eocene, perhaps 55 million years ago. What environmental pressure brought about this adaptation can only be guessed at. Thomas F. Barth of the University of Chicago has suggested that the advent of the widely successful order of rodents in the early Eocene may have led to the displacement of the primates from the shrub strata to the upper three strata of the forest canopy. In any case little is known about the form of the primates that made this transition.

In *Proconsul*, of the early to middle Miocene of 20 million years ago, the fossil record discloses a fully developed tree-climbing primate. His hand was clearly prehensile. His thumb, however, was imperfectly opposable. Functionally this hand is comparable to that of some of the living New World monkeys.

True opposability appears for the first time among the living primates in the Old World monkeys. In these animals the carpometacarpal joint shows a well-developed saddle configuration comparable to that in the corresponding joint of the human hand. This allows rotation of the thumb from its wrist articulation. Turning about its longitudinal axis through an angle of about 45 degrees, the thumb can be swept across the palm, and the pulp of the thumb can be directly opposed to the pulp surfaces of one of or all the other digits. This movement is not so expertly performed by the monkeys as by man. At the same time, again as in man, a fair range of movement is retained at the metacarpophalangeal joint, the site of pseudo opposability in the tarsier.

The hands of anthropoid apes display many of these anatomical structures but do not have the same degree of functional capability. This is because of certain specializations that arise from the fact that these apes swing from trees by their hands. Such specializations would seem to exclude the apes from the evolutionary sequence that leads to man. In comparing the hand of monkeys with the hand of man one must bear in mind an obvious fact that is all too often overlooked: monkeys are largely quadrupedal, whereas man is fully bipedal. Variations in the form of the hand from one species of monkey to the next are related to differences in their mode of locomotion. The typical monkey hand is rather long and narrow; the metacarpal, or "palm," bones are short compared with the digits (except in baboons); the terminal phalanges, or finger-tip bones, are slender and the tips of the fingers are consequently narrow from side to side. These are only the most obvious differences between the foot-hand of the Old World monkey and that of man. They serve nonetheless to show how too rigid an application of Frederic Wood Jones's criterion of morphological similarity can mislead one into assuming that the only important difference between the hands of men and monkeys lies in the elaboration of the central nervous system.

It seems likely that the terrestrial phase of human evolution followed on the heels of *Proconsul*. At that time, it is well known, the world's grasslands expanded enormously at the expense of the forests. By the end of the Miocene, 15 million years ago, most of the prototypes of the modern plains-living forms had appeared. During this period, apparently, the hominids also deserted their original forest habitats to take up life on the savanna, where the horizons were figuratively limitless. Bipedal locomotion, a process initiated by life in the trees and the ultimate mechanism for emancipation of the hands, rapidly followed the adoption of terrestrial life. The use of the hands for carrying infants, food and even weapons and tools could not have lagged far behind. As Sherwood L. Washburn of the University of California has suggested on the basis of observations of living higher primates,

tool-using must have appeared at an early stage in hominid evolution. It is a very short step from tool-using to tool-modifying, in the sense of stripping twigs and leaves from a branch in order to improve its effectiveness as a tool or weapon. It is an equally short further step to toolmak-

ing, which at its most primitive is simply the application of the principle of modification to a stick, a stone or a bone. Animal bones are a convenient source of tools; Raymond A. Dart of the University of Witwatersrand in South Africa has advanced the hypothesis that such tools

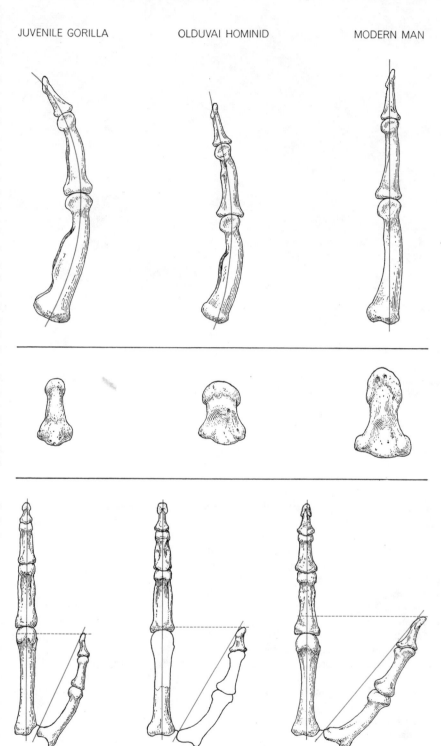

JUVENILE GORILLA   OLDUVAI HOMINID   MODERN MAN

HAND BONES of juvenile gorilla, Olduvai hominid and modern man are compared. Phalanges (*top row*) decrease in curvature from juvenile gorilla to modern man. Terminal thumb phalanx (*middle row*) increases in breadth and proportional length. Third row shows increase in length of thumb and angle between thumb and index finger. Olduvai bones in outline in third row are reconstructed from other evidence; they were not found.

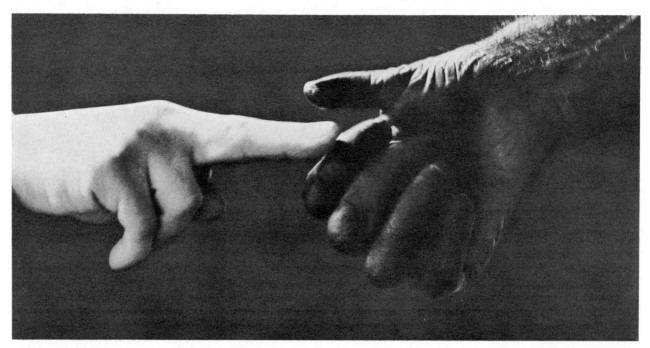

CHIMPANZEE, attempting to grasp experimenter's finger, uses an inefficient precision grip. Because animal's thumb is so short in pro-portion to the digits, it is compelled to bend the digits forward and grasp the object between the sides of index finger and thumb.

were used by early man-apes as part of an "osteodontokeratic" (bone-tooth-hair) culture.

The tools from the pre-*Zinjanthropus* stratum at Olduvai Gorge are little more than pebbles modified in the simplest way by striking off one or more flakes to produce a chopping edge. This technology could not have required either a particularly large brain or a hand of modern human proportions. The hand bones of the pre-*Zinjanthropus* individuals uncovered by the Leakeys in their more recent excavation of Olduvai Gorge are quite unlike those of modern *Homo sapiens*. But there seems to be no reason, on either geological or anthropological grounds, for doubting that the tools found with them are coeval. Modern man must recover from his surprise at the discovery that hands other than his own were capable of shaping tools.

At this point it may be useful to return to the analysis of the manual capability of modern man that distinguishes the power and the precision grip. When compared with the hand of modern man, the Olduvai hand appears to have been capable of a tremendously strong power grip. Although it was a smaller hand, the relative lengths of the metacarpals and phalanges indicate that the proportion of digits and palm was much the same as it is in man. In addition, the tips of the terminal bones of all the Olduvai fingers are quite wide and the finger tips themselves must therefore have been broad—an essential feature of the human grip for both mechanical and neurological reasons. The curvature of the metacarpals and phalanges indicates that the fingers were somewhat curved throughout their length and were normally held in semiflexion. Unfortunately no hamate bone was found among the Olduvai remains. This wristbone, which articulates with the fifth metacarpal, meets at a saddle joint in modern man and lends great stability to his power grip.

It seems unlikely that the Olduvai hand was capable of the precision grip in its fullest expression. No thumb metacarpal was found in the Olduvai deposit; hence any inference as to the length of the thumb in relation to the other fingers must be derived from the evidence of the position of the wristbone with which the thumb articulates. This evidence suggests that the Olduvai thumb, like the thumb of the gorilla, was set at a narrower angle and was somewhat short-er than the thumb of modern man, reaching only a little beyond the meta-carpophalangeal joint of the index finger. Thus, although the thumb was opposa-ble, it can be deduced that the Olduvai hand could not perform actions as pre-cise as those that can be undertaken by the hand of modern man.

Nonetheless, the Olduvai hand acti-vated by a brain and a neuromuscular mechanism of commensurate develop-ment would have had little difficulty in making the tools that were found with it. I myself have made such pebble tools employing only the power grip to hold and strike two stones together.

The inception of toolmaking has hitherto been regarded as the milestone that marked the emergence of the genus *Homo*. It has been assumed that this de-velopment was a sudden event, happen-ing as it were almost overnight, and that its appearance was coincidental with the structural evolution of a hominid of es-sentially modern human form and pro-portions. It is now becoming clear that this important cultural phase in evolu-tion had its inception at a much earlier stage in the biological evolution of man, that it existed for a much longer period of time and that it was set in motion by a much less advanced hominid and a much less specialized hand than has pre-viously been believed.

For full understanding of the subse-quent improvement in toolmaking over the next few hundred thousand years of the Paleolithic, it is necessary to docu-ment the transformation of the hand as well as of the brain. Attention can now also be directed toward evidence of the functional capabilities of the hands of early man that is provided by the tools they made. These studies may help to account for the radical changes in tech-nique and direction that characterize the evolution of stone implements during the middle and late Pleistocene epoch. The present evidence suggests that the stone implements of early man were as good (or as bad) as the hands that made them.

8 Paleoneurology and the Evolution of Mind

by Harry J. Jerison
January 1976

Tracing changes in the relation between brain size and body size in various groups of fossil and contemporary animals sheds light not only on the evolution but also on the nature of intelligence

The mind evolved. Paleoneurology, which deals with the evidence of nervous systems in fossil animals, provides new clues to the nature of that evolution. Integrated with information about the variety of brains and behaviors in living vertebrates and knowledge of how neural tissue is packaged in brains, the fossil record can be interpreted to develop a coherent account of the evolution of intelligence over the 500-million-year span of vertebrate history. That account provides fresh perspectives on the nature of intelligence as a biological phenomenon.

The approach complements more traditional ways of studying the evolution of mind: ethological analyses, based on naturalistic observations of species-typical behaviors of living animals, and psychological studies, which measure the competence of various species in standardized laboratory tests. The traditional analyses build on evolutionary relations among living animals to reconstruct the evolution of behaviors; the variety of present patterns of behavior are projected backward in time to probable ancestral patterns. With paleoneurological data, on the other hand, the approach to the evolutionary history of the brain is direct, and the history is interpreted in the light of present-day relations between brain and behavior.

The strategy of the paleoneurological analysis of mind is to identify a morphological trait as a correlate of mind, or biological intelligence. If the mind evolved, certain trends in the evolution of that correlated trait should be evident in the fossil record. For example, since there are obviously different grades in the distribution of intelligence or mind in living animals, it should be possible to measure an increase and a diversifica-

tion of the morphological trait in successive geological periods.

Charles Darwin was the author of an early statement of one hypothesis that relates mind to morphology when he wrote in *The Descent of Man:* "No one, I presume, doubts that the large proportion which the size of man's brain bears to his body, compared to the same proportion in the gorilla or orang, is closely connected with his mental powers." Some 80 years later Karl Spencer Lashley's more careful and explicit statement of the hypothesis made it applicable to the analysis of morphological data: "The only neurological character for which a correlation with behavioral capacity in different animals is supported by significant evidence is the total mass of tissue, or rather, the index of cephalization... which seems to represent the amount of brain tissue in excess of that required for transmitting impulses to and from the integrative centers."

The index, a measure of relative brain size, is defined as total brain size divided by the two-thirds power of the body size. (The exponent has to do with the relation between surface and volume, as I shall explain further along.) The ratio can be visualized if one plots data reflecting the present diversity of brain and body sizes in vertebrates [*see top illustration on page 84*]. Lashley's statement is borne out by the clear differentiation between "lower" and "higher" vertebrates and by the fact that the arrays of points representing both groups are similarly oriented at a slope of 2/3. It is easy to imagine the higher vertebrates as having evolved from a lower vertebrate grade as the result of the vertical displacement of a set of points in brain : body space. That displacement is in effect what is measured by the index

of cephalization. This implies that brain size is determined by a "body-size factor" and an "encephalization factor." Lashley's hypothesis was that only the encephalization factor was involved in the evolution of mind. Birds and mammals are "higher" vertebrates because they are more highly encephalized and are higher on a scale of biological intelligence than reptiles and other "lower" vertebrates. That is an interpretation of the data that agrees nicely with our intuition.

To assemble paleoneurological evidence one needs to know the brain and body size of fossil vertebrates. The brain size is determined from the volume of endocranial casts, or endocasts, that are replicas of the cranial cavity. In some cases natural endocasts are found: mineralized remains of sand or other deposits that replaced the soft tissue in a fossil skull. In other cases the fossil skull serves as the mold for a man-made acrylic or plaster endocast. The body size of the fossil vertebrate can be measured or estimated from models based on skeletal reconstructions. To take one example, the Eocene ungulate (hoofed mammal) *Uintatherium anceps* weighed about 1,250 kilograms; its endocast indicates a brain weight of about 290 grams [*see illustration on page 85*]. By assembling and plotting a large body of data of this kind it is possible to derive "brain : body maps," minimum convex polygons that enclose a set of brain : body points and enable one to analyze major shifts in encephalization without numerical analyses or indexes [*see bottom illustration on page 84*].

In the illustration a polygon for living mammals defines the region in brain : body space associated with the present mammalian grade of encephalization.

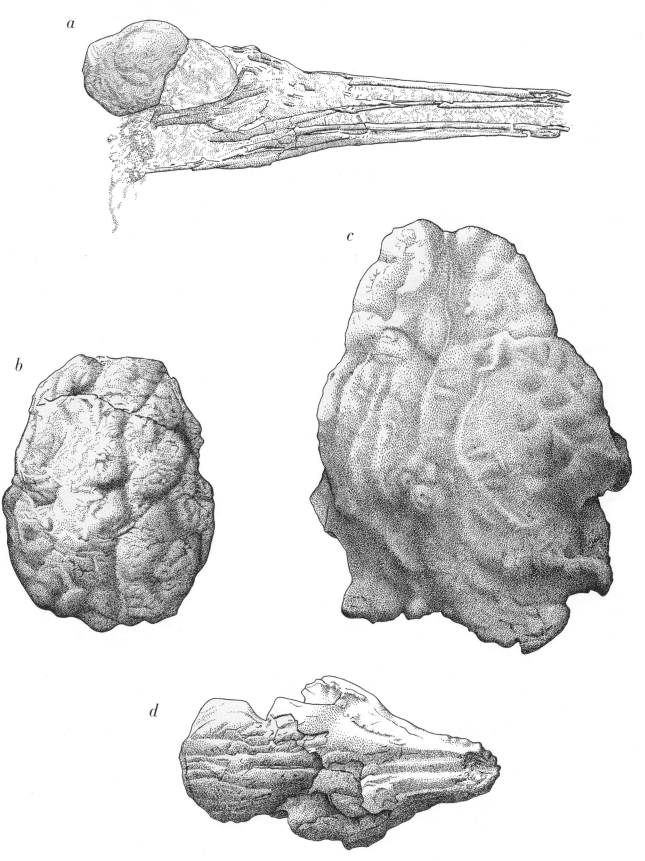

ENDOCASTS, casts of the inside of the cranial cavity, are displayed for four fossil animals. These endocasts, each drawn about one and a quarter times actual size, are natural ones: mineralized remains of sand or other deposits that replaced the soft tissues in the fossil skulls. The endocast of *Numenius gypsorum,* a bird of the Eocene (about 40 million years ago), is still in place in the long-beaked skull (*a*). The endocasts removed from their craniums are of *Potamotherium* (*b*), an otterlike Oligocene carnivore (about 25 million years ago), and *Dicrocerus* (*c*), an ungulate, or hoofed herbivore, from the Miocene (15 million years ago). The endocast of *Cainotherium* (*d*), an Oligocene ungulate, has parts of the skull in place. Casts are from the Muséum d'Histoire Naturelle in Paris.

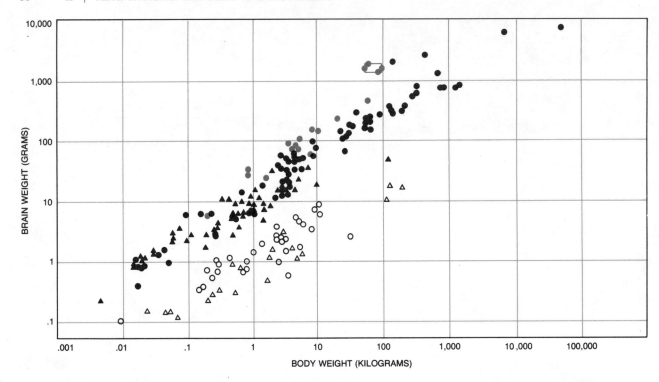

BRAIN SIZE is plotted against body size for some 200 species of living vertebrates. The data were collected by George W. Crile and Daniel P. Quiring some years ago. The four colored points connected by a rectangle represent the extreme measurements reported for man, indicating that variation within a species does not loom large in comparison to the distinctions among species. Data fall into two clearly delimited groups, which may be considered to be the lower and the higher vertebrates. In both cases the data fall along a line with a slope of 2/3 on log-log coordinates: brain size varies with the 2/3 power (which is the cube root of the square) of body size.

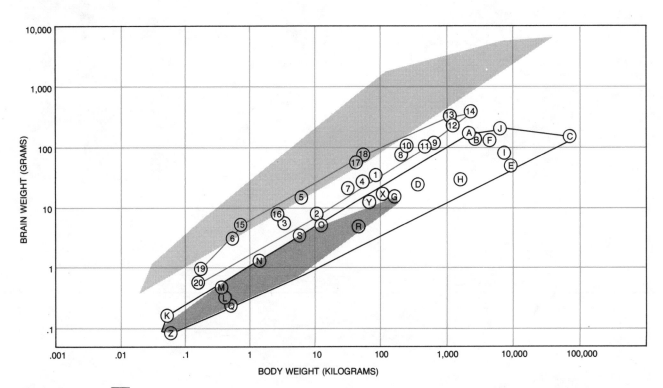

BRAIN : BODY MAPS are the minimum convex polygons that can be drawn to enclose a set of points representing brain size plotted against body size. Here the data for living mammals and reptiles have been taken from the illustration at top of page. Numbered points are for archaic ungulates and carnivores; oldest is *Triconodon* (20). Fossil reptiles are dinosaurs (*A—J*), pterosaurs (*K—O*) and mammal-like reptiles (*Q—S*); two amphibians (*X, Y*) and a fish (*Z*) are included.

The brain: body data of early fossil mammals are shown in the archaic-mammalian polygon. ("Archaic" designates taxonomic orders that are now extinct, having been replaced in their ecological niches by species from "progressive" orders.) The geologic time of the map for archaic mammals extends from the upper Jurassic period represented by the earliest well-preserved mammalian endocast (*Triconodon mordax*) to the late Eocene epoch with forms such as *Uintatherium,* or a time span of from about 150 to 40 million years ago. The polygon indicates that a single archaic mammalian grade persisted for the entire period. In the reptilian polygon the fossil data on dinosaurs, pterosaurs (flying reptiles) and therapsids (mammal-like reptiles) are entered along with the sample of living reptiles. Two fossil amphibians and one fossil fish are included, so that the reptilian polygon is a reasonable picture of the original lower grade of brain: body relations in jawed vertebrate species, the grade from which birds and mammals advanced when they first evolved.

The brain: body maps lead to a number of interesting conclusions:

1. Living reptiles have not departed significantly from their ancestral condition. A single brain: body map includes all the fossil and living species for which we have data, showing that the adaptive radiation of the reptiles and other lower vertebrates was accomplished without any major advances in relative brain size.

2. Adding fossil reptiles extends the reptilian polygon while maintaining its orientation in brain: body space. This fact and the data on mammals suggest that a fundamental biological process determines the orientation at a slope of about 2/3. Because the coordinate system is logarithmic the slope can be interpreted as providing exponents for the measures of body and brain (2 for the square units of a body surface and 3 for the cubic units of a brain volume) and thus for the exponent of 2/3 in the index of cephalization. That is a straightforward application of principles of physics. It makes sense biologically because of the way sensory and motor "surfaces" of the body are projected on the many structures that add up to the volume of the brain; maps of the cerebral cortex of mammals are often presented as grossly distorted but recognizable body surfaces drawn on a diagram of the brain.

3. Dinosaurs were not unusually "small-brained." The 10 dinosaurs that contributed data on fossil reptiles were

FOSSIL ANIMALS are reconstructed on the basis of their skeletal remains in order to provide an estimate of body size to compare with the brain size derived from an endocast. The body can actually be modeled and the model's volume can be determined, or the size can be estimated from a drawing. This animal is *Uintatherium anceps,* an archaic ungulate of some 50 million years ago. It weighed about 1,250 kilograms and its brain (*color*) weighed about 290 grams. Data are plotted as point No. 12 in archaic-mammal map on opposite page.

the source of the inference about the stability of the reptilian, or lower vertebrate, condition and the uniformity of slope for both fossil reptiles and living ones. The 10 dinosaurs were clearly normal reptiles with respect to brain size.

4. Encephalization and the evolution of intelligence occurred independently in birds and in mammals, which evolved independently from two different subclasses of reptiles. (The record on fossil birds is not considered in detail in this article; the 150-million-year-old fossil *Archaeopteryx,* the earliest-known bird, was intermediate between the reptilian grade and the grade of living birds.)

5. The reptilian map and the two mammalian maps are similarly oriented with a slope of about 2/3, and there are vertical displacements that indicate the progressive encephalization of the mammals.

6. Although the step from reptiles to mammals required a certain amount of encephalization (approximately a fourfold increase in relative brain size to transform a reptilian polygon into an archaic mammalian polygon), mammalian encephalization did not progress immediately but remained at a steady level for at least 100 million years. The earliest and the latest of the archaic species of mammals were all at about the same grade of relative brain size, enclosed within a single rather narrow polygon. That stability for such a long period of time suggests a successful response to the selection pressures of a stable new ecological niche.

7. Progressive evolution of encephalization within the mammals came late in their history, in the last 50 million

years of a time span of about 200 million years. That evolution transformed the archaic mammalian map into the map of living mammals by another four- or fivefold increase in relative brain size for the average mammal.

The final steps in encephalization are revealed in the fossil record of mammalian endocasts of the present geological era: the Cenozoic, which began approximately 65 million years ago. Those steps are best measured with a modern version of the index of cephalization: the encephalization quotient (E.Q.), which is the actual size of the brain divided by its expected size for an average living mammal. The expected size is determined by an equation, which states that the brain size equals the 2/3 power of the body size multiplied by a constant (.12) that represents the index of cephalization for an average living mammal [*see illustration on next page*].

In order to analyze the progressive evolution of encephalization in the mammals, I have computed average E.Q.'s and standard deviations for samples of fossil and living ungulates and carnivores and plotted them as a set of normal curves [*see top illustration on page 87*]. The curves have equal areas and can therefore be treated as normal probability distributions, that is, each curve shows the probability of the presence of species of the indicated degrees of encephalization in each assemblage. There was evidently a steady advance in average E.Q. during the Tertiary period (from about 65 million to about three million years ago), and there was also a diversification of E.Q. that was more or less proportional to the average E.Q. Significantly, the distributions overlap in the lower range of encephalization, indicating that some relatively small-brained species persisted in all the fossil groups and that they still persist among living mammals.

These are results that one would expect if encephalization (and intelligence) evolved as other traits have. Evolution involves morphological and behavioral adaptations to a variety of niches and to the invasion of new niches. The adaptive zones occupied by successive species of carnivores and ungulates during the Tertiary period must have included many niches in which there were selective advantages for species further encephalized. As more of these niches were invaded there would have been diversification, which would in turn have affected the frequency distributions of

E.Q.'s [*see bottom illustration on opposite page*].

The gradual change in encephalization throughout the Tertiary period has thus far been demonstrated only for the ungulates and carnivores among the orders of mammals. Although that may be owing simply to a lack of enough data on other orders, the limited evidence on other mammals suggests a different pattern of evolution: When an adaptive zone was entered, the succession of species in that zone attained a particular grade of encephalization rapidly and then maintained it. For example, the earliest-known insectivore endocasts, from about 35 to 40 million years ago, were already at the same grade of encephalization as average living species of insectivores, namely somewhat above the archaic level. Similarly, endocasts and body sizes for "dolphins"

of about 18 million years ago, the earliest in this group of cetaceans, show that they had already reached the E.Q. level of comparable living species such as the harbor porpoise. Dolphins, as large-brained mammals, may therefore represent an evolutionary picture very different from the more or less equally encephalized human species. As we shall see, the evolution of the hominid brain to its present size is a relatively recent phenomenon, having been completed only within the past million years.

The fossil evidence on the primate brain is not as complete as that for carnivores and ungulates, but it is orderly and involves several striking features. The primates have always been large-brained mammals. Early Tertiary prosimians had values of E.Q. ranging from .55 to 1.75. This can be contrasted with the range for all other early Tertiary progressive land mammals on which I have data, which is from .19 to .92. Living monkeys and apes are about twice as encephalized as other living land mammals (a mean E.Q. of 2.1 for simians as compared with 1.0 for average mammals). Primates of about 50 million years ago were also about twice as encephalized as their land-mammal contemporaries.

Encephalization within the hominids is measurable by cranial capacity rather than E.Q. (The comparison of cranial capacities is equivalent to the analysis of encephalization in species that are similar in body size, since it amounts to scaling vertically in brain : body space.) An australopithecine grade—a volume of about 500 milliliters—had been achieved in the early Pleistocene epoch, two or three million years ago, and probably as early as the Pliocene, five million years or more ago. A pithecanthropine grade has been recorded in a unique 800-milliliter endocast from Lake Rudolph in Kenya that is almost three million years old; the better-known true pithecanthropines are about a million years old and their cranial capacities range from about 750 to 1,250 milliliters [see "The Casts of Fossil Hominid Brains," by Ralph L. Holloway; SCIENTIFIC AMERICAN; Offprint 686].

The earliest endocast for *Homo sapiens* is about 250,000 years old, and man's present cranial capacity ranges from about 1,000 to 2,000 milliliters. The important evolutionary fact is the rapidity and recency of the increase in encephalization in the hominid lineage. There is no evidence of a change in encephalization in any other mammals in

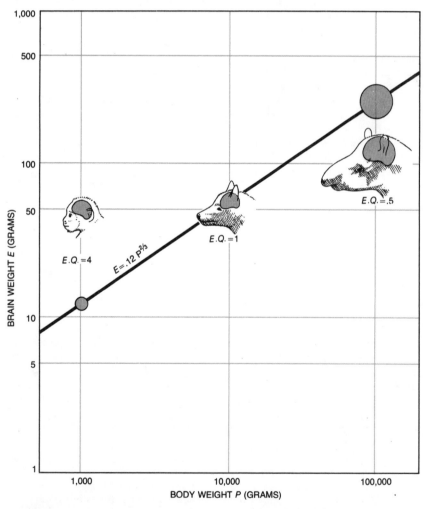

ENCEPHALIZATION QUOTIENT (E.Q.) is the ratio of an animal's actual brain size to its "expected" brain size. The expected sizes, represented by the diagonal, are given by an equation: brain size (E) equals .12 (the index of cephalization for an average mammal) times the 2/3 power of the body size (P). A hypothetical "smart" monkey has a brain four times as large (48 grams) as the expected size (12 grams) for its body size; its E.Q. is said to be 4. An "average" dog has an E.Q. of 1 and a "stupid" tapirlike animal has an E.Q. of .5.

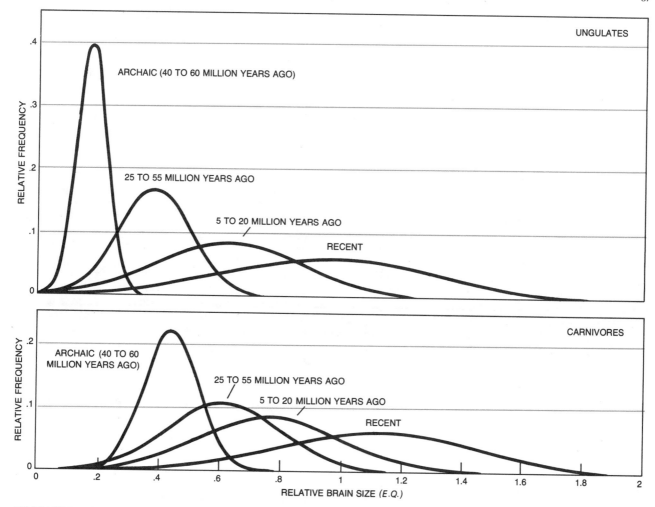

CHANGING DISTRIBUTIONS of relative brain size (E.Q.) are plotted for ungulates (*top*) and carnivores (*bottom*). Each curve gives the distribution of brain sizes for various species during specified periods; recent species are those living today. There is a steady increase in brain size and a concomitant diversification, with small-brained species persisting, shown by the flattening of successive curves. Note that these are "between species" curves. Differences within a species are usually not reflected in behavior.

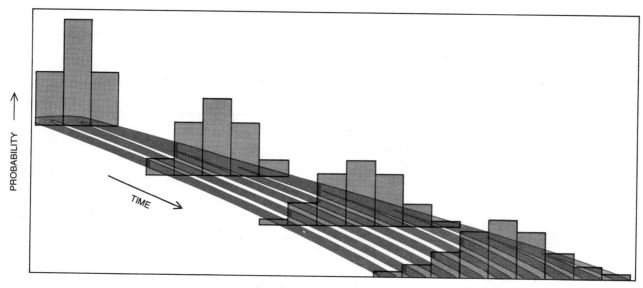

DIVERSIFICATION IN BRAIN SIZE reflects an adaptive radiation with respect to encephalization. The diverging pathways (*color*) suggest the increasing number of ecological niches making diverse demands on the brain. Niches appeared that required more encephalization, and other niches were preserved that made smaller demands. Animals adapted to a wider range of "encephalization niches" were able to survive. The bars reflect schematically the fact that distribution depends on discrete numbers of niches.

the past five million years [*see illustration below*]. In the remainder of this article I shall discuss some critical events in the history of the vertebrates that provide clues to the causes of encephalization and also to the nature of biological intelligence.

Vertebrates do not live by brains alone. That is the message from the insignificant encephalization in the many successful species of lower vertebrates and from the stability of relative brain size in the archaic mammals for at least 100 million years. The advance from fish to amphibian about 350 million years ago illustrates the conservatism in brain evolution. Here was an invasion of terrestrial niches, possibly the most demanding new adaptive zone in vertebrate history, yet according to the paleoneurological evidence there was no increase in encephalization. That was possible because the earliest amphibian required only minor alterations in the patterns of neurological and behavioral organization of its immediate ancestor among the bony fish. Sir James Gray has

shown that even the adaptations for movements on land could be "conservative." The legs of an amphibian served the same function as the inertial force of water for a swimming animal, providing a fulcrum that enabled early amphibians to be little more than fish that swam on land.

In the same sense, I believe, the earliest mammals were probably only slightly modified from their reptilian predecessors: they were "reptiles" that were active at night. This view, which explains the evolution of enlarged brains in early mammals, is consistent with the general evolutionary history of reptiles of the late Paleozoic and early Mesozoic eras, about 250 million years ago. The mammal-like reptiles called therapsids were the dominant forms during the late Paleozoic era, and they were replaced during the early Mesozoic by the ruling reptiles, notably the dinosaurs. The dinosaurs and therapsids were clearly in a kind of competition for the normal niches of land reptiles, and the mammal-like reptiles lost the competition, becoming completely extinct by the

mid-Mesozoic era, about 150 million years ago. The niches for which the therapsids competed and lost were for large diurnal animals that used vision as the normal sense for receiving information about events at a distance. The earliest mammals were a persistent remnant of only slightly modified therapsids, and they survived as small, nocturnal animals. They had improved auditory and olfactory systems and their visual systems were modified, with rod cells taking the place of cone cells. Both changes were appropriate for animals that were active at night.

The evolution of hearing and smell to supplement vision as a distance sense is sufficient reason for the evolution of an enlarged brain in the earliest mammals. The reason is to be found in the way neural elements are packaged in vertebrate sensory systems. In the visual system many of the circuits are in the retina, which contains an extensive and complex neural network that allows elaborate analysis of visual information. The corresponding neural elements of the auditory and olfactory systems of

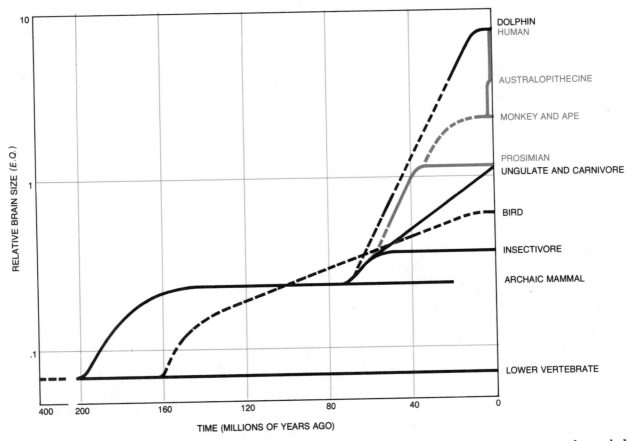

RATE OF EVOLUTION of grade of encephalization varied in different vertebrate groups and at different times. There was actually variability of brain size within groups too, so that the curves are somewhat arbitrary. In the case of cetaceans the highest grade attained is plotted to emphasize that the cetaceans that reached the dolphin grade did so long before the primates reached even an australopithecine grade; the recent and rapid evolution of the hominid brain is notable. The broken lines indicate gaps in data.

living vertebrates are in the brain proper [*see illustration on next page*].

In quantitative terms the effect of different modes of packaging is enormous. A small, highly visual lizard such as the American chameleon *Anolis* has in its retina at least a million sensory cells (cones) and about 100,000 ganglion cells, where the fibers of the optic nerve originate. Still other retinal neurons are involved in higher-order processing in that lizard's retina, and they must be almost as numerous as the sensory cells. The ear of *Anolis*, on the other hand, has only a few hundred sensory cells and other neural elements external to the brain. An auditory system analogous to the visual system would presumably have to have about as much integrative circuitry as there is in the retina, so that an "auditory" animal the size of *Anolis* would need space for almost a million integrative neurons and their dendritic fields to analyze inputs from the ears. There is no space for these in the middle and inner ears; the obvious place to package the additional material is in the brain itself, and solving the packaging problem would therefore require the enlargement of parts of the brain involved in audition. A similar argument would apply to an "olfactory" animal: its increased dependence on the sense of smell would require expansion of the forebrain systems that contain the integrative neurons of the olfactory system.

The brain of an early mammal with the body size of *Anolis* would have handled distance information in a reptilian way but with audition and olfaction as the receptor systems in place of daylight vision. Such a mammal's brain would have had to be enlarged compared with a normal Mesozoic reptilian brain in order to have space for the new neural networks that evolved to analyze nonvisual information. And so we see that the first expansion of the vertebrate brain may have been primarily a solution of a packaging problem and that it may only incidentally have resulted in the evolution of intelligence. Let us see how this crucial incidental result might have been attained.

The introduction of encephalized and finely discriminative audition and olfaction demands new ways of encoding neural information. Visual information is encoded at a retinal level with a structurally determined spatial code: the optics of the eye and the arrangement of retinal elements provide a grid that labels the location of stimulated cells. No such code is possible for sound or odors.

It is difficult for us, the least olfactory of land mammals, to imagine how spatial information could be encoded with the olfactory system, but consider how such coding is accomplished by the auditory system. Animals that use echolocation to identify the source and shape of distant objects do it by translating spatial information into a temporal code. Some such localization of environmental sounds in space must have been accomplished by the early mammals if audition was for them a precise distance sense. In the evolution and functioning of the neural apparatus necessary for such behavior two dimensions of sensory experience had to be encoded by the brain: space and time.

More was needed. Imagine an early mammal coping with life at twilight, sensing stimuli from distant sources. The stimuli are recorded by reptilelike vision (modified toward the mammalian retina with rod cells for night vision), mammalian hearing and mammalian smell, all providing information from the same environmental source. It would obviously be adaptive if the information received from the different sense modalities were given a common code, or label. The integrating code would work, in all likelihood, by the labeling of stimuli in the different modalities as coming from the same object in space at a particular time. And so we have the basic constructs of human conscious experience: objects in space and time. The conscious experience is essentially a construction of nervous systems for handling incoming information in a simple, consistent way.

This leads me to a few simple propositions. Reality, or the real world we know intuitively, is a creation of the nervous system: a model of a possible world, which enables the nervous system to handle the enormous amount of information it receives and processes. (That view is similar to one presented by the English psychologist K. J. W. Craik on the nature of explanation.) The "true" or "real" world is specific to a species and is dependent on how the brain of the species works; this is as true for our own world—the world as we know it—as it is for the world of any species. (That view was made familiar by the German biologist Jakob Johann von Uexküll, who described the "perceptual worlds" of animals and men.) The work of the brain is to create a model of a possible world rather than to record and transmit to the mind a world that is metaphysically true.

Biological intelligence, then, is a mea-

sure of the quality of the particular real world created by the brain of a particular species. The world as we know it ourselves, with the self as perhaps its most complex object, represents the human grade of biological intelligence. Different worlds are presumably constructed by different species. A very simple construction of a world may be characteristic of the lower vertebrates. As a matter of fact, no transformation of neural information—no construction—at all may be required in the lower vertebrates. Their behavior is tightly bound to specific stimuli by fixed-action patterns of response, in contrast to an "intelligence" system in which varied patterns of stimuli are transformed into invariant objects. The birds seem to be a special case. In birds the fixed-action pattern is the typical behavioral mode, and biological intelligence may be a little-used capacity. Yet experimental procedures showing that birds are well within the mammalian range of competence in performing standardized—albeit "unnatural"—learning tasks seem to affirm the basic validity of the judgment, based on relative brain size, that birds and mammals are at comparable grades of biological intelligence. Intelligence, in biological perspective, is clearly only one of several dimensions of behavior and is not the most important one for birds.

The further encephalization of the mammals beyond the archaic level involved new and peculiarly mammalian adaptations to niches that became available as a result of the extinction of dinosaurs and other ruling reptiles. These were daytime niches, and the mammalian response to them occurred in two stages. The first stage was an adaptive radiation that did not involve encephalization, in which effects were toward increases in body size; this was the archaic mammalian radiation. The second stage was a response to the new availability of daytime niches. It was inevitable that the visual system would evolve in some mammalian species and that those species would be at an adaptive advantage in daytime.

The new mammalian visual system would not be a simple retrogression to reptilian vision. Mammals had by then lost many of the reptilian visual adaptations; their normal vision was nocturnal, based on a new system of receptors (rod cells rather than cone cells) and a different analysis of the information from the receptors. Their vision must also have been much more encephalized than reptilian vision if it needed to be integrated with auditory and olfactory information. Lost adaptations do not reappear, and it is clear from the record of the mammals that their daytime vision was based on mechanisms different from those of reptiles. Mammalian vision is represented at forebrain levels, in the thalamus and the cortex, whereas reptilian vision has its most significant central representation at the level of the midbrain and is accomplished to a great extent at a retinal level.

What was the nature of the encephalized daylight visual system in progressive Tertiary mammals? A conservative approach would assume that such a system would be modeled after the other encephalized systems of the brain—the auditory, olfactory and night-vision systems—and would therefore involve the

temporal encoding of spatial information and also object formation. Because the peripheral information would already have been encoded spatially, the temporal code might result in the creation of "mental images," or memories of objects and of their spatial location. That construction would also involve other sense modalities, and so integrative systems for associating information contributed by the various sense modalities would have to be expanded. That called for still further morphological encephalization.

The step to man can be analyzed in a similar way. Several unusual selection pressures may help to explain the peculiar hominid adaptations and encephalization. It is currently accepted that the early hominids were nonarboreal pri-

mates that had invaded a niche comparable to that of predatory carnivores and that they had become modified for life in such a niche. What special problems may have faced such progressive primates, assuming that, like their living relatives, they were noisy social animals with reduced olfactory systems? Their geographic range would have been considerably more extensive than that of any other living primate, and they would have had to cope with that range with a much diminished sense of smell. Olfactory labels, with which wide-ranging social carnivores such as wolves mark a territory and map a perceptual world, would not have been available to the hominids. The development of adequate labels to mark the range may have required the further development of audi-

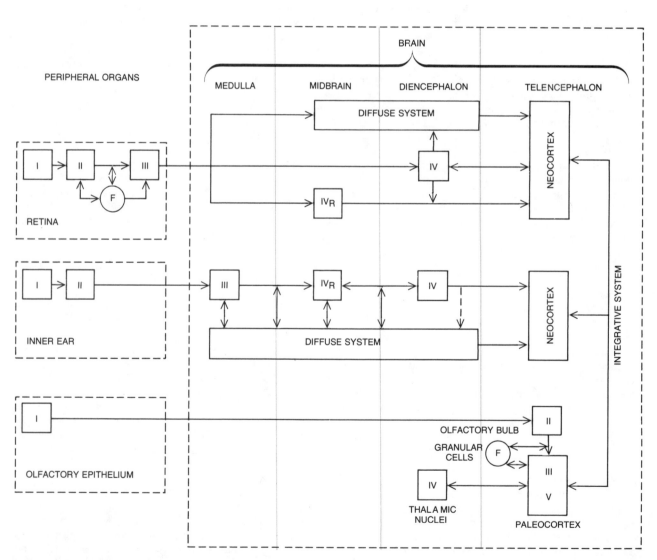

VISUAL, AUDITORY AND OLFACTORY SYSTEMS of living mammals are diagrammed in simplified schematic form to show that a much more significant fraction of the visual system than of the other two systems is packaged not in the brain but in a pe-

ripheral organ. The arrows show the flow of information through successive orders of nerve cells (*I through V*); stages labeled IV_R are parts of reflex control systems; *F* marks feedback loops. The integrative neocortical system (*right*) is specifically mammalian.

tory, visual and particularly vocal capacities—the last of these acknowledging the fact that primates are noisy animals. Such a development is consistent with the evolutionary model used to analyze encephalization in other mammals. In the case of the primates it suggests the evolution of primitive language as a further solution to the problem of creating a real world that provides an adequate model for the sensory events encountered during an animal's life.

A primary model of that type postulates language as a sensory-perceptual development. The availability of vocal labels can obviously result in the capacity to communicate linguistically, but I perfer to separate the role of language in perceptual activity from its role in communication; this makes it much easier to justify the evolution of language. If there were selection pressures toward the development of language specifically for communication, we would expect the evolutionary response to be the development of "prewired" language systems with conventional sounds and symbols. Those are the typical approaches to communication in other vertebrates, and they are accomplished (as in birds) with little or no learning and with relatively small neural systems. The very flexibility and plasticity of the language systems of the human brain argue for their evolution as having been analogous to that of other sensory integrative systems, which are now known to be unusually plastic, or modifiable by early experience. (Benjamin Lee Whorf and Edward Sapir pointed out many years ago one of the maladaptive features of this flexibility of the language system, which enables different societies to develop different languages and hence different realities, often with catastrophic effects on the interactions of human communities.)

I am proposing here that the role of language in communication first evolved as a side effect of its basic role in the construction of reality. The fact that communication is so central to our present view of language does not affect the argument. It is, in fact, theoretically elegant to explain the evolution of an important novel adaptation in a species by relating it to the conservation of earlier patterns of adaptation. We can think of language as being merely an expression of another neural contribution to the construction of mental imagery, analogous to the contributions of the encephalized sensory systems and their association systems. We need language more to tell stories than to direct actions. In the telling we create mental images in our listeners that might normally be produced only by the memory of events as recorded and integrated by the sensory and perceptual systems of the brain.

Mental images should be as real, in a fundamental sense, as the immediately experienced real world. Both are constructions of the brain, although it is appropriate to encode them in order to distinguish image from reality. The role of language in human communication is special because we have the vocal and manual apparatus to create spoken and written language. In hearing or reading another's words we literally share another's consciousness, and it is that familiar use of language that is unique to man. The point, however, is that it was necessary to have a brain that created the kind of consciousness communicated by the motor mechanisms of language.

That new capacity required an enormous amount of neural tissue, and much of the expansion of the human brain resulted from the development of language and related capacities for mental imagery.

The vertebrate brain evolved to control the normal range of behavior within each vertebrate species. It is reasonable to identify the brain of lower vertebrates as being adapted to control fixed-action patterns in response to specific patterns of stimulation, with few requirements for plasticity or flexibility. In the higher vertebrates, the birds and the mammals, plasticity and flexibility are evident in all living species. Yet the birds developed in their own direction, perfecting the fixed-action pattern as the basic behavioral response to environmental requirements. It is among the mammals that more flexible patterns of behavior have been the rule. If one defines intelligence as the capacity to construct perceptual worlds in which sensory information from various modalities is integrated as information about objects in space and time, the evolution of intelligence is most evident in mammals. That capacity was most elaborately developed in the primates, a group of mammals adapted toward adaptability. In the primates skeletal specialization was minimal and adaptations were more completely determined by the enlargement of the brain and the development of learned-behavior mechanisms than they were in any other vertebrates. The trend culminated in man, and we know it as the capacity for imagery, for language and for culture.

III

ADAPTATIONS: SPECIAL STRUCTURES

And in no other field does nature allow herself such sport; with the weapons of animals she has made a game—dividing into branches, for instance, the horns of stags; assigning simple horns to others . . . spreading others into palms and making fingers shoot out of these . . . [giving] to the bulls horns for attacking—indeed in this class she has also bestowed horns on the females, although in many she only gives them to the males; [and] to the crook-horn [she has given] upright horns twisted with a coil of wrinkles and sharpened at the end into a smooth point, so as to make them suitable for lyres.

Pliny
NATURAL HISTORY, XI, xlv

III ADAPTATIONS: SPECIAL STRUCTURES

INTRODUCTION

Vertebrates see and are seen. Their bodies may be camouflaged or brightly colored or marked. Pigments such as melanins (blacks, browns, some reds) or lipochromes (yellows, oranges, reds) are the common means of generating color. Melanin granules, for instance, are transferred from the sites of their synthesis to the epidermal cells of a feather or hair and become permanent features of those structures when their cells die. Some colors do not arise from colored pigments; instead, physical properties of crystalline or granular inclusions in cells or tissues generate "structural color" by diffracting and reflecting light. The blues of bird feathers or fish scales, the shimmering iridescence of a hummingbird's throat patch, or the silvery sides of a fish, all stem from structural phenomena. Not infrequently, structural color and pigment color are combined to yield a third color. For example, superficial yellow pigment and underlying blue structural color generate the green of most parrot feathers!

In "Reflectors in Fishes," Eric Denton summarizes the many ways that crystals of guanine are used by fish to produce color or pattern. He also shows how crystals are employed as mirrors to increase the efficiency of the visual system in sharks and bony fish, so that stray photons that may have passed through the visual pigment layer of the neural retina are reflected back and given a second chance to activate the visual process. Mirrors are also used to channel light (produced by luminescent bacteria that grow within "culture chambers" in the body) to the surface of certain fish as a means of camouflage or as a warning device.

Color and pattern are not the sole recognition signals of vertebrates. In the article "Horns and Antlers," Walter Modell describes antlers as sex-related recognition signals that have evolved in some male mammals. Horns, on the other hand, are usually found on both males and females of a species and may be used for defensive or offensive purposes. Horns, which are composed of dead keratinized cells of the sort that make up surface epidermis or hair, are not present at birth. They grow at a very rapid rate, however, of the order of 5 centimeters per month in the Dorset ram, for example. That rate slows in adults, though the horns of some buffalos do become very large indeed. Interestingly, the growth of horns is from the inside; that is, the oldest layers of horn are those on the exterior, whereas new cone-shaped layers of horn are added at the core and base of the horn.

Antlers grow anew each year upon some male mammals in response to changing levels of androgens. An extraordinarily rich blood supply to the site of antler bone formation supports growth at rates of up to 1 centimeter per *day*—this is perhaps the greatest rate of bone growth in the animal kingdom. If a deer is castrated after the antlers have started to grow during the summer season, antler growth will continue but full maturation will not be attained; that is, the skin covering of the antler (called velvet) will not be shed, and the antlers themselves will become permanent fixtures upon the deer's head (they will only be lost in cold climates where chilling may kill the cells). Normally, antlers are shed each year after the mating season because of the action of osteoclasts; osteoclasts are a type of bone cell that is also mobilized to digest bone and raise calcium levels in the plasma (see "Calcitonin," by Howard Rasmussen and Maurice M. Pechet, *Scientific American* Offprint 1200). One of the most intriguing aspects of antlers is the remarkable variation in their shapes. We cannot yet discern what selective pressures have operated to establish such bizarre branching patterns or huge sizes.

The last article in this Section, "The Head of the Sperm Whale," treats quite a different sort of head structure, the spermaceti organ of the sperm whale. Malcolm R. Clarke argues that sperm whales may use these oil-filled reservoirs to vary buoyancy during deep dives. Other students of marine mammals have found evidence that the spermaceti organ is an important component of the sound-generating system used by whales in echo-location and perhaps communication. The spermaceti organ may provide an example of how an animal uses one specialized organ for a number of independent processes.

The sperm whale, in its use of oil in the spermaceti organ to alter buoyancy, is not the only aquatic vertebrate to employ fatty materials for that purpose. Wax- and fat-filled swim bladders are common among fishes that inhabit very deep water. Still other fishes that may lack swim bladders lower the specific gravity of their bodies by including much oil in their bones. The fact that "oil is lighter than water" was recognized by vertebrates long before humans came into the scene!

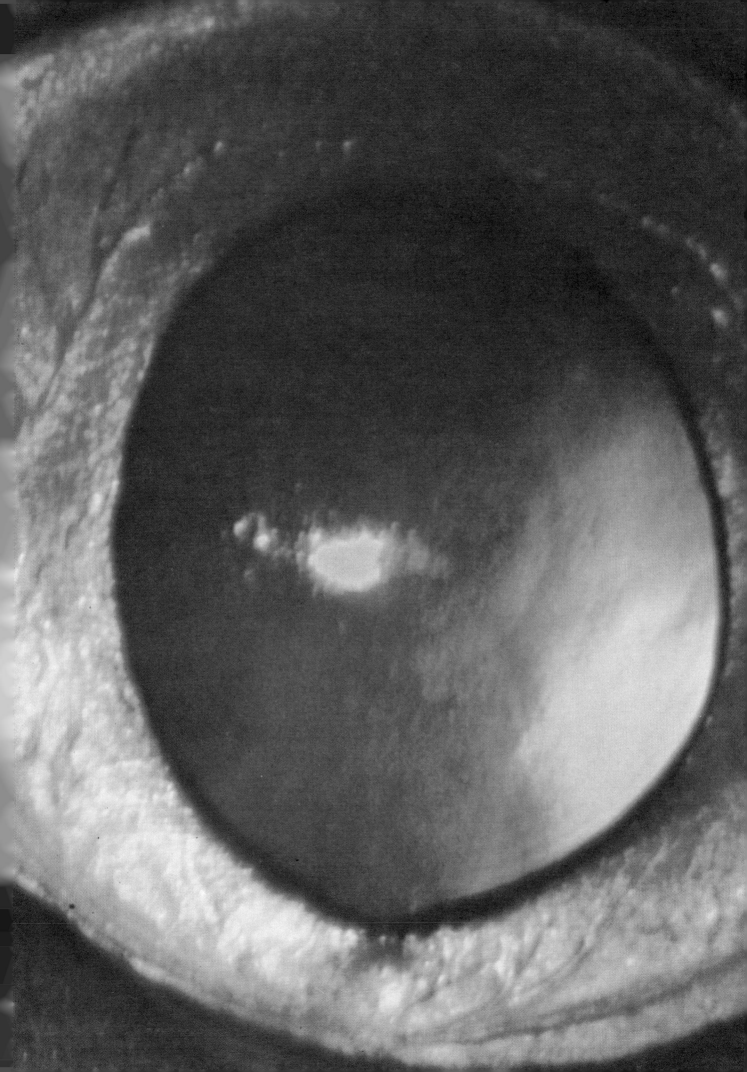

Reflectors in Fishes

by Eric Denton

January 1971

Many fishes are able to secrete nitrogenous compounds that form mirror-like layers of reflective crystals. In the scales and the skin the layers are displays or camouflage; elsewhere they aid vision

The silvery sides of many kinds of fish are so familiar as to be commonplace, but they are surely among the most beautiful objects in the animal kingdom. What makes the silvery sides silvery? It is the ability of these fishes to secrete certain nitrogenous compounds that form thin crystals. Arrays of the crystals, reflecting light like tiny mirrors, are found in the scales and the skin of the fish. Each array is so small that it takes about a million of them to silver a square centimeter of surface. Such crystals are not, however, restricted to the scales and the skin. In some fishes they form a reflecting layer in the eye, and in luminescent deep-sea fishes they act as reflectors that enable the fish to project the light it produces in special directions, sometimes even into a beam resembling the beam of an automobile headlight.

For a number of years J. A. C. Nicol of the University of Texas and I have been investigating the reflecting layers and their functions. The crystals of the reflecting layer range in shape from oblongs to needles [*see middle illustration on next page*]. For many years they were thought to consist only of the nitrogenous compound guanine ($C_5H_5N_5O$), which is named for the guano deposits from which the substance was first isolated. Recently L. M. Greenstein and his colleagues at the Mearl Corporation in New York have shown that the crystals also contain a closely related compound: hypoxanthine ($C_5H_4N_4O$) [*see illustration on page 99*].

When the individual crystals are immersed in water, they reflect about as well as glass does in air. Within the fish's cells, however, they are highly reflective because they are arranged in special stacks consisting of alternating layers of crystal (with a refractive index of about 1.8) and cytoplasm (with a refractive index of 1.33). Theoretical studies have shown that at any given wavelength of light the highest reflectivity will be found when each crystal and each cytoplasmic space between crystals has an optical thickness (thickness times refractive index) of a quarter of that wavelength. For light in the green part of the spectrum, for example, the ideal thickness of each crystal is seven millionths of a centimeter.

As the number of layers in a single array, or platelet, increases, so does the reflectivity of the platelet, and an ideal quarter-wavelength stack with only five crystals will reflect about 75 percent of the light that strikes it over a range of wavelengths that covers more than a third of the visible spectrum. M. F. Land of the University of California at Berkeley and I have found that platelets with dimensions that approximate the ideal quarter-wavelength characteristics are often secreted by marine animals, and also that platelets are often specialized by having the appropriate thicknesses of crystals and spacing to reflect particular colors. Electron-microscope studies that reveal very thin lamellar structures in reflectors have also been made by S. Kawaguti of Japan and J. M. Bassot of France. These stacks of thin plates behave quite differently from stacks in which the plates are more than a few wavelengths thick. A stack of five such "thick" plates reflects only 20 percent of the light that strikes it, but its reflectivity is fairly uniform over the entire spectrum.

Since many reflective crystal units reflect specific colors, the reader may wonder why a silvery fish is essentially colorless. It is because the platelets are usually arranged in several overlapping layers. A single scale from a herring viewed in white light, for example, displays several bright regions of contrasting color. In the fish, however, the reflecting areas of the scales overlap, and below the scales there are other reflecting layers. On the lower flanks of the herring the skin has a superficial golden layer, and below this layer is yet another layer of thin crystals that is called the argenteum, after the Latin word for silver. Since the colors of individual platelets arise from the interference, and not the absorption, of light, a wave band that is not reflected by the first layer of platelets it encounters can penetrate farther and be reflected by a second, a third or an even deeper layer. The total reflection built up in this manner includes all wavelengths and is therefore silvery [*see bottom illustration on page 101*].

It has long been realized in a general way that the reflective surfaces of such fishes as the herring, the mackerel and the salmon helped to camouflage them; some early underwater photographs taken by Francis Ward demonstrated how difficult it is to see some silvery fishes in their natural medium. Thanks to recent studies of the distribution of light underwater, particularly the investigations conducted by John E. Tyler and Rudolph W. Preisendorfer of the University of California at San Diego and by Nils G. Jerlov and his colleagues in Den-

GLOWING EYE of a small shark, the spiny dogfish, is seen enlarged in the photograph on the opposite page. Some of the light entering the eye is reflected back through the pupil by mirror-like crystals arrayed in a membrane behind the retina that is known as the tapetum lucidum. Any vertebrate whose eyes glow in the dark has a similar reflective membrane.

SILVERY SHEEN of the skin and scales of a smelt is produced by the reflections from arrays of crystals like those found in the eyes of the sharks. The crystals embedded in the skin of fishes are long and narrow; those on the inner surface of scales are wider.

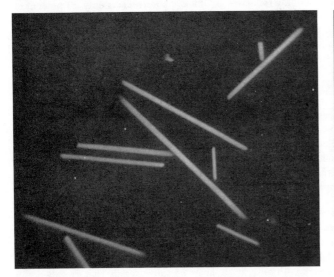

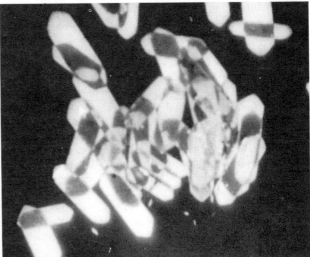

CRYSTALS OF TWO SHAPES are shown in these micrographs. They are from the skin of a herring (*left*) and the scales of a lantern fish (*right*). Many of the broader crystals are overlapping. The ends of the crystals are poorly defined because they taper.

SINGLE CRYSTAL taken from the scale of a herring is shown magnified 4,000 times in this electron micrograph. The crystal has been replicated in carbon and shadowed with metal to show how the two ends are tapered. Its thickness at the center is .07 micron.

mark, it is now possible to discuss how reflection serves as camouflage in more detail.

Light entering water is affected in three significant ways. First, as it passes from one medium to another it is refracted toward the vertical. Second, the precise angle of refraction is not the same everywhere because the surface of the water is not precisely horizontal but is disturbed by ripples and waves. Third, the light is absorbed and scattered in the water. As a result the greater the depth, the more underwater illumination comes to depend on absorption and scattering. Whether the sky is sunny or cloudy, the illumination underwater is almost constant in the horizontal plane, and regardless of the sun's position it is brightest almost directly overhead. The intensity of the light does, however, decrease rapidly with increasing depth.

The fact that underwater light has a fairly constant angular pattern of intensities simplifies the problem of camouflaging fishes. Consider for a moment how a human designer might undertake to camouflage an imaginary silvery fish with a square cross section. The rule of thumb for invisibility is that any light reflected from the fish to the eye of an observer must have the same intensity as the natural background light.

If the sides of such a fish are perfectly reflecting mirrors, they present no camouflage problem, since they will always reflect light identical in intensity with the light an observer would receive if the fish were not present. If the back—that is, the top—of the square fish is to be camouflaged, however, it must reflect only a fraction of the bright light impinging on it from above, and the reflected fraction must be identical in intensity with the dim light being scattered upward around the fish from below. This camouflage can be achieved by darkening the back of the fish so that it reflects only the appropriate amount of light.

The belly of the square fish presents the most difficult problem. To an observer looking up from below, this surface is always seen against the background of the bright light above. Moreover, all that illuminates the belly is the dim light from below, which is far too weak to keep the fish from being silhouetted against its bright background. Only two courses, or a combination of them, seem to be available to the designer. One is to abandon the square cross section and flatten the fish from the sides, giving it a narrow back and belly.

The other is to make the belly of the fish luminous, so that it will be as bright as the background above it. As will be seen, some fishes with the combined design have actually evolved. The expedient of producing light is only possible, however, for animals living deeper in the ocean, where the penetrating daylight is relatively dim.

A flattened shape, although it is good for camouflage, is not the shape that enables a fish to swim best. In point of fact most fishes have a shape that represents a compromise between attributes that favor one or another desirable quality, such as inconspicuousness, speed or maneuverability. The fastest fishes are cigar-shaped in profile and not greatly flattened sideways. How much camouflage is possible for them?

The desirable attribute of a dark back is achieved in fishes such as the mackerel and the herring through a special orientation of their reflective platelets. In the horse mackerel, for example, the platelets do not lie so much parallel to the surface of the fish as they lie parallel to one another. As a result an observer looking down from above sees the platelets edge on. Instead of being reflected by a silvery surface the light passes between the platelets and falls on areas of dark pigment in the skin. To an observer looking from the side the flank of the horse mackerel is well camouflaged by its array of parallel platelets. The silvery surface reflects light that has about the same intensity as the background illumination.

All the reflecting platelets in the horse mackerel are embedded in the skin, several layers of differently colored reflecting platelets overlying a silvery argenteum. The salmon and the herring have two sets of reflectors: long, thin crystals that form the layer of argenteum in the skin and other crystals that lie on the inner surface of the scales. The scale platelets are precisely organized and do not lie parallel to the surface of the fish, as a

simple experiment demonstrates. When a scale that has been detached from the lower flank of one of these fishes is placed among the scales on the upper flank without changing its orientation with respect to the fish, it will be seen to reflect light best in directions that are quite different from those in which its neighbors do.

Divers' reports testify to the effectiveness of the camouflage produced by silvering; a school of silvery fish can sometimes come quite close to a diver before he notices them. Only an observer looking up from below finds them easy to see.

A pattern of camouflage that is suited to one environment obviously may not be suitable in another, and fishes that migrate between one environment and another often show marked changes in appearance. A striking example is afforded by the Atlantic salmon. The parr, or immature salmon, lives in freshwater streams. Its back is marked with a disruptive pattern of bars and red spots that enables it to blend in with the stream bottom. As the young salmon approaches the time for its downstream migration to the sea it develops a herring-like camouflage: its back becomes dark and its sides very silvery.

The use of silvery reflecting layers for camouflage is found in marine animals that are themselves virtually transparent. An animal that consisted entirely of almost transparent tissues would of course be very well camouflaged in the diffuse light underwater. Many tissues, however, cannot function properly and still let light pass. The eye, for instance, could not form an image if light entered it from all directions. Red muscles, the liver, the kidney and such specialized organs as the squid's ink sac are other examples of tissues that cannot function and be transparent. Nature's solution is to treat each of these organs as if it were an entire animal: the sides of the organ

NITROGENOUS COMPOUNDS that comprise the reflecting crystals in fishes are guanine (*left*) and hypoxanthine (*right*). The molecules are alike except for an amino group (*color*).

are silvered and the top is darkened. The stratagem is evident in the almost transparent larvae of flatfish such as the plaice and the sole. The only readily visible parts of these tiny fishes are their eyes and digestive tract; these organs are darkened on top and silvered elsewhere.

The same kind of camouflage is used on the opaque parts of the otherwise translucent squid: its eyes, digestive tract and ink sac.

Surface reflectors do not always serve to provide camouflage. The bright surface markings of such fishes as the neon tetra, an aquarium favorite, are clearly intended for display. Leaving such special cases aside, any silvery fish can be either hard to see or conspicuous, depending on its orientation. Diving friends tell me that on occasion a fish will roll over and swim on its side for a

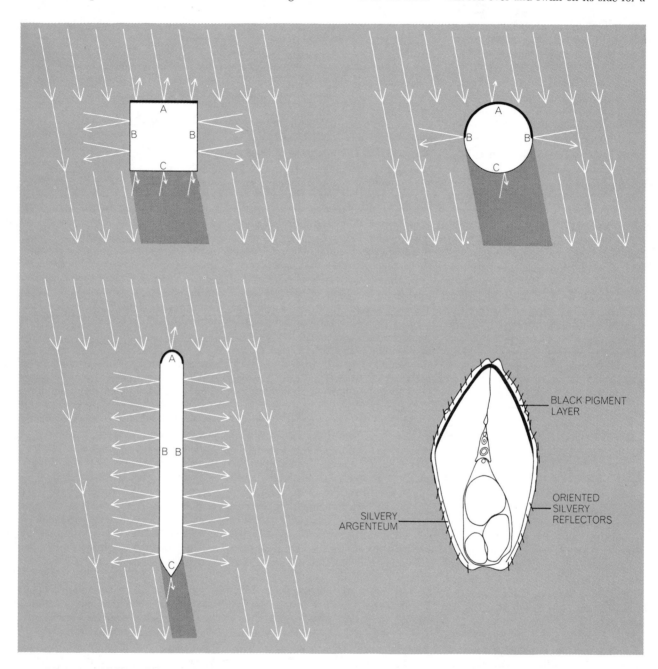

IMAGINARY AND REAL FISH demonstrate how silvery surface reflectors render many fishes inconspicuous when viewed from above or the side. A square fish (*top left*) is exposed to light of differing intensity: the strongest comes from above (*A*), weaker light comes from the sides (*B*) and the weakest comes from below (*C*). The back of the square fish will be invisible to an observer who looks down from above if most of the light from above is absorbed and the amount that is reflected is only as intense as the weak light from below. Similarly, its sides will be invisible if they act as mirrors and reflect back to an observer light that is equal in intensity to the light from the background. Under most circumstances, however, the belly of the square fish will remain conspicuous to an observer looking up from below because the very dim

light that the belly reflects is far less intense than the background light. In effect the fish casts a shadow, and a fish that is circular in cross section (*top right*) casts a much larger shadow than a fish that is equal in area but flattened in shape (*bottom left*). Only a few fishes, however, are so flattened; most fast swimmers are more nearly round. A cross section of a real fish (*bottom right*) shows how nature solves the camouflage problem. The reflecting crystals (*black*) are generally parallel to one another, rather than lying parallel to the surface. This makes the flanks of the fish mirror-like when viewed from most positions to one side. From above only the edges of the reflectors are visible and the dark pigment below them absorbs the descending light. The fish is the horse mackerel; it is well camouflaged except to observers looking up from below.

while; when a silvery fish does this, it becomes very conspicuous indeed. Schools of juvenile fish swimming near the surface can look like rivers of silver. The reason is that the near-surface zone is one where the symmetry of light distribution is least perfect and random motions are most likely to produce eye-catching reflections.

It has been suggested that the loss of camouflage, because of changed orientation, may at times serve a useful purpose. When a school of silvery fish is attacked by a predator, bright flashes are sometimes seen as individual fish in the school twist and turn. It is possible that the flashes distract the predator.

Among the fishes that are not shiny are the cartilaginous fishes: sharks, rays, dogfishes, angelfishes, devilfishes and the torpedo fish. Although such fishes have no crystal platelets in the skin, almost all of them have platelets in the eye. The platelets lie behind the retina, where they form a silvery reflecting layer known as the tapetum lucidum, or "shining carpet." The tapetum is not restricted to fishes; when a cat's or a dog's eyes gleam in an automobile's headlights, the reflection comes from a similar kind of silvery layer behind the retina.

The function of the tapetum is to improve the eye's efficiency. In vertebrates the eye resembles a camera: it has a lens that forms an image of the outside world on a sensitive surface, the retina. Visual pigments in the retina are transformed by the absorption of light, and this is a first step in the process that leads to a message being sent along to the brain. Any light that passes through the retina without being absorbed by a visual pigment is simply wasted. In an eye with a tapetum light that is not absorbed during its inward passage through the retina is reflected back and so has a second chance of being absorbed.

Possession of a tapetum would also have certain potential disadvantages. The reflecting platelets might scatter the light reaching them to other parts of the retina and in so doing make the image of the outside world less sharp. The reflection might also make the animal much more conspicuous, like a cat caught in the headlights. It has been learned, however, that neither of these potential disadvantages actually holds.

The tapetum of fishes was first examined in detail in the early decades of this century by the German zoologist V. Franz. One of the curious features he discovered was that not all the arrays of reflecting platelets behind the retina were, as might be expected, parallel to

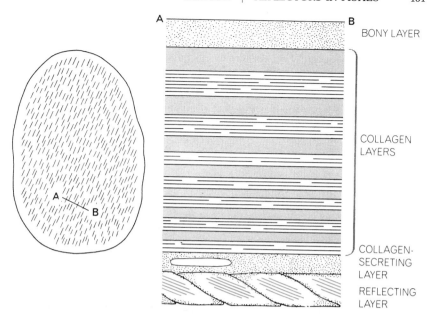

SCALE OF A JUVENILE SPRAT is seen as viewed from the inside and enlarged 60 times (*left*) and in a cross section (*right*). Its actual length is about a millimeter. Dashes that cover almost its entire surface mark the long axes of its numerous arrays of crystal platelets that comprise the reflective surface. The line of cross section is indicated by *A–B*; the bony surface layer of the scale is at the top of the section. Below it are several layers of collagen, a collagen-secreting layer and finally the reflecting layer itself, where individual arrays of platelets, each containing about five crystals, lie at an angle to the plane of the scale.

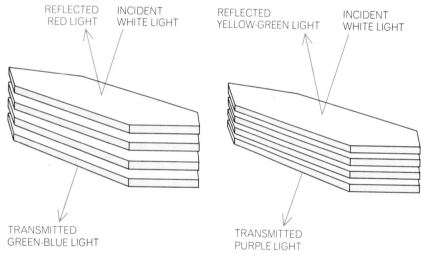

STACKS OF CRYSTALS reflect different light wavelengths depending on the thickness of the cytoplasmic layers between the crystals. When the cytoplasm is about as thick as the crystal (*left*), the array reflects light of the longer wavelengths in the visible spectrum; when the cytoplasm is only half as thick (*right*), reflectivity shifts to middle wavelengths.

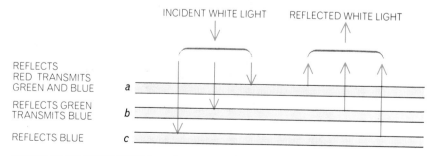

OVERLAP OF REFLECTORS produces the bright sheen characteristic of silvery fishes. If the first layer (*left*) reflects red light (*a*), the second green (*b*) and the third blue (*c*), the reflected increments (*right*) will reach an observer's eye as light of full spectral range.

the surface of the retina. In many parts of the tapetum they were inclined at a considerable angle to the surface. Franz also discovered that when some fishes with a tapetum were exposed to bright light, a black pigment soon appeared in the reflecting layer and covered up the platelets.

The reason many of the platelets lie at an angle to the retinal surface is now clear. In the eye of the spiny dogfish,

for example, once the way the iris restricts the entering light is taken into account it becomes evident that all the platelets in the tapetum are approximately perpendicular to the rays that strike them. Any light that does not interact with the visual pigments before or after reflection either escapes from the eye through the pupil or is absorbed by the black inner surface of the iris. The arrangement of the platelets prevents

their scattering light to other parts of the retina and thereby diminishing the sharpness of the visual image [see upper illustration below].

It has been thought that the function of the appearance of black pigment in the tapetum in response to bright light is to reduce the rate at which the visual pigment is bleached. Since, for some of these animals, the brightest light to which they are exposed in life is prob-

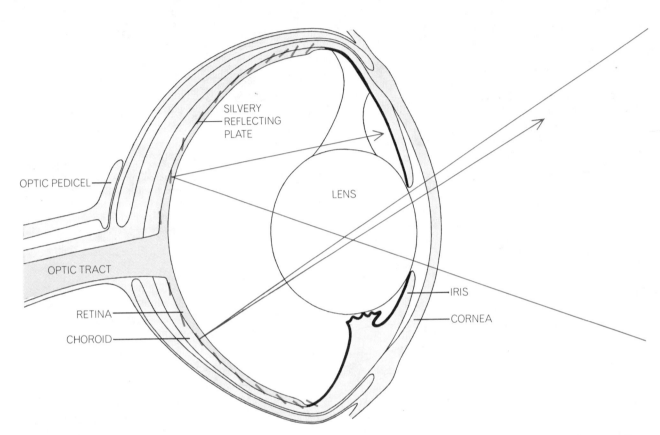

REFLECTORS IN TAPETUM of shark's eye (color) are oriented so that when light rebounds from them, it is absorbed either by the retinal rods or by the black inner surface of the iris, or leaves the eye through the pupil. Diagram omits refractive role of lens.

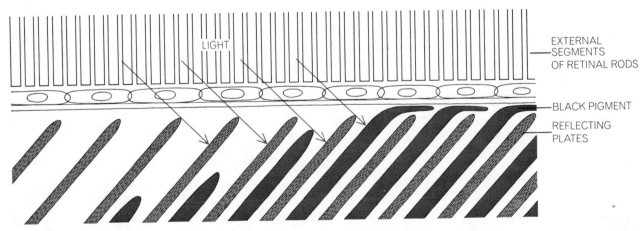

PIGMENT IN TAPETUM advances between the reflector arrays when the eye is exposed to bright light and soon halts further reflectance; progress of the advance is shown from left to right in this drawing. The mechanism makes the animal less conspicuous.

ably 100,000 times brighter than the light at which they can just see, and the maximum pigmentation of the tapetum can only reduce the absorption of light by about 40 percent, this can scarcely be true. It is much more likely that the function of the pigment is to reduce the risk that eyeshine from the tapetum would betray an otherwise inconspicuous fish to predators or prey.

We also found that an eye with a tapetum usually contains about half the visual pigment per unit area found in the eye of an animal that lives in the same environment without the benefit of a tapetum and, in accord with this, that the parts of the receptor cells that contain the visual pigment are only about half as long. The greater efficiency of an eye with a tapetum enables it to absorb about the same amount of light as an eye without a tapetum in spite of its having a smaller amount of visual pigment.

Many deep-sea fishes have light-producing organs, and in some of them the light is aimed in a particular direction. Such aiming is achieved by reflecting mirrors and by lenses. E. R. Gunther reported observing an unusual fish of this kind on a voyage of the research vessel *William Scoresby* in the 1930's. The fish was silvery and eel-shaped and from nine to 12 inches long. Gunther wrote: "From a pair of luminous organs in the orbital region the fish...emitted a beam, of varying intensity, of strong blue light which shone directly forwards for a distance of about two feet."

A more sophisticated device is found in some fishes that live in the middle depths of the ocean. During the voyage of the Royal Research Ship *Discovery* in 1969 J. B. Gilpin-Brown of the Marine Biological Association in Plymouth, P. G. Wright of Makerere University College and I studied such fish of the genus *Pachystomias*. We found that large light-producing organs that are situated close to the eyes and face outward in the same direction are covered with a tissue that could only transmit orange and red light. In one of the fishes these organs were seen to emit flashes of orange-red light.

Almost all the deep-sea fishes have eyes that are sensitive to light in the blue-green region of the visible spectrum because it is light of these wavelengths that penetrates most deeply into the ocean. One consequence of this adaptation is that deep-sea fishes are generally very insensitive to red light. *Pachystomias*, however, has retinal pigments that are particularly sensitive to red light. In

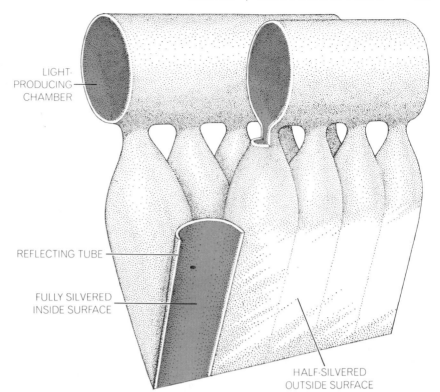

BIOLUMINESCENT ORGAN of one species of the deepwater hatchetfishes has a pair of cylindrical tubes with reflective linings (*top*) that feed light into two arrays of wedge-shaped reflecting tubes that emit the light downward along both flanks of the fish. In the drawing parts of the structures have been cut away to reveal the reflective inner linings (*color*).

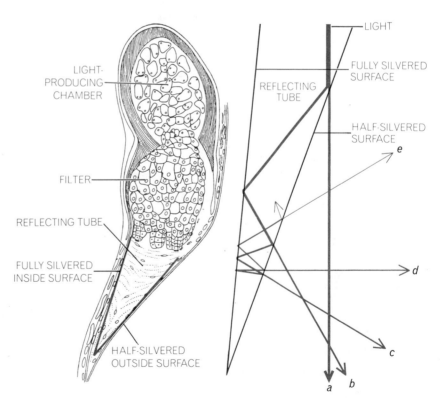

SINGLE UNIT from a light-emitting organ of a related hatchetfish species is seen in cross section (*left*). The reflecting surfaces of its wedge-shaped lower structure are outlined in color. The light path through such a structure is shown schematically (*right*); the outer wall is only half-silvered, so that some of the light entering from above escapes (*a*), and the balance is progressively reflected and emitted at diminishing intensities (*b—e*). This is exactly the angular distribution of light needed to match underwater light conditions.

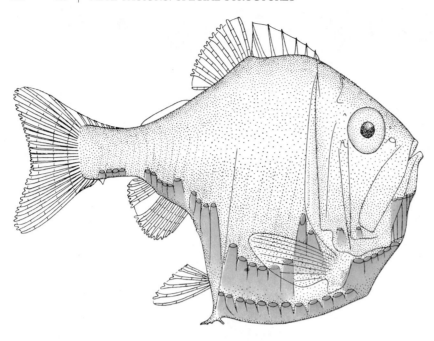

HATCHETFISH of the species *Argyopelecus aculeatus* is one of those the author and his colleagues studied aboard the research ship *Discovery*. The fish is about seven centimeters long; an array of down-pointing lights runs from chin to tail, mainly on the ventral surface.

other words, fishes of this genus are like a sniper armed at night with an infrared "snooperscope." They can see their prey without being seen themselves.

Pachystomias also has photophores, or light-producing organs, aimed downward and located along the belly. These photophores have filters that confine the emission of light to the blue part of the spectrum. Downward- and sideways-directed blue lights are characteristic of many fishes, such as the hatchetfishes, which live in the middle depths of the sea. The purpose of the downward-directed illumination has been suggested by William Dixon Clarke of the General Motors Defense Research Laboratories: it eliminates the "shadow" the fish casts with respect to an observer looking up from below. Fishes with such photo-

phores, unlike all the others, are able to camouflage their belly. The blue-transmitting filters on their photophores ensure that the wave band of the emitted light matches the wave band characteristic of the attentuated daylight in the depths where they live.

On the *Discovery* Gilpin-Brown, B. L. Roberts and I examined in detail the photophores of certain hatchetfishes, *Argyopelecus aculeatus* in particular. We found that the organs had ingenious features that enabled the fishes to match their background regardless of the angle of view. Sets of photophores in *A. aculeatus*, for example, are located fore and aft along the fish's belly [see illustration above]. The light is produced in chambers that share their light among groups of reflecting tubes. Sometimes re-

flecting tubes on both sides of the fish receive light from a common chamber. The interior of such a chamber is highly reflective, and the light can leave it only through small holes in its floor that lead into the reflecting tubes.

The reflecting tubes are typically wedge-shaped and are lined with reflecting material of two kinds. The lining of the inner surface of a tube consists of a dense array of long, thin crystals with their long axes running from top to bottom. The crystals on the outer surface of the tubes are quite different; they are broad and few in number, so that the part of the organ in contact with the surrounding water is only "half-silvered" and can emit light.

When light from above strikes the half-silvered outer surface of the tube, some of the light passes into the water and some is reflected. The reflected light is then returned to the front surface by the highly silvered inner surface of the tube but this time at a different angle. Because of the wedge shape of the photophore each double reflection bends the remaining light increasingly upward in the direction of the surface. As a result the brightest emitted light is directed downward and the emission becomes progressively dimmer as the angle of view is changed from the bottom to the side. This is precisely the pattern of illumination that is needed to match the natural distribution of daylight in the deep ocean. We have shown that some fishes do in fact emit light whose angular distribution is close to that required to match the natural distribution of daylight in the sea at the depths where these animals live. In effect the silvered interior surfaces that reflect and diffuse the luminescence of the deep-sea fishes are performing the same function as the external reflectors of other fishes: both kinds of silvering help to make the animals hard to see.

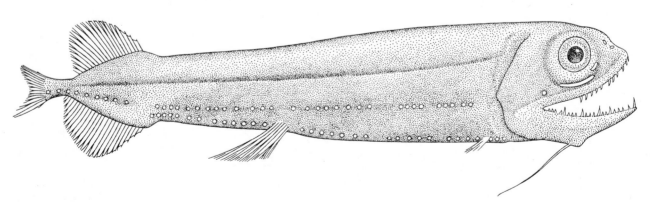

DEEP-SEA PREDATOR of the genus *Pachystomias* produces bright flashes of light in the red part of the spectrum from organs near its eyes. Deep-sea fishes are largely insensitive to red light, but this predator is not. Thus it can see prey and yet remain unseen.

Horns and Antlers

by Walter Modell
April 1969

They are commonly believed to be rather alike but in actuality they are quite different. Among other differences, the material of horns is related to skin and the material of antlers to bone

Some years ago a curator at the New York Zoological Society received a shipment of antelopes from South Africa to which he had been looking forward eagerly. It included rare specimens that were hardly known to Americans except in crossword puzzles: the hartebeest, the eland, the waterbuck, the impala and others. When the antelopes were unloaded from the ship, the curator was stricken with shock and horror. They had all been dehorned. In order to save space and prevent injury the shipper had polled the animals' horns with a cattle dehorner, assuming that they would regrow their "antlers" after settling down in their new home. Unfortunately the shipper had made a serious zoological error, failing to distinguish between horns and antlers. All the animals in the shipment had horns, which as every dairy farmer knows do not regenerate after polling. The curator rejected the mutilated animals as being unsuitable for exhibition in a zoo, and since no one else wanted them they were slaughtered and given to the large cats, which for the first time in a long while ate as they had been brought up to eat.

The hapless animal dealer who confused horns with antlers was not particularly ignorant but was a victim of a common misapprehension. Many people and even reference books are not entirely clear on the differences between horns and antlers. The differences are fundamental and complex, and they present interesting problems in biochemistry, physiology, animal behavior and evolution.

True horns, antlers and similar cephalic adornments are found today only in five families of ungulates (animals with hooves): (1) the Rhinocerotidae (rhinoceros), characterized by one or two permanent midline nasal horns; (2) the Bovidae (cattle, sheep, goats and antelopes), characterized by a pair of symmetrical permanent horns; (3) the Antilocapridae (pronghorn antelope), which annually renew their pair of symmetrical horns; (4) the Cervidae (moose, caribou, elk and deer), which annually renew a pair of antlers, and (5) the Giraffidae (the giraffe and the okapi), whose cephalic protuberances are permanent and paired but are neither horns nor antlers.

In functional terms horns make sense: they serve their possessors as effective weapons. The knobs on the giraffe's head are less understandable; borne some 18 feet above the ground, they are hardly in a position to attack anything except perhaps a low-flying airplane, and in any case they are short, blunt and cushioned with a tuft, so that they cannot inflict much damage. The function of antlers is even more mysterious; apart from giving the animal a noble appearance these headpieces have little utilitarian justification; indeed, they are an encumbrance. Antlers are too delicate to serve as weapons. When antlered animals really fight, they use their hooves and not their antlers. For several months of the year, between the annual shedding and the regrowth, the animal does without antlers and seems not to miss them at all.

The only observable function served by antlers is that during the mating season the males use them to tilt with other males in winning a harem. This unique application often ends unhappily; it is not uncommon for the two contestants to lock antlers (not horns!) so that both are immobilized and perish, and the herd loses the genes of what may well be its two best stags. As we shall see, the antler is a strange and uneconomic experiment of nature, extravagantly costly to its possessors in several ways, and it seems destined eventually to disappear.

In distinguishing antlers from horns we note to begin with that they are composed of entirely different materials. Horn consists mainly of keratin, the protein that is the principal constituent of hair, nails, hooves, scales, feathers, claws and other tough structures derived from epidermal tissue. Like hair and nails, horn is not a living, sensitive tissue: it has no nerves or blood supply and hence is insensible to pain and does not bleed when it is cut. Horns grow slowly and attain their definitive size and shape by extension from their source in an epidermal layer surrounding a bony core, the *os cornu* at the base of the frontal bone of the skull, as the animal grows to adulthood. If the horny material is cut off, it is not regenerated.

Antler, in contrast, is living tissue that resembles true bones of the body in physiology, chemical composition and cellular structure. During the antler's development it is covered with a hairy skin called velvet, which has a rich supply of blood vessels and nerves. While the antler is in velvet, it bleeds profusely when injured, and the skin is sensitive to touch and pain. At this stage the antlers are not only tender but also fragile. As the season progresses the antlers become ossified, the velvet is shed and the bare tines of bone are exposed. At the end of the mating season (usually in December in temperate regions of the Northern Hemisphere) the animal sheds its antlers, and four to five months later (in April or May) it begins to grow a new set. In short, the antler is a deciduous organ that is cast off and renewed annually like the leaves of a tree.

Horns and antlers differ significantly in their architecture. Antlers, at least those sought as trophies, tend to be large, complex and ornate. Antlers are paired,

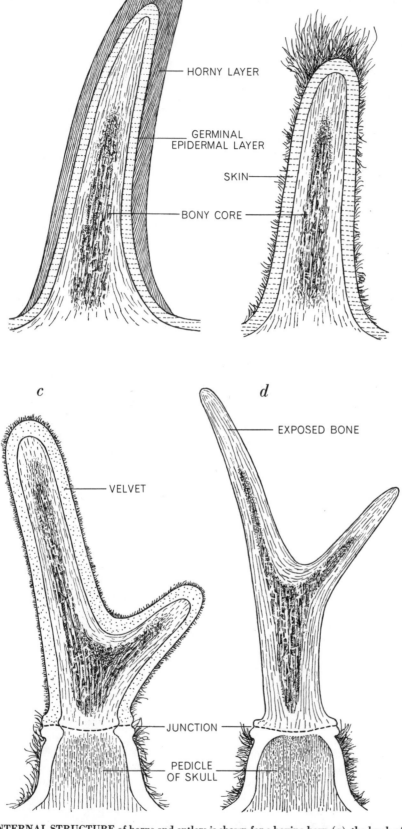

a

HORNY LAYER

GERMINAL
EPIDERMAL LAYER

SKIN

BONY CORE

b

c

VELVET

d

EXPOSED BONE

JUNCTION

PEDICLE
OF SKULL

INTERNAL STRUCTURE of horns and antlers is shown for a bovine horn (*a*), the knob of a giraffe (*b*) and the antler of a deer, with its "velvet" covering (*c*) and with the bony structure exposed (*d*) after the velvet has been shed. The junction with the skull where growth begins in the spring and where the antler breaks off at the time of the annual shedding is indicated. Developing antlers have blood vessels and nerves, whereas horns entirely lack them.

although in some species they are not symmetrical. In certain animals, such as the reindeer and the moose, antlers have a winglike, palmate structure—like the palm of a hand with extended fingers. From year to year after a deer has reached maturity and until it reaches its prime each new crop of antlers becomes larger and more elaborate, adding branches and "points" that provide a measure of the animal's age and vigor. The number of points is also a measure of the magnificence of the deer; in Scotland, for example, a deer with 12 points is called a royal stag. In some species of Cervidae now extinct the antlers weighed more than the animal's entire internal skeleton. At the other extreme the tiny pudu of the Chilean Andes, a reddish deer only about a foot high, has simple pygmy antlers consisting of almost invisible spikes two or three inches long.

The horns of the Bovidae, although they too can be magnificent, are clearly structured for use rather than ornamentation. They exist as symmetrical pairs (an exception is the four-horned antelope with two symmetrical pairs) in a rich variety of forms: curved, twisting, coiled, helical and zigzag. They all end, however, in a single strong spear capable of impaling or tossing an adversary. (Testimony to the formidable power of horns as weapons has been erected outside the bullring in Madrid in the form of a statue of Sir Alexander Fleming, the discoverer of penicillin, whose wonder drug has reduced the death toll among gored matadors.) It also seems significant that in almost all species of Bovidae the female as well as the male possesses horns (although they are often smaller in the female), whereas among the Cervidae antlers are secondary sexual characteristics of the male (the only known exception being the reindeer and its identical American version, the caribou). This again suggests that the antler, unlike the horn, did not evolve primarily as a weapon of defense.

Let us examine the biological and evolutionary distinctions between horns and antlers in more detail. Horn is made up of filaments that closely resemble hair, and these filaments arise from papillae in the skin that are much like hair follicles. It is clear, however, that horn is not, as it has sometimes been said to be, simply a mass of agglutinized hair. On microscopic examination it can be seen to be made of distinct hollow filaments, whereas hair fibrils are solid. Furthermore, horns originated much earlier (probably at least 50 million years earlier) than hair, which apparently developed only after the arrival of mammals.

The simplest and most primitive horn of our day is the horn of the rhinoceros. It is made up of tubular, filamentous secretions from the skin that are cemented together to form a projection from the animal's nose. Having no bony core, the horn consists of solid keratin. As it grows it becomes cemented to the nasal bone, but if a dead rhinoceros's head is skinned, the horn often comes away with the skin, to which it is firmly attached.

In cattle and other animals of the bovid family the horns are hollow, as one can see in an antique powder horn. The horn is mounted like a shoe directly on the spikelike *os cornu,* which projects from the frontal bone of the skull. Part of the frontal sinus can often be found in the center of the *os cornu.* Horns arise from an inner epidermal layer (the *stratum germinativum*) immediately covering the *os cornu.* This layer lines the developing horn and produces its slow growth by continuing to secrete filaments. The reason a horned animal cannot regrow a polled horn is that the operation destroys the essential filament-secreting epidermal tissue.

The only horned animal that sheds its horns periodically is the pronghorn antelope (*Antilocapra americana*), a native of North America that was once numerous but was almost wiped out by hunters before conservationists took measures that have effectively preserved the species. Each year the pronghorn develops a set of true horns that consist of keratin and grow on an *os cornu.* The prong is an extra spike on the horn.

This animal's annual shedding process is not even remotely related to that of antler replacement. The new horn grows while the old one is still in place and pushes the old shoe off the *os cornu* as it achieves full development, so that the animal is never without a horn. Although such an experiment is not recorded, it is probable that, if the old horn were cropped at the base, a new one would not grow. The pronghorn's horn differs from the horns typical of the Bovidae in that it is often covered with a considerable growth of hair. Yet the pronghorn antelope is so like members of the family Bovidae in all its obvious physical features that I believe it belongs in that family.

In contrast to horn, antler is a unique anatomical object. It is far and away the fastest growing postnatal bone known. My interest in this unusual tissue goes back to my days as a second-

year medical student, when I began a study on seasonal changes in the elk at the laboratories of the New York Zoological Society. I showed one of my histologic sections of growing antler to my professor, James Ewing, who was then the world's outstanding student of malignant growths. On examining it under the microscope, he described the tissue, which was extraordinarily rich in mitotic figures and gave other signs of fierce growth, as a sample of malignant bone sarcoma. When I told Ewing that it was actually a slice of a normal growing antler, he urged me to pursue the study of the tissue because of its simulation of malignant growth.

Unfortunately neither I nor any other investigator since has been able to discover the nature of the mechanism that controls the exuberant growth of antler cells. Although under the microscope the

actively growing antler tissue cannot be distinguished from that of malignant neoplasm of bone, it is clear that its development is under rigid control. Instead of spreading out in all directions the cells produce a sharply defined structure growing away from the head. In the course of a few months the original cells grow into a large, bony tree of great complexity. Then, at the end of the mating season, special mechanisms of the body in effect cleanly amputate this structure and it is discarded. No surgeon has yet achieved such success in removing a bone sarcoma. It is noteworthy, however, that like a malignant neoplasm the antler recurs (every year until the stag becomes very old) because a few primordial cells are left behind.

Antlers have intrigued and mystified naturalists since ancient times. Until the beginning of the 19th century they were

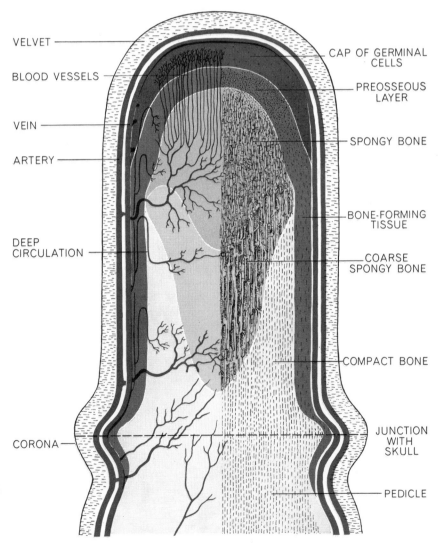

BLOOD SUPPLY of growing antlers is provided through the elaborate structure of arterioles and veins depicted at left. The major structural zones of the antler are shown at right.

a   *b*   *c*

ARCHITECTURE OF HEADPIECES is suggested by five horned animals and the giraffe. The horned animals are (*a*) the mouflon, (*b*) the springbok, (*c*) the wildebeest, (*d*) the rhinoceros, whose horn is made up of tubular filaments as other horns are but, unlike

generally thought to be composed of wood; indeed, French naturalists named them *bois,* and they attached the term *écorce,* meaning bark, to the velvet that came off when the antlers matured. The discovery that they were actually bone did not come until some 19th-century investigator boiled antlers and identified gelatin in the residue.

Not all naturalists took so naïve a view

as the French. Since antlers are a male characteristic, some early investigators sought to relate them to male physiology. Aristotle correctly noted: "If stags are castrated before they are old enough to have antlers, these never appear, but if castrated after they have antlers, their size never varies nor are they subject to annual change." In recent years it has been shown that injections of the male

hormone testosterone can cause a castrated male deer to develop antlers. The male hormone can also stimulate a spayed female deer to produce knobs, covered with skin resembling velvet, at the sites on the skull where antlers normally grow in the male. I have observed (as have others) elderly female deer with similar knobs after menopause.

How is the antler formed? Investiga-

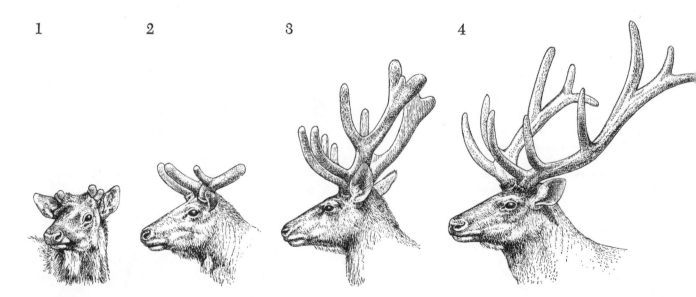

1   2   3   4

ANNUAL EVOLUTION of antlers, belonging in this case to the wapiti elk, begins (*1*) with the appearance of the velvet-covered buds of new antlers in April, about six weeks after the former antlers have been shed. Within two weeks (*2*) the characteristic branched pattern has appeared. By the end of May (*3*) the antlers are well developed and fully covered by velvet. During this

d   *e*   *f*

them, is solid rather than hollow and (*e*) the pronghorn antelope, which is the only horned animal with horns that are deciduous, or periodically shed. The giraffe's head (*f*) has growths that are not true horns but rather are protuberances covered with a hairy skin.

tors many years ago, using ordinary stains (hematoxylin and eosin) to study by microscope the intimate details of the growth of the tissue, concluded that antlers, like the long bones of the body, developed through an intermediate cartilage stage. I examined the microscopic details of antler development with special silver staining techniques and found that, on the contrary, the bone of antler,

like the bone of the skull, is formed by direct ossification of the framework of fibrous tissue that develops first. It becomes cancellous (spongy) bone with an internal cavity that is continuous with the cavity of the skull and shares its blood supply. Curiously, unlike other spongy bone in the body, such as the sternum and the pelvis, the bone of the antler does not manufacture blood, al-

though there is considerable blood (and a little fatty marrow) in it.

Under the influence of a hormonal rhythm the male deer begins to grow its antlers in the spring (late April or May in the U.S.). The developing antler bulges out of the velvet that covers the pedicle, a bony platform atop the frontal bone of the skull, soon after the old antler separates from it. From a few

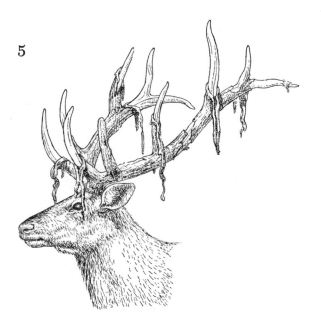

5

6

stage the animal is careful to avoid hard objects. By August (4) the antlers are mature. Growth has ceased and the velvet has begun to dry at the tips. When the bony material of the antler has become fully hardened, the velvet dies (5) and peels off in ragged shreds. After the antlers are mature and have lost their velvet (6) the wapiti, which is an American elk, becomes sexually aggressive.

ANCIENT HORNED ANIMAL was *Triceratops*, a reptile with three horns and also a horny shieldlike structure. Horns apparently originated with the early large reptiles.

ANCIENT ANTLERED ANIMAL, much more recent than *Triceratops* but also extinct, was the giant Irish elk. Its massive antlers weighed more than its entire internal skeleton.

fibroblasts (embryonic cells) left behind when the former antler was cast a mass of fibroblasts develops below the skin, and the velvet cover grows to conform with the developing antler. The fibroblasts rapidly form the armature of the antler, much as a branch and twigs grow out of a tree. Meanwhile osteoblasts (bone-forming cells) begin to stream into this framework and to lay down the bone-forming calcium. The resulting bone has a thinner cortex and consequently is not nearly as strong as the bone of the skull, but apart from the fact that it does not form blood it is indistinguishable from the cancellous bone of the skull and other spongy bones of the skeleton.

By September the antler has grown to full size and has firmly united itself to the pedicle. The bone at the base becomes progressively denser and eventually cuts off the flow of blood from the skull to the antler's interior. Some blood is still supplied, however, by arteries coursing through the velvet covering the antler. Soon, as a result of some mechanism that has not yet been satisfactorily explained, the velvet proceeds to degenerate. It dries up and is shed by the stag, coming off in strips when the animal rubs its antlers against trees or shrubs.

At this stage the antlers have no blood supply or nerves and are insensible to painful stimuli. The bare bone and sharp tines of the antler are exposed, and for a few weeks in the fall the many-spiked antler might be usable as a weapon. It is an awkward weapon at best, however, and its development is not particularly well timed, because by that season the fawns born of the stag's preceding matings are fleet enough to escape predators and no longer need the sire's protection. As I have noted, when stags tilt with each other to gain a harem for the new mating season, the antlers may be dangerous to both adversaries. Furthermore, how is one to explain the curious fact that sometimes the harem winners are stags (called hummels) that for some unknown reason have failed to develop antlers? Could they be better fighters because they lack antlers?

The antler's implausibility as a weapon suggests that it did not actually evolve for this function. One interesting current hypothesis is that antlers serve the deer and other Cervidae as a cooling device during the summer. The velvet covering the antler provides an admirable means of radiating body heat, because of its considerable surface area and its exten-

TYPICAL HORNS AND ANTLERS are shown in front and side views. At top are the horns of the kudu ram, a large antelope that is found in much of Africa. At bottom are the antlers of the caribou, which is the North American equivalent of the reindeer.

POSSIBLE EVOLUTIONARY TREND in antlered animals may be represented by the Chinese water deer (*left*) and the musk deer (*right*), which are the only animals of the family Cervidae that do not possess antlers. The two species, however, have evolved tusks.

sive apparatus of peripheral blood vessels and sebaceous glands. The fact that the velvet dies at the end of the summer and the antlers are shed not long afterward lends some support to this surmise about the antler's primary reason for being. It is difficult to imagine any other plausible reason why nature should have endowed animals with an elaborate superstructure that is used only during a few weeks of the year and is discarded annually after being produced at great metabolic expense. (To the question of why females lack these radiators the supporters of the hypothesis respond that females stay in the shade.)

In the late fall some of the cement holding the base of the antler to the pedicle is resorbed, the connection weakens and the antler is cast—not as an active process but by breaking off when it happens to strike something. The break occurs neatly at the junction with the skull where the antler started to grow in the spring; there is a little bleeding, but the blood promptly clots. The process is painless. A new growth of hairy velvet (rather than the hairless, nonsweating scar tissue that forms in the normal healing of wounds) quickly covers the wound on the exposed skull, and the organs and cells involved in the formation of antler go into a dormant period, lasting from late December until the fol-

lowing April or May. An injection of testosterone during this period, however, can trigger antler growth.

Notwithstanding the peculiarities and apparent frivolity of antlers, they cannot really be called a freak of nature. As a headpiece they are one version, albeit an exotic one, of a phenomenon that goes back to the early history of land animals. Horns are at least 100 million years old. In the Cretaceous period there were dinosaurs and crocodile-like reptiles with horns. Some of these growths beggar the imagination of science fictionists. *Triceratops* had three horns: a huge spike rising from the nose and one above each eye. *Styracosaurus* had an upright nasal spike nearly two feet long and a neck shield with six spikes thrust out from its edges. Skull protuberances have also been found among fossils of some early mammals, notably the elephantine ungulates. The horns were formidable weapons in some early mammals, such as *Arsinoitherium*, but in others, such as *Uintatherium* and *Brontotherium*, they were blunt extensions from the skull and were probably covered with skin. By the end of the Eocene epoch some 40 million years ago the rhinoceros, its naked horn as menacing a weapon as the horns of the ancient reptiles, had begun to appear.

The rhinoceros is unusual in several ways. It is the only odd-toed animal that has a horn. The Rhinocerotidae include the only nonmythical unicorn, but several surviving species of rhinoceroses have two horns, both positioned on the midline one behind the other. The rhinoceros's horn is also exceptional in that it is solid keratin, not a hollow shoe. In prehistoric times the animal must have made considerable use of its horn as a weapon, but today the rhinoceros's massive size and armor are sufficient protection to discourage attack by its less massive contemporary competitors.

Paradoxically, in our era the horn has been the rhinoceros's undoing. For more than 1,000 years man, now the animal's principal enemy, has been hunting down the rhinoceros for its horn. In China ground rhinoceros horn has long been prized for its supposed values as an aphrodisiac and a medicine for various ailments, and in the medical markets of China and Africa today the horn is said to be worth half its weight in gold. A thriving trade is also conducted in dried rhinoceros blood and in rhinoceros hide for use as a warrior's shield. A few rhinoceroses manage to survive in Asia by keeping out of sight, and in some parts of Africa the animal is protected from hunters by law.

During the Miocene epoch, beginning

some 25 million years ago, horns and antlers developed among many species of two important families of ungulates, the Bovidae and the Cervidae. These two families are so remarkably alike in many ways that they may well have had a common ancestry. The great differences between horn and antler, however, indicate that the two forms of head appendage had different origins. The Bovidae revived the keratinous horn growth that was already a 100-million-year-old carryover from the age of reptiles. The Cervidae introduced the antler as a basically new growth. According to the available fossil record, the first antlered cervid was *Dicrocerus,* an ungulate of the early Miocene that grew a very simple antler in the form of short spikes. Because antler is so skimpily constructed that it deteriorates about twice as fast as the skull bone under ordinary exposure, it may well be that its ancient history is not as well preserved in the paleontological record as that of skull and skeletal bone. The record does show, however, that in the Pleistocene epoch antlers became common among the Cervidae and some of them grew to monstrous size. Probably the most impressive antlers of all time were borne by the great stag (*Cervus megaceros*) of the Ice Age; its pair of antlers had a spread of three meters and weighed about 70 kilograms (154 pounds)! Among the living members of the family Cervidae today all but two species have antlers. The exceptions are the Chinese water deer and the musk deer [*see illustration on preceding page*].

There are reasons to believe the giraffe and the okapi are closely related to the Cervidae and evolved from the same group of ancestors. The permanent knobs on the head of the giraffe and the okapi are not made of keratin; hence they are not true horns. The giraffe's knobs are formed of a bony core extending straight up from the skull and are covered with a hairy skin that gives the knobs their tufted appearance. In the okapi a bit of bare bone is exposed at the tip. The knobs of these two strange beasts seem closer to antlers than to horns and perhaps are best compared to antlers permanently in velvet.

In a curious way horns and antlers are invariably associated with certain other apparently unrelated anatomical features. Headpieces of one kind or another (horns, antlers or knobs) are possessed by virtually all even-toed ungulates with a four-chambered (or true ruminant) stomach; this includes the Bovidae, the Cervidae, the pronghorn, the giraffe and the okapi. With the single exception of the rhinoceros no odd-toed animal (which includes the horse, the ass and the zebra) has a headpiece, and headpieces are also absent in all even-toed ungulates with false ruminant, or three-chambered, stomachs (camels, llamas and others), as well as in those with a single-chambered stomach (pigs, peccaries and hippopotamuses). What connection can there be between horns or antlers and a perfect ruminant stomach and even-toed hooves? By what odd quirk of evolution did these seemingly unrelated characteristics come to be associated with one another, if indeed the association is not mere coincidence?

Be that as it may, this enigma is much less intriguing than the mystery of why antlers evolved and why they have persisted so long. The evolutionary success of horns, which have proved their value over 100 million years of trials and are now firmly incorporated in a great number of species of ungulates, is quite understandable. The antler, on the other hand, is obviously an encumbrance with a very limited, if not entirely questionable, usefulness. Perhaps the main surprise is that it has lasted for upward of 25 million years and is still retained by almost the entire family Cervidae. There is evidence, however, both in the fossil record and in the hunting history of man, that antlers as well as the antlered animals are on the decline. Their often suicidal tilts with their antlers, by killing off some of the best and strongest stags, may have played a part in this decline. The preservation of the Cervidae is now aided by the U.S. program for the protection of wildlife, but this has led to a proliferation that has made them a true pest and necessitates large-scale seasonal hunting of the animals to spare them from starvation and our forests from ruination by debarking. Overprotection of the animals, with its consequent disturbance of ecology, may not be in our interest.

It may be that in the long run natural selection can save the Cervidae by eliminating those with elaborate antlers, so that the surviving members of the family come to be like the two nonantlered species, the Chinese water deer and the musk deer. In any case, it appears that antlers, if not the antlered animals themselves, are doomed by evolution, and that the eight arboreally ornamented reindeer of Santa Claus, like the handsome unicorn, will one day become strictly a legend.

11

The Head of the Sperm Whale

by Malcolm R. Clarke
January 1979

It can represent a quarter of the animal's length and a third of its total weight. The oil-filled spermaceti organ housed within it may keep the whale neutrally buoyant during dives

Among the great whales the sperm whale is most clearly recognizable in having a head that seems disproportionately large. The sperm whale's head can make up more than a third of the animal's total weight (50 tons for the average adult male) and more than a quarter of its total length (an average of 60 feet). There is good reason for this apparent disproportion. The sperm whale's skull accounts for perhaps 12 percent of the weight of the head. The other 88 percent consists mainly of a peculiar anatomical feature located in the whale's snout above the upper jaw: the spermaceti organ. The organ is a complex mass of muscle and oil-filled connective tissue. The oil is what gave the sperm whale its name; in a large male the organ may hold four tons of spermaceti oil.

Such a mighty organ clearly must have a very important function in the life of the sperm whale. The purpose of the great "case," as whalers often call it, has long been a subject of speculation. Even the structure of the spermaceti organ was scarcely known until the past decade or so. Before then guesses about its function could rest only on generalities and on what was known about similar but far smaller organs in a few other toothed whales. Today enough is known about the organ to make it possible to suggest its main purpose: to enable the whale to remain neutrally buoyant when it is submerged.

Why were the anatomical details of the organ not known much earlier? Part of the answer is that pioneers in the field gathered their data by dissecting sperm-whale fetuses. It happens, however, that both the skull and the snout of the fetus differ greatly in proportion from the same components in the head of the adult whale. Another part of the answer is that it is no small task to dissect an adult sperm whale. Without such commercial whaling facilities and tools as flensing platforms, steam winches, five-meter steam-driven saws and razor-sharp flensing knives the dissection would be quite impossible. Even with such aids and the cooperation of the commercial whalers this is not a trivial task.

On the flensing platform half-ton masses of fat, flesh and fiber are cut, rolled and pulled off this way and that until the observer's sense of orientation is easily lost. It is no wonder that even in this decade a book has been published that presents the principal structures of the sperm whale's head upside down. Only after watching and photographing many sperm whales being cut up on the flensing platform could one hope to clarify the anatomy of the adult whale's snout, and that is the task I undertook. My work was greatly advanced by the capture of one small adult whose head the commercial whalers cut for me in a series of transverse sections 20 centimeters thick. I was then able to photograph the sections and measure them in detail.

Many functions for the spermaceti organ have been suggested. Among them are that it is a means of generating and focusing sounds (and for receiving them), a means of moving air between the whale's lungs and its nostrils when the whale is deep underwater, a means of opening and closing the whale's long nasal passages, a means of absorbing nitrogen from the bloodstream in the course of deep dives and even a means of attack and defense.

Certainly this complex organ may have more than one function. Variations in the chemical composition of the spermaceti oil in different parts of the organ suggest that one function may well be the channeling (or focusing) of sound generated by the whale. At the same time it is hard to accept the suggestion of one worker that such focusing can concentrate sound intensely enough to stun the squids that are the sperm whale's main prey. More than one of the other suggestions, however, seem reasonable, if difficult to demonstrate.

Although an understanding of the structure and proportions of the head of the adult sperm whale is a necessary preliminary to studying the function of the spermaceti organ, one must also know something about the biology and behavior of sperm whales, in particular how these toothed whales differ from other whales that do not have a large snout. Moreover, the large quantity of oil contained in the spermaceti organ suggests that the oil itself must serve some special function; one needs to know something of the physical properties of the oil to understand what role it plays in the life of the sperm whale.

Let us begin by reviewing the biology and behavior of the sperm whale in search of clues to the function of the animal's snout. One immediately apparent fact is that sperm whales are unusual, although not unique, on the roster of toothed whales in being distributed worldwide. They have been hunted for centuries in every ocean from as far as 60 degrees north latitude to 40 degrees south. Since the development of modern whaling techniques the sperm whale has also been pursued to the high latitudes of the Antarctic. Together with its baleen-whale cousin the right whale, the sperm whale was of particular importance when the whaling industry depended on hand harpooning from open boats: unlike many other whales, these two stayed afloat after being killed.

The food of the sperm whale consists almost entirely of that Concorde of the snail family, the squid. To catch these speedy denizens of the depths the sperm whale dives deep and stays down for long periods. A large sperm whale is typically submerged for 50 minutes of an hour-long diving cycle. During a 10-minute surface interval between dives the whale will take 50 to 60 breaths of air. Dives longer than 50 minutes have been observed; the record dive is somewhat more than 80 minutes.

Sperm whales not only stay submerged for long periods but also frequently go deeper than 1,000 meters. One sperm whale, watched by sonar, was observed to go below 2,250 meters; even deeper descents can be inferred from the presence of bottom-dwelling

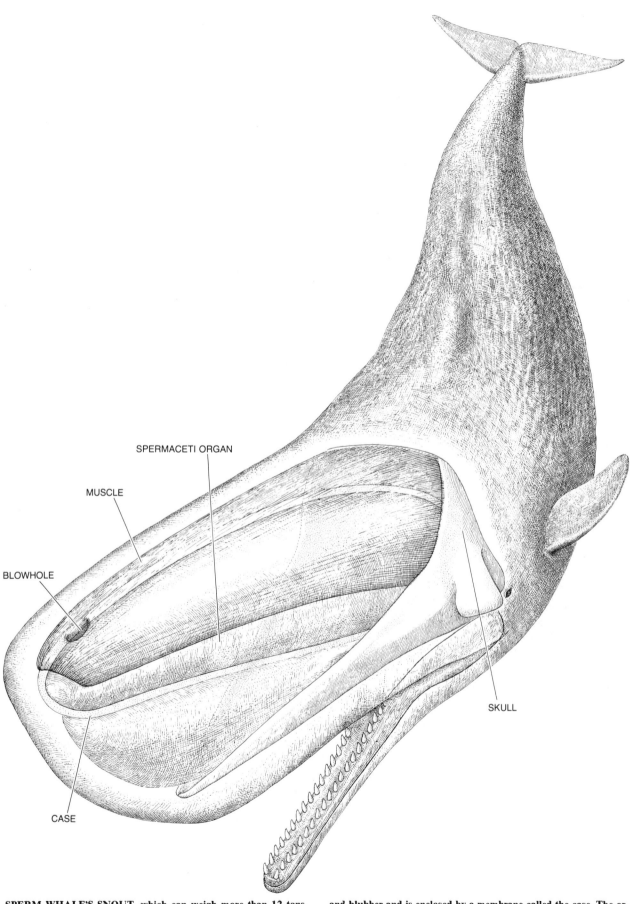

SPERMACETI ORGAN

MUSCLE

BLOWHOLE

CASE

SKULL

SPERM WHALE'S SNOUT, which can weigh more than 12 tons, is largely occupied by the spermaceti organ, a complex oblong mass of oil-filled connective tissue that is surrounded by layers of muscle and blubber and is enclosed by a membrane called the case. The organ is cradled in a long conical depression largely formed from the upper jawbones of the sperm whale's elongated skull (*light color*).

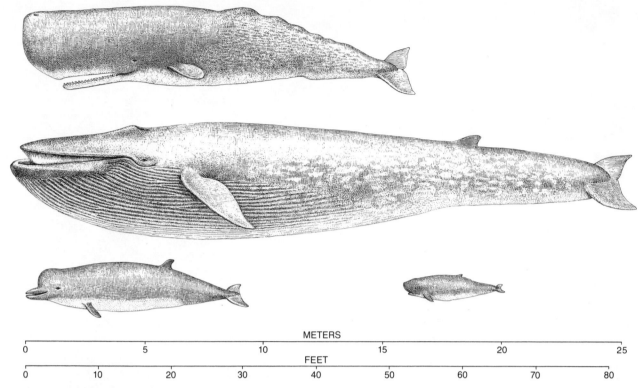

THREE TOOTHED WHALES that possess spermaceti organs are compared in this drawing with the largest of all whales. They are the sperm whale *Physeter catodon* (*top*), the pygmy sperm whale *Kogia* (*bottom right*) and the bottle-nosed whale *Hyperoodon* (*bottom left*). **The largest whale is the blue, or sulfur-bottom, whale *Balaenoptera musculus*, one of the suborder Mysticeti: the toothless baleen whales.**

sharks in the stomach of a sperm whale captured in an area where the bottom depth was greater than 3,000 meters.

Why should a sperm whale need to dive so deep? Squids are found at all depths in every ocean. They are a principal food of many other air-breathing vertebrates: seabirds as well as such marine mammals as seals and lesser toothed whales, including porpoises. Perhaps the reason for the sperm whale's diving behavior is that the deeper a squid-eater can descend, the farther it can outdistance its competitors and the greater the stocks of squids within its reach are. Certainly the sperm whale can catch deep-living squids that are beyond the reach of seabirds, seals or porpoises. The whale can also reach bottom in the zone where the sea floor drops steeply to the abyssal plain from the edge of the continental shelf. Here on the continental slope, at depths of between 200 and 3,000 meters, many squids lay their eggs and can be found in large numbers and dense concentrations. Few air-breathing animals other than the sperm whale can hope to reach such easy pickings.

A peculiarity of the sperm whale's diving cycle is that the animal frequently surfaces within a few hundred yards of the point where it began its dive. The reason is not that its underwater time has been occupied by a slow descent and ascent. The whales are known to descend at a speed of about four knots (120 meters per minute) and to ascend at a speed of about five knots. Hence a round trip to a depth of 1,000 meters would not take more than 15 minutes. As we have seen, the duration of a deep dive is some three times longer. Both the duration of the dive and the fact that the sperm whale's place of emergence is close to its place of submergence suggest that when the whale comes to the bottom of its dive, it must sometimes lie almost still in the water.

Many, although by no means all, squids are fast swimmers over short distances. One may therefore wonder why a sperm whale would lie still at depth instead of actively pursuing its swift prey. The whale's lower jaw is long and narrow. It is the jaw of a snapper; even when it is open, it offers little water resistance. Perhaps the sperm whale's hunting strategy relies less on active pursuit and more on silent hovering followed by a quick pounce into a passing shoal of squids. Little or no daylight penetrates these hunting depths, but most of the squids on which the whale preys are luminescent. In its efforts to catch these speedy invertebrates a still, silent whale may well have the advantage over a swimming one.

A whale can lie still underwater only by being very nearly neutrally buoyant, that is, by having the same density as the surrounding water. In the older classes of man-made submersibles that displaced from 1,600 to 2,000 tons buoyancy had to be controlled within 40 liters of water (between two and three hundred-thousandths of the displacement weight) to enable the vessel to lie still in the water for listening purposes. Many water-dwelling animals can also achieve buoyancy within very fine limits and so can lie still at the depth where they live. For example, some fishes counter the sinking effect of those body tissues that are denser than water by storing low-density fats; other fishes manage neutral buoyancy by means of an air-filled swim bladder. Many squids do the same by replacing the dense sodium ions in their body with less dense ammonia ions.

In its biological and behavioral aspects, then, the sperm whale exhibits some unusual features. Among them are migratory behavior that takes the males from equatorial waters to polar ones, deep dives of long duration, the ability to lie still when submerged and the property of floating when dead. No one feature is unique to the sperm whale, but only the sperm whale is known to combine them all.

Of these four features two concern buoyancy and the other two—great range both horizontally and vertically—involve changes in ambient water conditions that are accompanied by changes in buoyancy. Although the sperm whale could carry the right amount of fat or air to be neutrally buoyant at a particular geographical location and depth, if the whale depended on such a static system for the control of buoyancy, the different water densities at other depths and geographical locations would push it either down or up.

Is this the clue we are seeking? Is the spermaceti organ a device for controlling buoyancy over a wide range of conditions? If it is, the organ must be able to vary its density. How could it do so? Only one substance is present in the organ in large quantities that is also able to undergo a substantial change in density: the spermaceti oil itself. In point of fact this oil has long been known to have properties different from those of other whale oils. When the liquid oil is dipped out of a dead whale's head and exposed to ambient air temperatures, it soon loses its clarity and becomes a soft crystalline solid.

Temperature probes of freshly killed sperm whales show that when the whales are resting on the surface, the temperature of the spermaceti oil is 33 degrees Celsius (90 degrees Fahrenheit). The oil begins to crystallize, or congeal into a solid, when its temperature drops below 31 degrees C. Unlike the crystallization of water, which is almost instantaneous at the freezing point, the crystallization of spermaceti oil is a gradual process that is not completed until the temperature drops several degrees. When spermaceti oil freezes, it becomes denser and therefore occupies less volume. And occupying less volume, it displaces less of the ambient seawater and is less buoyant.

If the temperature of the spermaceti oil could be varied, then, the changes in density that accompany changes in temperature might be enough to let the sperm whale control its own buoyancy. Are such changes in temperature physiologically possible? The problem is of course one of the loss and gain of heat. With that in mind let us consider the anatomy of the spermaceti organ.

The tissues that house the spermaceti oil in the whale's snout have a dense network of capillaries supplied with blood by large arteries that enter the snout at the rear. The circulation of the arterial blood is therefore the principal means of conveying heat to the oil. The same circulation at the capillary level is also the main distributor of heat within each block of spermaceti tissue; when the tissue is cooled locally, the movement of blood through the capillaries helps to spread the cooling effect.

In addition the larger arteries and veins in the snout of the sperm whale lie side by side; this countercurrent system assists the cooling of the spermaceti tissue by the exchange of heat between the warmer incoming arterial blood and the cooler outgoing venous blood. The arteries that supply blood to the snout are

surrounded, particularly at the point where they pass through the skull, by a dense network of veins carrying cooler blood away. The heat exchange can maintain a sharp difference in temperature between the snout (where the blood is normally below 34 degrees C.) and the rest of the whale's body (where the blood is normally above 37 degrees C.).

The sperm whale can lose heat through the surface tissues of the snout, either by "passive" conduction through the blubber and skin or by "active" heat transport: circulation of the blood to the papillae, minute fingerlike structures within the skin. As we shall see, passive conduction by itself is much too slow a process for the achievement of neutral

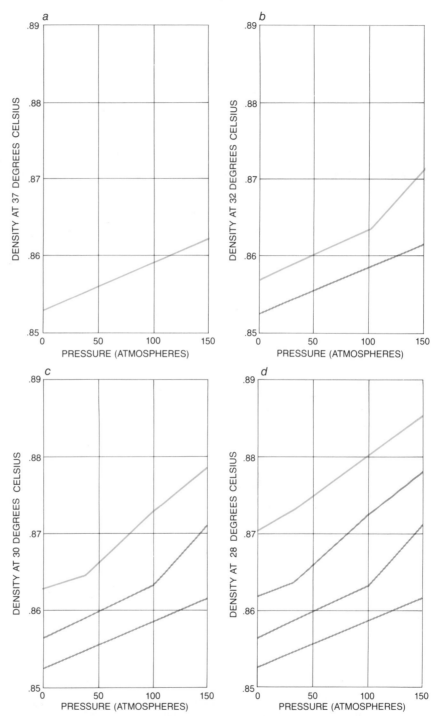

SPERMACETI OIL varies in density according to its temperature. These graphs show the density of the oil at progressively lower temperatures: at 37 degrees Celsius (a), at 32 degrees (b, *with the 37-degree reading included for comparison*), at 30 degrees (c) and at 28 degrees (d). Pressure also affects the density of the oil. Each 10 meters of additional depth adds one atmosphere (14.7 pounds) of pressure; thus at a depth of 500 meters the pressure is 50 atmospheres. The slopes of the four graphs trace the increasing density of the oil up to the equivalent of 1,500 meters. The findings are from a study made at the British National Physical Laboratory.

buoyancy in the time available. Active heat transport is a more promising possibility.

The anatomy of the sperm whale's head suggests still another means of heat loss: the sperm whale's nasal passages. Asymmetrical nasal passages are characteristic of toothed whales, but no other toothed whale has passages like those of the sperm whale. The simpler of its nasal passages, the left one, runs back from the cavity under the animal's single blowhole, curves to pass the left side of the spermaceti organ and enters the skull just in front of the brain case. The sperm whale breathes through this muscular tube, which can be expanded until it becomes circular in cross section.

The right nasal passage is totally different. Although it too begins in the cavity under the blowhole, it at first runs forward and is a tubular passage of small diameter. Then it widens and flattens out as it passes downward to form a broad, flat chamber, the vestibular sac, located at the front of the snout just under the blubber. A broad horizontal opening in the rear wall of the sac, often called the "monkey's mouth," gives entry to the continuation of the nasal passage, a wide, flattened tube that runs back the length of the snout to a point just in front of the brain case. There the wide passage narrows as it enters the skull and meets the left nasal passage in a common cavity. Just forward of this narrowing the right nasal passage opens upward to connect with a second sac, the nasofrontal sac, located above the central part of the skull crest.

The right nasal passage is thus often more than a meter wide; its roundabout course through the snout is five meters or more long. Not only does it pass through the core of the spermaceti organ but also its two sacs cover the front and rear ends of the organ.

The interior of the right nasal passage is lined with a delicate layer of black tissue. Under the black layer are two layers of white tissue, first an elastic layer and then a fibrous one; the three layers together form a wall that is between .6 millimeter and one millimeter thick. Spermaceti tissue lies directly in contact with the wall of the nasal passage on all sides, and capillaries from the spermaceti tissue enter the white elastic layer.

The intimate relation between the sperm whale's right nasal passage and the spermaceti organ is such that if seawater enters the nasal passage, the spermaceti oil will be markedly cooled. One can calculate the rate of oil cooling for any given temperature of seawater (on the basis of heat exchange between the blood in the capillaries and the cooler seawater) from the total area and the thickness of the nasal-passage wall. One can go on to calculate the rate of heat loss via the skin of the whale's snout, via the right nasal passage or via both heat-exchange areas combined and so determine the time required for the whale to reach neutral buoyancy over a range of selected depths in either an Antarctic environment or an equatorial one.

In either environment, with minimum values in the calculation, the right nasal passage proves to be a slower heat exchanger than the skin of the whale's snout. The calculated difference between the two surfaces, however, is probably not significant; the nasal passage is elastic, and its wall could be expanded to present a greater heat-exchange area than the minimum I have calculated.

My calculations show that a 30-ton

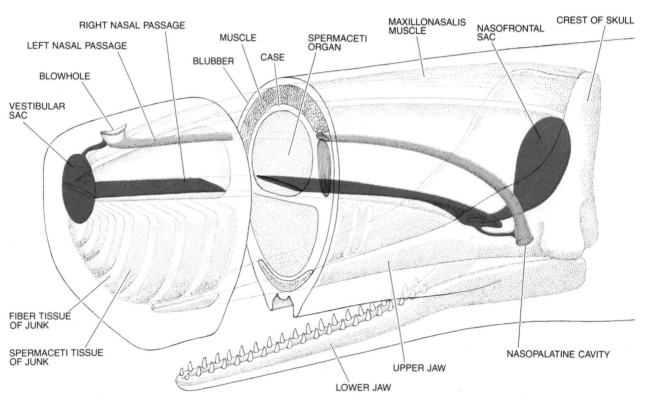

NASAL PASSAGES of the sperm whale are not only asymmetrical but also intimately associated with the spermaceti organ. The left nasal passage (*dark gray*) is the simpler of the two. Beginning in a cavity under the blowhole, the passage curves to pass along the left side of the spermaceti case and terminates in the nasopalatine cavity of the skull. The right nasal passage (*color*) also begins under the blowhole; it then runs forward and widens to form a vertically oriented sac, the vestibular sac, at the forward end of the spermaceti case. A narrow horizontal opening at the back of the sac provides for the continuation of the passage rearward through the interior of the spermaceti organ until the passage approaches the scooplike crest of the whale's skull. There the nasal passage gives rise to a second vertically oriented sac, the nasofrontal sac, located at the back of the spermaceti case, before narrowing to enter the skull cavity that is shared in common with the left nasal passage. The complex route followed by the right nasal passage is some five meters long, and the passage itself is in places more than a meter wide. Over most of its length vessels from the network of capillaries in the spermaceti-oil tissue extend into the wall of the nasal passage. A major muscle of the snout, the maxillonasalis, runs from the crest of the skull to the forward half of the spermaceti case.

sperm whale, exploiting both heat-exchange areas simultaneously, could adjust to neutral buoyancy in less time than it normally takes to swim to a depth of 500 meters. In dives from 200 to 1,000 meters deep, exchanging heat via the snout skin alone, the whale would reach neutral buoyancy within five minutes of attaining the desired depth. If both heat-exchange areas come into play, the interval would be shortened to three minutes.

In the Antarctic, because lower water temperatures mean greater buoyancy at and near the surface, the calculated time needed to reach neutral buoyancy at a depth of 100 meters, with either heat exchanger in play alone, comes to about 20 percent of the total submersion time of even a prolonged dive. At first this may appear to be too great an investment of time to make the attainment of neutral buoyancy worthwhile. There is, however, a counterbalancing factor: swimming in the cold surface water of Antarctic latitudes, the whale may maintain the temperature of its spermaceti oil at a level lower than the 33 degrees C. characteristic of equatorial waters. If it does, cooling to neutral buoyancy would be quicker. In addition, when the whale was submerged during a dive, its resorting repeatedly to the nasal heat exchanger could hasten the achievement of neutral buoyancy. My calculations show that below 200 meters a filling of the nasal passage with seawater twice rather than once would be more than enough to exchange the required quantity of heat.

Just how can seawater be drawn into the whale's right nasal passage? Surrounding the outer wall enclosing the spermaceti tissues are large muscles. They run from the front half of the "case" to attachments on the crest of the skull. Their contraction would suffice to raise the front end of the case, thereby lifting the upper half of the nasal passage. The same contraction would also open the front end of the nasal passage and draw water in from the cavity under the blowhole. The relaxation of the muscles would expel the water.

How far might the seawater travel along the right nasal passage? There are additional muscle fibers in the floor of the passage and still other fibers that run forward within the spermaceti tissue from the front wall of the nasofrontal sac. The contraction of these fibers would hold down the bottom half of the right nasal passage, ensuring that the water would travel at least as far as the nasofrontal sac.

One cannot exclude the possibility that seawater also reaches the spermaceti organ by a different route: the left nasal passage. Once drawn into this shorter tube from the cavity under the blowhole, the water could be pumped to the cavity where both nasal passages meet;

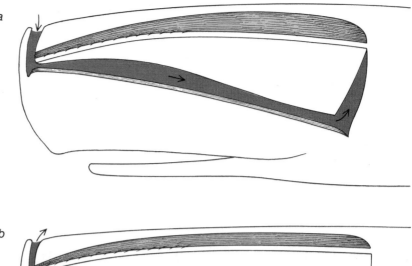
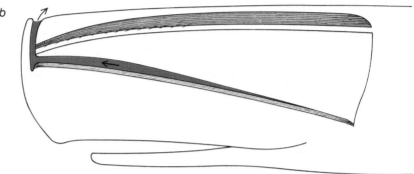

WATER INTAKE through the right nasal passage (*a*) may be accomplished by contraction of the major snout muscle (*dark gray*). Muscle action would widen the passage (*color*) so that water could enter; small muscles within the spermaceti tissue would aid the process. Relaxation of these muscles would allow the passage to narrow again (*b*), thereby expelling the water.

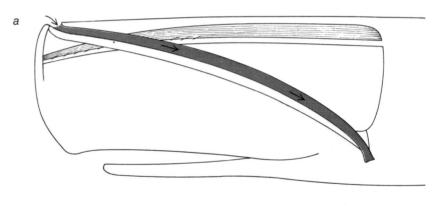
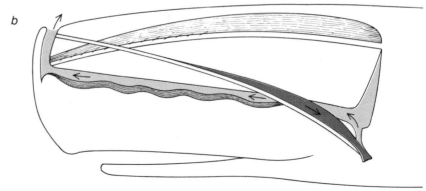

ALTERNATE INTAKE ROUTE is the left nasal passage (*color*). Contraction of a minor muscle, the nasal-plug muscle, would widen the left passage and draw water into the nasopalatine cavity (*a*). Thereafter (*b*) the action of muscle fibers in the floor of the right nasal passage (*gray*) could pump water forward through the spermaceti organ and out through the blowhole.

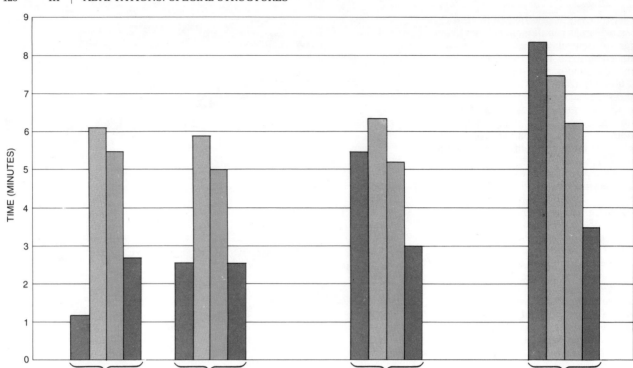

THREE ALTERNATIVES for cooling spermaceti oil when a whale is diving at the Equator take different lengths of time, as is apparent in this graph. The calculations are for a 30-ton whale diving with full lungs. The first bar (*gray*) in each of the four sets shows the time needed by the whale to swim to the indicated depth at a speed of five knots. When the dive is to 200 meters, even maximum heat exchange via the right nasal passage and the skin of the snout combined (*color*) must continue for more than an additional minute before neutral buoyancy is achieved. Heat exchange via the right nasal passage (*light gray*) or via the snout skin (*light color*) would have to continue even longer. With deeper dives the trend favors achievement of neutral buoyancy during the time the whale is swimming to depth. In a dive to 500 meters the time required to achieve neutral buoyancy through maximum heat exchange is equal to the time of descent. In a dive to 1,500 meters even the least efficient form of heat exchange, via the nasal passage alone, achieves neutral buoyancy during descent.

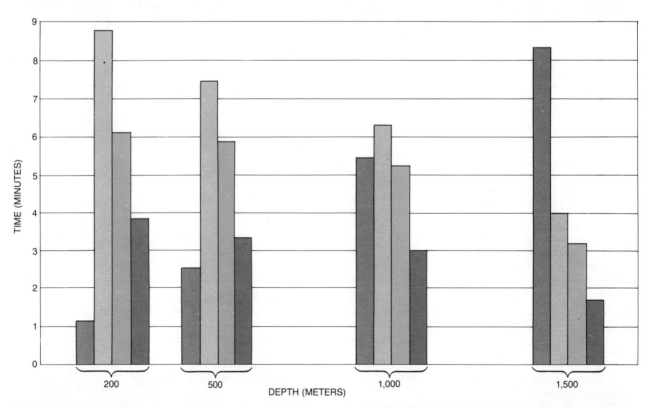

SAME ALTERNATIVES are calculated for a dive in polar waters; again the whale weighs 30 tons and is swimming down with full lungs at a speed of five knots. Maximum exchange of heat (*color*) almost achieves neutral buoyancy during a descent to 500 meters. Heat exchange via the snout skin alone (*light color*) achieves neutral buoyancy during a descent to 1,000 meters. Even the least efficient form of heat exchange, via the nasal passage only (*light gray*), achieves neutral buoyancy four minutes before the whale reaches 1,500 meters.

the water could then enter the rear of the right nasal passage and be moved forward by the contraction of the muscle fibers in the passage floor. If that is the case, one wonders why a relatively major role has been assigned to these few fibers and a relatively minor one to the very large muscles attached to the case. Nevertheless, on the two occasions when dissection revealed the presence of seawater in a sperm whale's right nasal passage, I also found water in the left passage, and so this alternate route cannot be ruled out.

To find out what changes in the density of spermaceti oil accompany changes in temperature and pressure I asked members of the staff of the British National Physical Laboratory to conduct density measurements over a range of temperatures and pressures. They found that whereas changes in density are mainly a function of temperature, pressure also has a marked effect, particularly near the onset of freezing.

At this point in my study I faced two related questions. First, what changes in buoyancy would a sperm whale encounter over its range of diving depths in both a polar and an equatorial environment, supposing it had no means of controlling buoyancy? Second, is the quantity of spermaceti oil in the head of an adult sperm whale large enough to achieve neutral buoyancy by means of density changes over the same range of depths and environments?

Turning to the first of these questions, the factors that influence sperm-whale buoyancy include the density of the seawater, the effect of pressure on both the liquid and the solid components of the whale's tissues and the effect of pressure on the volume of the whale's lungs. Variations in these factors all follow established physical laws, and therefore their effect on a whale of any particular size can be calculated. The calculations, of course, depend on reasonably accurate estimates both of the proportions of liquid, solid and gas in the diving whale and of the buoyancy of the whale when it is floating on the surface. With such a large animal it is not easy to make precise estimates of this kind. Those used in our study have therefore included minimum and maximum values as well as mean values. The mean values are used in this discussion, but the conclusions remain valid even when the extreme values are used instead.

In moving on the surface between equatorial and polar latitudes a sperm whale experiences a change in surface-water temperature of as much as 26 degrees C. Moreover, if the whale dives to a depth of 1,000 meters at the Equator, the change in temperature may be as much as 23 degrees C. These changes in temperature are accompanied by changes in the density of the seawater. Density is mainly dependent on temperature,

but salinity and pressure also enter into the calculations. For example, if a whale that is neutrally buoyant on the surface at the Equator were moved to an Antarctic latitude such as 55 degrees south, the increased density of the surface water would give the whale an increased buoyancy amounting to a little more than .3 percent of its body weight. The increase in water density the whale would encounter during a deep dive, whether at the Equator or at 55 degrees south latitude, would also increase its buoyancy. Because of the greater change in water density with increasing depth at the Equator, however, the increase in the whale's buoyancy there would be about five times more than it would be in the Antarctic: at a depth of 2,000 meters it would amount to almost .5 percent of the animal's body weight.

So much for water density. What about water pressure? Its effect on the water content of the whale's tissues proves to be very slight, as is its effect on the oil content. As for the solid components of the whale's tissues, they are almost incompressible, and so their mass gives the whale added lift as the pressure increases. At a depth of 1,000 meters the lift would equal about .1 percent of body weight. Lumping the effects of water pressure on all three tissue components, the sperm whale at a depth of 1,000 meters would have extra lift equivalent to .07 percent of its body weight.

Water pressure also acts on the gas in the whale's lungs during a dive and so affects the whale's buoyancy, particularly at depths of up to 200 meters. When a freshly killed sperm whale is floating on the surface in the Tropics, the volume of its body above the water roughly equals the volume of air in its lungs. By the same token, when a sperm whale exhales on the surface, it comes close to attaining neutral buoyancy. Sperm whales may sometimes exhale before diving; under these circumstances the reduction in buoyancy should facilitate the dive. If instead the whale dives with its lungs full, by the time it reaches 200 meters the effect of the water pressure on the gas in the lungs will have brought about the same reduction in buoyancy as an exhalation at the surface would have. Below 200 meters there is no great difference between diving with full lungs and diving with empty ones.

If one considers the combined effects of seawater density and pressure on the whale, one can calculate the lift or the downthrust the sperm whale will experience at various depths and in various environments. For example, the whale is very positively buoyant on the surface at the Equator. If the animal dives with full lungs, the rapid decrease in lung volume causes a downthrust equal to about .1 percent of its body weight at 100 meters. By the time the whale reaches 200 meters the change in the water density

will more than counterbalance this first effect, and the animal will have positive lift. The lift will increase as the whale descends until at 2,000 meters it will equal about .2 percent of the body weight. In Antarctic latitudes, since the cold surface water is denser than the warm surface water at the Equator, the sperm whale begins its dive with greater buoyancy and so does not experience much downthrust in the early part of its descent.

Calculations indicate that throughout its geographical range the sperm whale must adjust its overall density by an amount equivalent to .2 percent of its body weight in order to achieve neutral buoyancy below 200 meters. I have suggested that such an adjustment could be made by a "freezing" of the whale's spermaceti oil and have also described the heat exchangers available to cool the stored oil to a solid. But is the freezing effect adequate to the task? Among the known facts that are useful in answering this question are the findings of the National Physical Laboratory with respect to the density of spermaceti oil at various temperatures and pressures. It is also known that a 30-ton whale will have nearly 2.5 tons of spermaceti oil in its head and that at the surface in equatorial latitudes the oil will be at a temperature of 33 degrees C.

Starting with these facts, we can go on to calculate the spermaceti-oil temperature required to counterbalance the whale's natural lift at successive depths. The calculation shows that below 200 meters the temperature of the oil need be lowered by only a few degrees (and never below 29 degrees C.) to attain the required densities. As an example, if a sperm whale in Antarctic latitudes exhales before diving, the temperature of the spermaceti oil need not fall below 30 degrees in order to counterbalance the whale's natural lift. When the temperature and the density are plotted together, the slopes of the temperature lines and the slopes of the lines indicating the required densities are strikingly similar.

With both the anatomical and the physical data in hand one can consider how the dissipation and the regaining of heat might be timed during the sperm whale's diving cycle. To begin with an example in the equatorial environment, one finds that the heat generated during a diving cycle probably cannot be dissipated during the 10 minutes of each cycle that the whale spends at or near the surface. The reason is that at any depth shallower than 100 meters the difference between the water temperature and the subcutaneous temperature of the whale is less than 2.6 degrees C. Even if the active loss of heat at the surface by vasodilation (that is, an expansion of the blood vessels in the skin) is combined with passive heat loss at depth by conduction, only part of the

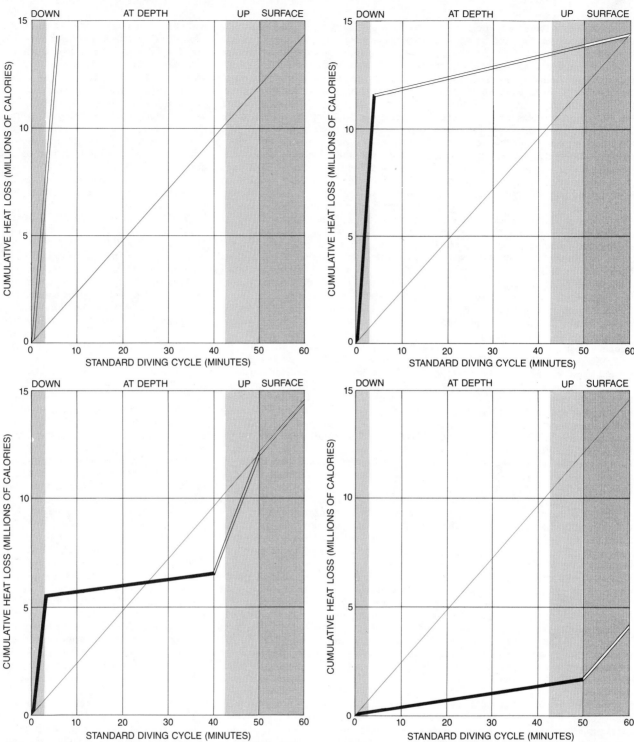

HEAT BUDGET OF A DIVING CYCLE calls for a loss of nearly 15 million calories over a 60-minute period that is divided into some three minutes for descent to a depth of 500 meters (*colored band at left*), some 40 minutes of stalking prey at that depth, some seven minutes for ascent to the surface (*second colored band*) and 10 minutes of rest at the surface (*gray band*). The loss of heat is required to balance the body-heat gain resulting from exertion during the diving cycle. Each of the four graphs plots a steady rate of heat gain (*colored diagonal line*) and the rate of heat loss that is achieved by one of four different strategies of heat exchange. Calculations are for a 30-ton whale diving to 500 meters with full lungs at the Equator. The open line in graph *A* represents the result of a dive with continuous heat exchange via body skin and snout skin until the spermaceti oil is cool enough to bring about neutral buoyancy. The heat budget would remain badly out of balance during most of the diving cycle. The line in graph *B* represents the result of a dive with heat exchange via the right nasal passage, body skin and snout skin (*solid portion of line*) until neutral buoyancy is achieved. Heat loss is then reduced by vaso-constriction. Such a strategy would produce a budget that starts badly out of balance but improves steadily thereafter. The line in graph *C* represents the result of a dive with heat exchange via the right nasal passage only during the descent; after that, heat loss is reduced by vasoconstriction until some 40 minutes have passed (*open portion of line*). Heat is then transferred from the body to the spermaceti oil until the oil temperature rises to 33 degrees C. Thereafter vasodilation allows further heat exchange via skin surfaces. This strategy would produce alternating net heat loss and net heat gain until the whale returned to the surface. Strategies involving loss of heat via the snout skin, alone or in combination with the right nasal passage, would give rise to similar curves. One of these three strategies is the likeliest to be adopted by the whale. The line in graph *D* represents a dive with heat exchange minimized by vasoconstriction until the whale returns to the surface. The slight heat loss following vasodila-tion at the surface in equatorial waters (*open portion of line*), together with limited heat exchange during the dive, would produce an out-of-balance budget; the whale could not dive again within 10 minutes.

heat the whale generates during the active 50 minutes of its diving cycle will be removed. Heat cannot be retained by means of vasoconstriction during the submerged period of the dive because some of the sperm whale's time in deep cold water must be spent in losing heat and achieving neutral buoyancy.

Looking further, we find that the greater part of the heat the whale generates during the active part of the diving cycle may either be lost slowly, by a controlled partial vasodilation that keeps pace with heat production, or be lost rapidly by more extensive vasodilation during one or more short periods in the course of the diving cycle. Brief but extensive vasodilation would not call for prolonged control and would at the same time allow the whale to make use of variations in spermaceti-oil density for buoyancy control. For example, as the whale descends, the oil would be cooled by heat exchange not only via the skin but also probably via the right nasal passage. When the whale has attained neutral buoyancy, soon after reaching the desired depth, any further active cooling of its oil would be halted by means of vasoconstriction. Minor adjustments in buoyancy thereafter could be made by admitting small amounts of seawater into the right nasal passage; that would eliminate any dependence on skin vasodilation for heat loss.

During most of the 50 minutes of the diving cycle spent at depth the body of the sperm whale, and its muscular mass in particular, rises in temperature. The whale's muscles need not increase in temperature more than two degrees C. in order to store all the heat needed to bring the solidified spermaceti oil back to its normal fluid temperature. Just before the whale begins its ascent to the surface the supply of blood to its snout would be increased, carrying heat from its warmer body to the spermaceti organ. Hence at the same time that the whale's body temperature is falling to its normal level the thawing spermaceti oil is rapidly rising to its normal 33 degrees C. Heating the oil of course decreases its density and increases its volume, and so the whale's buoyancy shifts from neutral to positive; this would help to lift the whale toward the surface even if the animal were exhausted. Any muscle-generated heat in excess of that needed to reheat the oil could now be lost by vasodilation.

Using as an example a 30-ton sperm whale diving to a depth of 500 meters in equatorial waters, one can now consider five alternatives with respect to the most efficient means of heat loss. As the first alternative let it be assumed that vasoconstriction does not take place over the entire skin surface until the whale has descended to the selected depth and achieved neutral buoyancy. During the descent heat will be lost over the entire skin surface; the total heat loss will approach the total heat gain as a result of muscular activity during the diving cycle. When the passive loss of heat through the whale's blubber is added to this active heat loss, the total loss possibly exceeds the total gain. This first alternative must therefore be eliminated.

Let it be assumed as the second alternative that vasoconstriction takes place over the whale's entire skin surface as soon as the descent begins, so that almost all the cooling of the spermaceti oil is a result of heat exchange in the right nasal passage alone. Under those circumstances the time required to reach neutral buoyancy will be greater than if heat exchange were simultaneously in progress through the skin. Nevertheless, neutral buoyancy is achieved in about six minutes (12 percent of the 50-minute dive time). The dilation of the blood vessels of the spermaceti tissue before the whale begins to ascend would then shunt heat from its body muscles to its snout until the oil temperature returns to its normal 33 degrees C. Any excess body heat generated during the dive could then be lost by vasodilation.

In the third alternative the assumption is zonal vasoconstriction. The blood vessels of the whale's body skin constrict immediately after submergence but the blood vessels of the snout skin remain dilated until the spermaceti oil is cooled to the point of neutral buoyancy at depth. Such a sequence would lead to neutral buoyancy even more rapidly than the sequence outlined in the second alternative. It requires the further assumption, however, that the sperm whale can dilate and constrict the blood vessels of its snout skin and body skin independently.

In the fourth alternative heat exchange through the right nasal passage and heat exchange through the whale's entire skin proceed simultaneously. This sequence leads to neutral buoyancy faster than either the second or the third alternative. Moreover, the amount of the body heat generated during the dive that remains to be lost to the seawater is smaller.

The fifth and most efficient alternative involves heat exchange via the right nasal passage and the snout skin only.

All except the first of these five alternatives are plausible. The fourth alternative has an advantage over the third and fifth in not postulating a capacity for zonal vasoconstriction and vasodilation. Before the third and fifth alternatives are dismissed on those grounds, however, it should be noted that such zonal dilation is similar to the human facial blush. It seems at least conceivable that sperm whales too can blush.

As evidence for heat exchange via the right nasal passage I have found seawater in this passage on two occasions. Therefore the second, fourth and fifth alternatives are all likely ones, and the fifth alternative, involving both the right nasal passage and the snout skin, is the most likely of all. In this connection, in several recently killed sperm whales the spermaceti tissue closest to the surface of the snout was found to be slightly warmer than more central tissue. This suggests that an efficient cooling system, namely heat exchange via the right nasal passage, had been operating in the center of the snout.

Changes in the density of the spermaceti oil would of course affect the sperm whale's center of gravity. As the oil increases in density during the whale's descent the center of gravity will shift forward; as the density decreases just before or during the ascent the center of gravity will shift backward. These changes would actually be advantageous during both descent and ascent because the whale often moves almost vertically down and up. The shift means, however, that the whale's fore-and-aft trim, to borrow a nautical expression, will not be the same when the whale is submerged as it is when the whale is on the surface. The main factor in the whale's trim while it is on the surface is not the spermaceti ballast but rather the considerable buoyancy provided by its air-filled lungs, so that under these circumstances the role of the spermaceti is trivial. Perhaps, however, the forward shift of the center of gravity at depth actually gives the whale its best trim during the time in the diving cycle when the submerged animal is neutrally buoyant and either lying still or swimming.

Sperm whales less than about 11 meters long have less spermaceti oil in relation to their size than larger ones. The smaller whales are also comparatively shallow divers and do not migrate to such high polar latitudes as the larger whales. As a result they encounter smaller variations in water density, and even though their supply of the oil is smaller, it would suffice to compensate for the smaller range of water densities.

For an air-breathing, warm-blooded, long-diving animal control of buoyancy has numerous advantages. At the surface extra buoyancy allows the whale to relax. At depth the animal's hunting success, which is dependent on silence and good hearing, is not jeopardized by the need to keep swimming in order to remain at a particular level. Moreover, if it were necessary for the sperm whale to keep swimming in order to remain at the same level during its 50-minute submergence, it would have to invest a larger amount of energy in its diving cycle.

These are not the only advantages derived from the control of buoyancy. Imagine for a moment a hypothetical sperm whale that was less buoyant on the surface than is actually the case. Over a part of its range of diving depths and in certain geographical locations this hypothetical whale would actually

SPERM WHALE'S HEAD is peeled of its blubber on the flensing platform of a factory ship in the Antarctic; teeth of the lower jaw are clearly visible. Upper-jaw teeth are usually absent.

depths made possible by oil-density control could hardly be matched by a system that depends on an exhaustible reservoir.

Loss of heat from the spermaceti oil as a means of buoyancy control imparts still another distinct advantage. In the Tropics the water temperature at the surface may be less than three degrees C. below the subcutaneous temperature of the whale. At depths shallower than 100 meters the body heat generated by a deep dive could not be lost during the usual 10-minute resting time at the surface between dives. Both the heat exchange that is involved in freezing the spermaceti oil during the descent phase of the diving cycle and the body-heat shunt involved in thawing the oil before and during the ascent phase help to overcome this equatorial handicap.

Few whales are known to dive as deep or for as long a time as the sperm whale. Two other toothed whales—the bottle-nosed whale *Hyperoodon* and the pygmy sperm whale *Kogia*—have spermaceti organs. Both are long-duration deep divers. In view of the complicated structure of their snout it seems probable that both also use the spermaceti organ for buoyancy control. This conjecture cannot be proved, however, until much more is known about the dimensions, anatomy and diving habits of these whales.

Some other whales, for example the fin whale *Balaenoptera physalus,* can make rapid excursions to considerable depths. The dives of the fin whale, perhaps to depths of some 300 meters, seldom last longer than 10 minutes; the whale swims continuously and surfaces some distance from its point of submergence. These baleen whales usually swim faster than the three deep-diving toothed whales. It is clear that they do not submerge for long periods of time or in order to feed as the three deep divers do; their prey is the shrimplike krill found in swarms at or near the surface.

As I have noted, many suggestions have been made regarding possible functions of the spermaceti organ other than buoyancy control. None, however, accounts as adequately for the size and structure of the organ. Such a complex anatomical structure, however, could certainly serve more than one function. Several of the other proposed functions are quite compatible with the buoyancy-control hypothesis I have proposed. Nevertheless, so many peculiarities of the sperm whale's snout and its cargo of oil can be explained by the buoyancy-control hypothesis that it is hard not to accept this as the organ's main function. Final proof of the hypothesis, however, must await measurement of the temperature or the density of the oil within the spermaceti organ in the course of a deep dive. This is a difficult and costly task but not an impossible one.

experience downthrust during the diving cycle. Downthrust, of course, is a hazard to any air-breathing animal that makes deep dives of long duration. By not only having extra buoyancy at all depths but also having control over its buoyancy the sperm whale has a failsafe diving system. Even if the animal became exhausted during a deep dive, it could still pop up to the surface for air. For example, a sperm whale that has exhausted itself, perhaps by short bursts of high-speed swimming after squids, can shunt the heat that is a by-product of its muscular activity into the spermaceti organ and again become positively buoyant. Such an exhausted whale is just as likely to reach the surface again as a fresh whale that can easily swim up.

What about alternative means of buoyancy control? For example, either collapsing the lungs or emitting air while submerged would affect buoyancy. On close consideration neither stratagem proves to be as efficient as control

by means of oil density. The degree of lung collapse depends entirely on water pressure, whereas buoyancy also depends on seawater density, which in turn is dependent on temperature. As diving depths increase, these factors work in opposite directions; only at two points in a deep dive do the two factors exactly match to yield neutral buoyancy.

As for the emission of air, that would be an ineffective means of controlling buoyancy over much of the whale's diving range. For example, below 600 meters even a small emission of air would greatly reduce the volume of air remaining in the whale's lungs for the return to the surface; in some geographical areas a whale that had emitted air at depth would experience considerable downthrust as it swam through the thermocline, the abrupt transition in water temperature that is encountered between 200 and 100 meters. Furthermore, once air is emitted it is lost. The many minor adjustments of buoyancy at different

IV

ADAPTATIONS IN
REPRODUCTION

Nightingales pour out a ceaseless gush of song for fifteen days and nights on end when the buds of the leaves are swelling—a bird not in the lowest rank remarkable. In the first place there is so loud a voice and so persistent a supply of breath in such a tiny body; then there is consummate knowledge of music in a single bird: the sound is given out with modulations, and now is drawn out into a long note with one continuous breath, now varied by managing the breath, now made staccato by checking it, or linked together by prolonging it, or carried on by holding it back; or it is suddenly lowered, and at times sinks into a mere murmur, loud, low, bass, treble, with trills, with long notes, modulated when this seems good—soprano, mezzo, baritone; and briefly all the devices in that tiny throat which human science has devised with all the elaborate mechanisms of the flute, so that there can be no doubt that this sweetness was foretold by a convincing omen when it made music on the lips of the infant Stesichorus. And that no one may doubt its being a matter of science, the birds have several songs each, and not all the same but every bird songs of its own. They compete with one another, and there is clearly an animated rivalry between them; the loser often ends her life by dying, her breath giving out before her song. Often younger birds practise their music, and are given verses to imitate; the pupil listens with close attention and repeats the phrase, and the two keep silence by turns: we notice improvement in the one under instruction and a sort of criticism on the part of the instructress.

Pliny
NATURAL HISTORY, X, xliii

IV

ADAPTATIONS IN
REPRODUCTION

INTRODUCTION

In this Section we consider three cases of reproductive specialization in vertebrates. In the first article, "The Phalarope" by Otto Hahn, reversal of sex roles in a species of bird is shown to be hormone-dependent and to involve changes in both male and female phalaropes. Here, as in other instances in vertebrates, both physical features (feather color, patterns of color, feather shape, brood patch formation, and so on) and behavioral activities (display, egg incubation, feeding of hatchlings, and so on) are controlled by hormones such as testosterone and prolactin. The situation is even more complicated in that other hormones may also be involved; thyroxine, for instance, may fluctuate in levels in the blood and cause patterns of pigment stripes upon the new breeding plumage of a male bird. Coordinated activity of hormones produced by the anterior pituitary, the gonads, the adrenals, and other endocrine organs is required to elicit the full range of structure and behavior required for successful reproduction.

Some fishes, too, are subject to sex reversals. There are certain "cleaner" fish (so called because they remove ectoparasites from other fishes) that live near the Great Barrier Reef and normally comprise small groups. A single dominant male and a harem of from three to six females will occupy a given territory. The females are hierarchical and have a dominant member that is usually the oldest (fittest?) of the sex. A fantastic event occurs if the dominant male dies. If foreign males do not invade the territory when this happens, the dominant female *changes sex*—in fact, within a few hours of the male's death, the dominant female may be showing "male" aggressive displays. Within a day or two, male courtship and spawning behaviors are evident! The transition to the point of releasing functional sperm takes longer—14 days or so—but that is surely a surprise in itself. D. R. Robertson, the Australian zoologist who has studied these fishes, suggests many advantages for this bizarre arrangement. For instance, the "male" genotype at any one time might be thought of as being the most highly adapted for local environmental conditions and survival, since that genotype is derived from the female genotype that survived longest in that locale. It is certainly intriguing, and may be quite appealing to some, that these cleaner fish live in a society in which "males are produced only when needed"!

The second article in this Section, "Mimicry in Parasitic Birds," by Jürgen Nicolai, demonstrates a marvelous set of adaptations resulting in precise matching of subpopulations of parasites and hosts. As background, we should recall that it is common among many types of fish that very large numbers of eggs should be fertilized, but that few offspring should survive to reproduce successfully. In general, birds and mammals produce far fewer offspring, though there is substantial variability between species. For instance, some large sea birds produce only a single egg every other year, whereas small song birds may lay ten or so during each of two periods in a single summer.

Parasitic birds have adopted the strategy of laying their eggs in the nests of other birds. The hosts both incubate the eggs and feed the hatchlings along with their own brood. In order to be successful, the widow birds of Africa must synchronize their reproductive activity precisely with that of their potential hosts. In addition, the parasite's eggs must match those of the host in size and color, and the hatched chicks must have appropriate size, markings, and behavior so as to fool the host adults into feeding and protecting a counterfeit brood. Some parasitic bird offspring simply kill host offspring so as to receive full attention from the host parents; such a strategy, if systematically pursued, is deleterious to survival of the host population. By precise matching of their young to host young, widow birds avoid that difficulty.

Perhaps the most fascinating aspect of widow bird parasitism concerns singing. Different subpopulations of widow birds have acquired slightly different markings so as to match subpopulations of host finches precisely. If a given widow bird female were to mate with a male from a subpopulation having different markings, then the hybrid offspring would likely be intermediate in character and so be rejected by host parents. To avoid this problem, widow bird females mate only with widow bird males that include, in their breeding season song, phrases of the songs sung by the appropriate subpopulation of host finches! Widow bird males learn these phrases when they are themselves young hatchlings in the nests of finch foster parents. Thus, song is used as a barrier to hybridization that might otherwise produce unsuitable offspring.

We can fully appreciate this unexpected use of song after considering the various sounds made by birds. A bird generates calls and songs with its syrinx, a specialized organ located at the bifurcation of the trachea into the two primary bronchi. A tympanic membrane that may vibrate and generate sound is found in the wall of each bronchus; and, in fact, different individual notes in a bird's song can be shown to come from either the right or the left tympanic membrane and syrinx. If one severs the nerve leading to one side of an adult bird's syrinx, then the notes from that side are eliminated permanently from the song; the intact side continues to "sing" its notes in correct order and timing, but now there are gaps left by the absence of notes from the side of the severed nerve. The bird cannot "learn" that its song is deficient and cannot perform the whole repertoire on the intact side. However, if such an operation is performed on a young bird that has not yet perfected its full song, the intact side can be used for the whole song. Obviously, the avian voice box adds a versatility in sound production that cannot be matched by mammals, including humans; it allows the fantastic diversity described so eloquently by Pliny and innumerable poets.

One of the main conclusions of recent work on bird song is that the slight variations observed between different birds of one species may be individual recognition markers. It is hard to generalize about as diverse a topic as bird song, but it does seem probable that the characteristics of the notes or the syllables of the song are varied by individuals, whereas the basic time spans between the notes are kept relatively constant in order to provide species-recognition information. Since pitch, frequency modulation, and amplitude can all be varied, it is not difficult to imagine how even minute differences may identify each bird, and so allow pair maintenance, recognition, and so on to operate.

As a result of a number of recent experiments, we are able to determine quite precisely the stage at which young birds "learn" or acquire their songs. The zoologist Peter Marler and his collaborators have investigated populations of white-crowned sparrows living north, south, and east of San Francisco Bay. Each population sings its own dialect of the basic white-crowned sparrow song. A subsong, or a full song unlike that of any dialect, will develop in cage-reared, isolated white-crowned sparrows. If a mixed group of young from the three populations is never allowed to hear parental songs, the birds will develop their own unique dialect, which can in turn be taught to their offspring. Normally a newly hatched bird hears its own dialect being sung for the first two or three months of life. Then in the fall and winter the older birds of the population stop singing. Several months later when spring time approaches, the maturing bird begins singing, and soon perfects the parental dialect.

After a series of experiments in which captured young white-crowned sparrows were raised in isolation, it was found that the period of learning the dialect occurs between about two weeks and two months of age. From then on, the bird "remembers" in some way what it heard early in life. As might be expected, a nestling from one population placed with another group acquires the new dialect and sings it the following spring. This is an extraordinary feat. The young nestling acquires the song early, by listening. It does not practice or receive any reward for learning the song. Then months elapse before singing starts; the bird practices and varies its song until it finally duplicates what it heard months before.

It is of interest, therefore, to investigate the relationship of hearing to dialect. A bird deafened at birth can at best sing only the species subsong. If a bird is deafened at two months of age, we might guess that it had heard the dialect for a sufficient length of time and that no interference with singing would occur. However, this is not true, because the dialect song has not been perfected. In fact, the experiments show that a bird must be able to hear during the weeks that it is perfecting its dialect song. Thus it appears that the bird must compare its own song noises with those that were heard months before, matching those it produces to the original sounds to produce the correct dialect. After the song has been perfected, deafening has little effect except over several months, when the song gradually drifts away from the original melody; apparently, some feedback from hearing acts to maintain the normal dialect song.

Perhaps the most surprising example of this type of learning has been observed in female white-crowned sparrows. Normally, they never sing, yet, like the males, they are exposed to dialect singing during youth. When a female is injected with testosterone (a male gonadal hormone), she soon breaks into song and before long is singing the dialect! Thus she too has retained the memory of early sounds, though as a rule she would never emit them. The utility of such retention is obvious if she can use it also to identify males of her regional population when pair formation and breeding take place.

With the foregoing as perspective we can now reconsider the use of song by parasitic widow birds. It appears that the young widow bird offspring listen and learn aspects of foster parent song, and then sing phrases of that song when they mature and mate. As a result subpopulations of parasites remain "homed in" on subpopulations of hosts. What a lovely example this is of adaptation.

One of the most fascinating variations on bird song is antiphonal singing. Some shrikes, warblers, and barbets use antiphony for pair identification and location under conditions of limited visibility, particularly in tropical habitats. As pointed out by W. H. Thorpe in "Duet-Singing Birds," *Scientific American* Offprint 1279, one of a pair of birds will begin by singing notes or groups of notes that are part of a song; then, the mate will break in, singing the missing notes; the process is repeated again and again as an antiphonal duet (see Figure 1).

It is probable that the use of antiphonal singing as a form of personal identification and location has arisen in species in which territoriality is not rigidly enforced. Mixing, both between and within species, is much freer in the tropical habitats than in more strict territorial regions of the temperate zones.

Figure 1. The duet pattern on a pair of shrikes recorded on the Mozambique Coast. This is a duet of intermediate complexity; the A and B birds alternate as shown. After repeating this duet a number of times the birds may suddenly change to another song in their repertoire. On rare occasions W. H. Thorpe has heard three birds participating in a trio, but the biological function of such behavior is a mystery. The scientific pitch shown in this illustration is 256 hertz; the bar length equals 1.5 seconds. [From W. H. Thorpe and Myles W. North, *Nature*, 208, 1965.]

Consequently, accurate pair identification signals may have become so important that antiphony evolved. Examples of an extreme form of the imitative abilities of antiphonal singers are seen in parrots and mynah birds, which have an amazing imitative repertoire. They too may have acquired such capabilities in order to have distinctive nuances in song for individual identification.

The final article in this Section, "The Social System of Lions" by Brian C. R. Bertram, concerns a special case of mammalian reproduction in which the males of a species have much shorter effective reproductive lives than the females have. Among the strategies that have evolved in male lions to increase their contribution to the gene pool of their species is the killing of young cubs in a pride that were fathered by previous males. From 3 to 12 female lions are the core of a lion's breeding unit, the pride; those females tend to be related genetically. Even though there is no strictly limited season for reproduction in lions, the females of a pride are likely to come into estrus (heat) together. This may help to reduce competitive fighting among males, since the common estrus makes it more likely that a female in heat will be available for each of the males. This is a somewhat different strategy than that displayed by humans (as is explained in "An Essay on Vertebrates" in the companion volume to this book, *Vertebrates: Physiology*). Female humans are not only asynchronous in mating behavior but also able to mate at times when ovulation does not occur. It will be interesting to discover whether the high frequency of matings in female lions that do not result in births is in part because of copulation in the absence of ovulation; if so, this would be an example of a behavioral modification of reproductive physiology possibly selected for as a means of reducing the injuries to males that might result from their fighting with each other.

BREEDING-SEASON PLUMAGE of the female phalarope (*bottom*) is more colorful than the plumage of the male (*top*). In this somewhat stylized rendering the birds are drawn to actual scale; the female is larger as well as more colorful than the male. These birds are Wilson's phalarope (*Steganopus tricolor*), a species that breeds in Canada and over much of the U.S. west of the Great Lakes.

The Phalarope

12

by E. Otto Höhn
June 1969

*The reproductive behavior of this aquatic shorebird is
most unusual. The female phalarope aggressively
courts the male; the male phalarope warms the eggs
and rears the young birds*

The difference or similarity in coloration between a male bird and a female of the same species usually reflects the behavior of the species in courtship and breeding. In species where the male bird is more colorful than the female, the male tends to play a dominant role in courtship and a minor one in rearing the young. If the breeding plumage of the two sexes is similar, the female usually participates actively in the events of courtship and the male has a large share in the care of the eggs and fledglings. Then there are a few species in which the female is more colorful than the male, and here the female takes the initiative in courtship and the male alone rears the young. In this last group is the small, long-necked shorebird known as the phalarope.

In the process of selecting a mate the brightly colored and (compared with the male) large female phalarope is the aggressor. Soon after the eggs are laid the female bird typically leaves the breeding ground. Then for 21 days the small, dull-colored male phalarope incubates the eggs and for 10 more days looks after the downy young. Shortly before the incubation period the male bird sheds feathers from its abdomen and the bare skin thickens and becomes engorged with blood. This "brood patch" enables the male to warm the eggs and the young birds.

In addition to its unusual reproductive behavior the phalarope has other singular characteristics. For one thing, in shallow water, where other shorebirds wade, the phalarope swims. Floating high in the water, it looks as if it were made of cork. The phalarope is an excellent swimmer, and associated with this ability are lobed membranes that extend from the sides of its toes. These flexible paddles are much like the ones found on the highly aquatic coot and grebe but are

considerably smaller. The structure of the phalarope's foot is responsible for the bird's name: the genus *Phalaropus* is from the Greek for "coot-footed."

As it swims about the phalarope daintily picks up bits of food from the surface of the water or searches for food below the surface by dipping its head into the water, in the manner of surface-feeding ducks. At times the phalarope engages in an unusual feeding maneuver: it rotates like a top, dabbing at the water with its beak as it turns. Specimens of Wilson's phalarope (a species named after the 18th-century American naturalist Alexander Wilson) have been observed spinning as fast as 60 revolutions per minute, spearing the water

with their beak once in each revolution. These rotations may serve more than one purpose, depending on environmental circumstances. In shallow water the spinning movement could stir up food particles lodged on the bottom. In deep, cold pools on Greenland, which are frequented by phalaropes, the spinning is believed to activate mosquito larvae, making them more conspicuous to the birds. In some instances the rotations may simply be due to an abundance of food. Turning to the side for a bit of food, the bird may see another morsel farther to the rear, and after snapping up the first it continues the turning movement to reach the second.

There are three species of phalarope;

ARCTIC PHALAROPES are the two other phalarope species. At top is the northern phalarope (*Lobipes lobatus*) in summer breeding plumage; at bottom is the red phalarope (*Phalaropus fulicarius*), which is also seen in breeding plumage. The birds are females.

they belong to the order Charadriiformes, which includes plovers and sandpipers. One species, the red phalarope (*Phalaropus fulicarius*), breeds on the tundra of the arctic zone in both the Eastern and the Western Hemisphere. The nest of the red phalarope is never more than a few miles inland, and the birds spend the nonbreeding season—the winter—at sea, often many miles from land. They feed on crustaceans and other small animals of the marine plankton, and they rest and sleep afloat. Thus apart from the summer breeding season they are true seabirds. The wintering areas of this species are off the east and west coasts of South America and the west coast of Africa [*see top illustration on page 134*]. On its migrations the red phalarope apparently flies over the sea; the birds are rarely seen on land south of their breeding grounds.

A second species, the northern phalarope (*Lobipes lobatus*), nests in the subarctic region of the Eastern and Western hemispheres. This species tends to build its nests farther inland than the red phalarope. It too winters at sea. Its wintering areas, which are nearer the Equator than the red phalarope's, lie off the coasts of Peru, Argentina, northwest Africa, southern Arabia and Indonesia.

Wilson's phalarope (*Steganopus tricolor*) is the most familiar of the three species. It is confined to the Americas. The breeding grounds of Wilson's phalarope extend from the Great Lakes westward across the prairie provinces of Canada into British Columbia and farther south over a large area of the western U.S. The main wintering area of the bird is inland on the pampas of Argentina.

All the phalaropes have a dull fall and winter plumage, gray above and white below. It is the same in both sexes. In the breeding plumage of spring and summer other colors appear, although more faintly in the male bird [*see illustration on page 130*]. The only other bird species in which the female is more colorful than the male are the tropical and subtropical painted snipes (*Rostratula*) and the quaillike hemipodes (*Turnices*) of the Eastern Hemisphere. The female is somewhat larger than the male in all three phalarope species. The average weight of Wilson's phalaropes in a group I collected during the breeding season was 2.4 ounces for the females but only 1.7 ounces for the males.

I have observed the reproductive behavior of Wilson's phalarope in southern Canada and the behavior of the other phalarope species along Hudson Bay and in Alaska. The process of pair for-

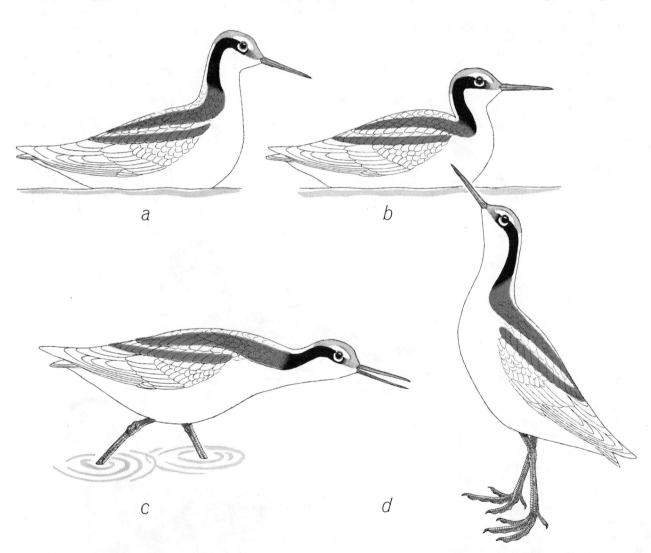

a

b

c

d

POSTURES OF THE PHALAROPE are associated with reproduction. During the period of pair formation the female Wilson's phalarope stays near a particular male, normally swimming with her head up (*a*). A threat posture is assumed (*b*) when another female phalarope approaches. While walking, the female bird assumes a different threat posture to ward off intruders (*c*). The head-up pose (*d*) is taken by a pair of phalaropes before their first mating. It probably signifies that no threat is intended.

mation is best known in Wilson's phalarope. In a group of the birds one sees a female keeping close to a particular male, following him as he swims here and there. When another female approaches the pair, the first female lowers her head and pulls it in until her long neck is scarcely visible. This posture is a threat gesture: it sets the stage for a forward lunge. With her head down the threatening female swims purposefully toward the intruder. When she is about a foot away from her, she takes wing and with her neck extended and her legs dangling flies at her. Usually the threatened female retreats, and after a brief chase the threatening bird returns to the vicinity of the male. If one female threatens another while the birds are walking on land, the threatening bird extends her head—sometimes with her bill partly open—toward the enemy. Actual fights, however, are rare. I have seen only one. The two birds hovered in the air beak to beak, each trying to stab the other as they fluttered with their legs dangling. The aerial fight lasted for only a few seconds; the intruding female gave up and flew away.

It appears, then, that the female phalarope chooses a particular male and wards off other females by threat and, if necessary, attack. Thus the male bird becomes accustomed to the company of a particular female and later directs his sexual advances toward her. Male phalaropes show very little aggressive behavior at any time.

Before two phalaropes mate they sometimes face each other with their heads raised, the bill of each bird aimed at a point above the other's head. For a few seconds the birds hold this pose. Since the phalarope threatens and fights with its beak, the posture apparently signifies that no threat is intended. The female then indicates readiness for mating by taking a crouching position. The male rises up like a miniature helicopter, hovers over the female's back and mating follows. The raised-beak display seems to be a prelude only to the first mating between two individuals. When the birds are more familiar with each other, the display is omitted; the male simply hovers aloft and lands on the female's back.

What is the biological significance of the reversed roles of the sexes in the phalarope? At the University of Alberta my colleagues and I have been studying the bird to learn the physiological factors that underlie its behavior. The greater aggressiveness of the female phalarope compared with the male suggests that in the earlier phases of the breeding cycle the female may secrete more androgen (male sex hormone) than the male does. There are many studies of animals that link aggressiveness to male sex hormones. It has been shown that among birds the administration of androgen increases the aggressive behavior of doves, female herring gulls and castrated young male chickens.

That androgens play a role in the production of the phalarope's breeding plumage has been shown by J. E. Johns of Montana State University. In his experiments with Wilson's phalaropes and northern phalaropes in their drab winter plumage he plucked feathers from the shoulder of the birds. He then found that if the plucked phalaropes were given the male sex hormone testosterone, they grew new feathers like those of the normal female in breeding plumage, regardless of whether the experimental bird was a male or a female. Phalaropes that received no hormone after plucking grew feathers of the dull winter plumage, similar to the feathers that had been plucked. Other experimental phalaropes were given a female sex hormone, namely estradiol, which in combination with the pituitary hormone prolactin induces formation of the brood patch. Whether administered alone or together, these hormones did not give rise to female nuptial plumage.

It remained to be shown that the female phalarope does in fact produce large amounts of androgen. My colleagues and I collected ovaries and testes from a group of phalaropes and, for comparison, from other bird species in which the male is the dominant sex: mallard ducks, domestic fowl and redwinged blackbirds. We isolated the individual hormones (by means of chromatography) from extracts of the birds' gonads and measured the amount of each hormone. In birds where the male is dominant the amount of testosterone in the testis was five to 10 times greater than the amount in the ovary. In phalaropes, however, there was about as much testosterone in the ovary as in the testis, and in some instances there was more testosterone in the ovary.

Evidently the brilliant plumage of the female phalarope is due to the high production of androgen by her ovaries. The same factor is probably responsible for the aggressive behavior of the female. We ruled out a possible involvement of hormones of the adrenal cortex, because the amount of androgen in this gland was slightly greater in the male than in the female both in phalaropes and in the control birds (mallard ducks).

How it is that the male phalarope and not the female develops a brood patch and incubates the eggs has also found an endocrine explanation. It has been known for some time that an estrogen, or female hormone, together with prolactin will cause finches of both sexes to develop a brood patch. (The brood patch normally appears only on the female finch.) In phalaropes we found that only a combination of androgen and prolactin

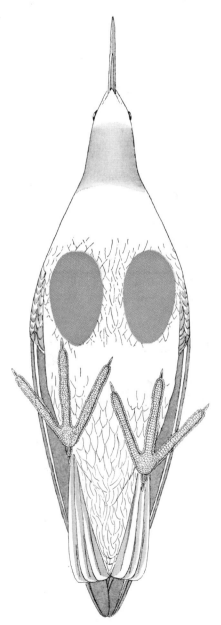

BROOD PATCHES appear on the abdomen of the male phalarope before the period of incubation. The exposed skin, which becomes engorged with blood after the feathers are shed, is brought in contact with the eggs and with the young birds. The female phalarope does not produce brood patches.

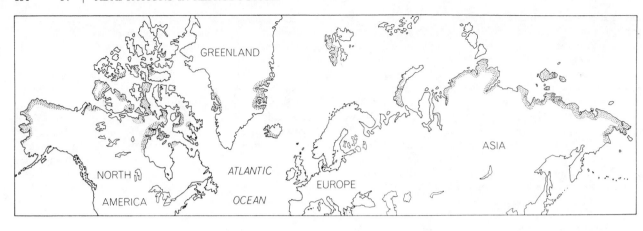

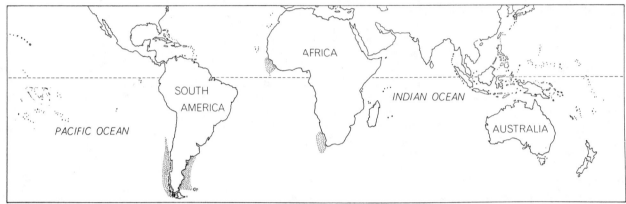

DISTRIBUTION OF RED PHALAROPE indicates the bird's arctic breeding grounds (*top*) and the areas where it spends the arctic winter (*bottom*). The nests of this species are never more than a few miles inland and their migratory flights are over the sea.

DISTRIBUTION OF WILSON'S PHALAROPE (which is named after the 18th-century American naturalist Alexander Wilson) shows the bird to be in North America during the breeding season (*left*) and in South America during the nonbreeding season (*right*).

could induce the formation of brood patches, both in females, which in nature do not develop these patches, and in males at a time of the year when they do not naturally form them.

Our earlier investigations had shown that the female phalarope is an efficient producer of androgen. The failure to develop brood patches could hardly be due to a lack of androgen, but it might result from insufficient production of prolactin. Prolactin is readily detected with a test employing the crop sac of the pigeon, which is a dilated section of the pigeon's gullet. In a nonbreeding pigeon the lining of the crop sac is thin and smooth, but in a breeding pigeon, just before the young hatch, the tissue of the lining proliferates into a wrinkled mass known as the crop gland. (The surface of the crop gland, sometimes called "pigeon's milk," is shed by the pigeon and eaten by its young.) This proliferation results from the increased secretion of prolactin from the pigeon's pituitary gland. The same effect can be achieved by injecting a sufficient amount of prolactin; therefore whether or not a given sample contains this amount of the hormone can be determined by injecting it into the skin over the crop sac and observing whether or not the crop gland develops.

With this technique we tested an extract prepared from the pituitaries of 39 breeding male phalaropes. Injected into a test pigeon over a four-day period, the extract gave rise to a response similar to one produced in another pigeon by 10 micrograms of prolactin. A similar extract from 40 female phalaropes did not affect the crop sac at all. Apparently the pituitary of the female phalarope secretes much less prolactin than the pituitary of the male. This result was confirmed by further experiments. We employed test pigeons that had been made more sensitive to prolactin by the injection of a subthreshold dose of the hormone; such a dose does not affect the crop sac, but it reduces the amount of prolactin required for a response. Tests with pigeons thus prepared indicated that during the breeding period the pituitary of the male phalarope contains about three and a half times more prolactin than the pituitary of the female. Evidently the female phalarope fails to develop brood patches because of an insufficiency of prolactin, one of the two hormones required for the formation of the patches. This lack of prolactin is almost certainly the reason female phal-

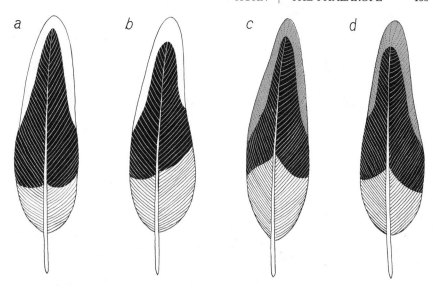

PLUMAGE IS TRANSFORMED when phalaropes receive testosterone (male sex hormone). The shoulder feather of a northern phalarope in dull winter plumage looks much the same whether the feather was grown in the wild (a) or in captivity (b). When a bird in this plumage is plucked and given testosterone, the new shoulder feather that is grown (c) resembles the shoulder feather of the female of the species in breeding plumage (d). The bright-colored female plumage is produced whether the phalarope is a male or a female.

aropes also have no urge to incubate their eggs.

One may speculate on how the partial reversal of the sexual roles of the male and female phalarope evolved. Suppose that in an ancestral phalarope both sexes had an androgen-dependent colorful breeding plumage, and that both sexes shared in rearing their young, as many living shorebirds do. Then suppose that a hereditary inability to secrete enough prolactin to form brood patches and to induce the urge to incubate eggs arose, perhaps as a mutation, in certain fe-

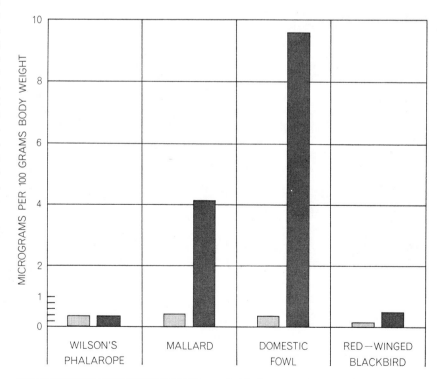

DISTRIBUTION OF TESTOSTERONE in male (dark gray) and female (light gray) birds is based on measurements of the amount of the hormone in their gonads. In the species compared with the phalarope the male is more colorful and more aggressive than the female.

136

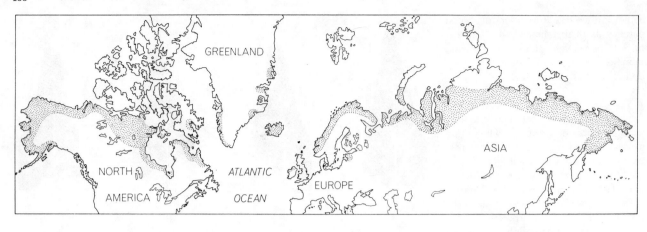

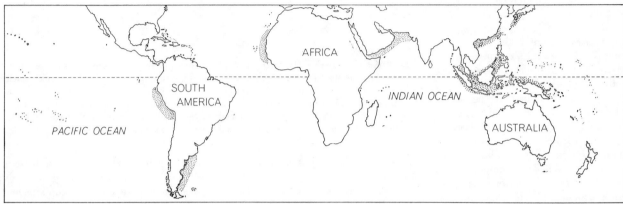

DISTRIBUTION OF NORTHERN PHALAROPE indicates that this species tends to nest south of the red phalarope (*top*) and to winter nearer the Equator (*bottom*). Like the red phalarope, the northern phalarope feeds on small animals of the marine plankton.

males. Pairs of birds in which the female was so affected could still reproduce if the male entirely took over incubation and the rearing of the young. Freed from the urge to incubate, these somewhat nonmaternal females would leave the breeding grounds earlier than normal females. Then they would be less exposed to predation, since they would not have to remain in the nest to incubate the eggs and rear the helpless young. As a result more of these females would survive to reproduce again.

The exposure of the male bird to predation while incubating the eggs and rearing the young would tend to favor individual males with an inconspicuous breeding plumage and thus with a low secretion of androgen at the time the plumage appeared. Brightly colored feathers would not, however, be a disadvantage to a female bird that had been liberated from the duties of the nest. On the contrary, a more colorful female would probably be more attractive to males and perhaps more dominant in encounters with females lacking a mate. Selection, then, would favor a female bird with an enhanced androgen secretion at the time breeding plumage is grown. The continuation of this increased androgen production into the early breeding season would account for the aggressiveness with which the brightly colored female phalarope courts the rather pallid male.

Mimicry in Parasitic Birds

by Jürgen Nicolai
October 1974

Various species of birds lay their eggs in the nest of another bird, which then incubates the eggs and feeds the young. The widow birds of Africa achieve this result by some remarkable feats of mimicry

The number of eggs a female animal produces is inversely proportional to the probability that the egg will give rise to a female that produces more eggs. All invertebrate animals and most lower vertebrates need to broadcast eggs by the hundreds and thousands in order to ensure the existence of subsequent generations. The number of eggs laid by a bird, although much smaller, also reflects the probability of reproductive success. Consider the greatest of ocean birds, the wandering albatross. It occupies breeding grounds on isolated islands of the Southern Hemisphere where it is completely undisturbed, and the female albatross lays a single egg every other year. In contrast, most species of European tits and both species of goldcrests respond to the stresses of severe winters and numerous predators by laying two yearly clutches of eggs, each numbering between eight and 12.

Any bird that is subject to severe environmental stress or predation pressure is at a reproductive disadvantage. It must incubate its eggs with the heat of its own body, and so the eggs must be kept all in one place, usually in a nest. A predator therefore needs to search out only a single target. One group of birds, the turkeylike megapods of Australasia and the Pacific, has surmounted this handicap by an ingenious stratagem. Instead of building nests they pile up mounds of plant material where they bury their eggs to be incubated by the warmth of plant decay [see "Incubator Birds," by H. J. Frith; SCIENTIFIC AMERICAN, August, 1959]. Still other bird species have discovered another solution to the problem. They deposit their eggs, one at a time, in the nest of another species; the eggs are then incubated and the hatchlings raised by the host.

Birds that practice this kind of parasitic parenthood are found around the world. Among them are various members of the cosmopolitan family of cuckoos, some species of American cowbirds, a black-headed duck (*Heteronetta atricapilla*) in South America, all species of honey guides and certain weaverbirds in Africa. I have investigated the behavior of this last group, observing in particular the several species of a subfamily, the Viduinae, within the family of weaverbirds. The birds in this subfamily are commonly called widow birds. My work has been conducted both in the field and in our laboratory at the Max Planck Institute for Behavioral Physiology in Seewiesen.

Although a parasitic bird has overcome the disadvantage of putting all its eggs in one nest, the very nature of its solution to the problem introduces certain other difficulties. For one thing, if parasitic parenthood is to be successful, the parasite's reproductive cycle must be synchronized with the cycle of the host species. For another, it would be fatal if the host rejected either the parasite's egg or the hatchling that emerged from it. With respect to the egg a strategy of mimicry has evolved. For example, parasitic cuckoos lay eggs with markings that closely resemble those of the host's eggs. Widow-bird eggs are unmarked, as are their host's; they differ from the eggs of the host species only slightly in size and shape.

With respect to the acceptance of the hatchling the parasite strategies that have evolved are various. Several species of cowbirds punch tiny holes in the host's eggs before depositing their own, thus ensuring that no host egg ever hatches. An African honey-guide hatchling emerges from its egg armed with formidable hooks at the tip of its beak; its host's offspring are all soon fatally wounded, leaving the parasite to consume all the food the host provides. Some cuckoo nestlings simply nudge their foster siblings out of the nest during the first few days of life; others mature so much more rapidly than their nestmates that the foster siblings' development is inhibited and they die prematurely.

In the long run, of course, strategies that result in the death of the host's young are unproductive. The parasitic parent needs to use the nests of several host pairs each year and a dwindling host population means fewer nests. The widow birds have evolved a strategy that circumvents this problem; like the egg strategy, it is imitative. Each widowbird nestling is indistinguishable from its host's nestlings in size, in color and markings, in gesture and in call. Thus the intruders can grow up among their alien nestmates with no risk of being rejected by their foster parents. Such a strategy ensures that the numbers of the host population do not diminish.

It seems surprising that the widow birds should have evolved this particular strategy. Their chosen hosts—various species of finches, the entire family Estrildidae—are noted for having mouths that are colored and marked in a complex and conspicuous manner [see *illustration on page 138*]. No two of the 125 species of estrildid finches have identical mouth markings. The color of the palate may be whitish, red, yellow or bluish. The palate markings may form a three-spot pattern or a five-spot pattern or may be only a fine horseshoe-shaped line; the spots themselves may be black or violet. Moreover, the fledglings' gape papillae vary in shape and color. They may be ivory white, cornflower

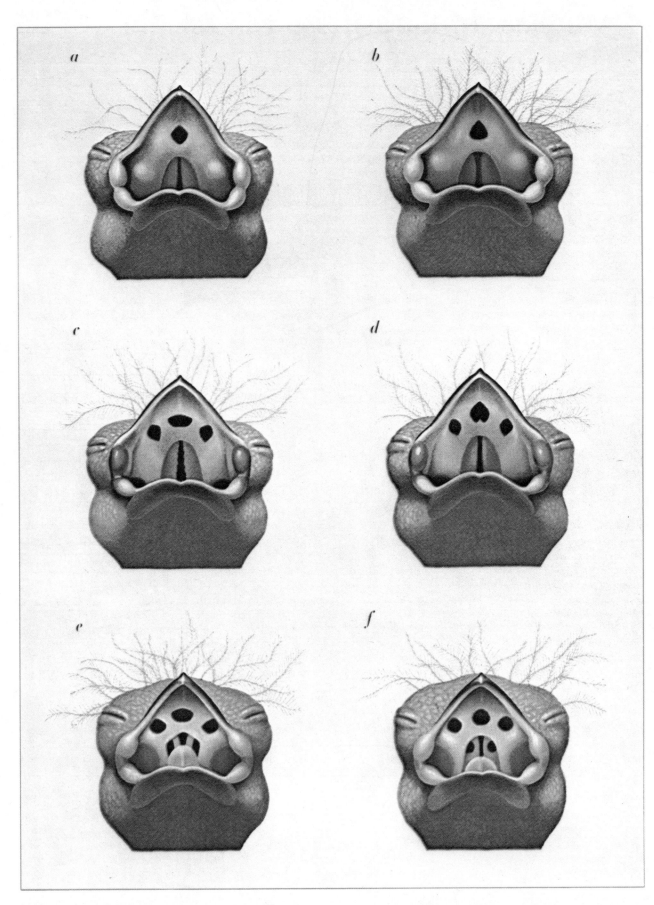

MOUTH MARKINGS of host nestlings, which stimulate the host parents' feeding response, are mimicked in detail by all the parasitic species of African widow birds. Shown here are the markings of a host-species nestling, the melba finch (*a*), and its parasitic sibling, the paradise widow bird (*b*), of a second host species, the purple grenadier (*c*), and its parasite, the straw-tailed widow bird (*d*), and of a third host-parasite pair, Jameson's fire finch (*e*) and the purple combassou (*f*). Many other mimicries of the type are known.

blue, yellow or faintly violet, and they may appear as simple thickenings or resemble small pearl-shaped warts.

In parallel with these elaborate mouth markings there has evolved among the estrildid finches a precise, genetically based "knowledge" of species-specific patterns and stimulus-receiving mechanisms that are correspondingly selective. By the simple experimental method of placing nestlings of one finch species in the nest of other species my colleagues and I have demonstrated that the parent birds will feed only the nestlings that display the appropriate species-specific markings. Even minor deviations in pattern are noted by the parent birds, and the deviants are ruthlessly weeded out by starvation.

A parasite nestling constantly interacts with its foster parents from the time of its emergence until it leaves the nest. The host birds are thus an essential part of the parasite's environment. The host sets a series of examples that instructs the parasite with respect to such matters as environmental standards and feeding habits. One instance of this process of parallel adaptation is the synchronization of the parasite's and the host's reproductive cycle.

Most species of estrildid finches breed during the rainy season. The finches' gonads begin to swell when, after months of drought, a series of showers brings the vegetation to renewed life. The timing is appropriate because the increase in the food supply that comes with the rainy season makes it easier to feed nestlings. Several finch species begin their breeding activities when the rains start; others do not begin until the middle of the rainy season or toward the end of it. A few species, such as the yellow-winged pytilia and the aurora finch of West Africa, delay their breeding until the dry season, a time when they find feeding conditions appropriate.

In each of these instances the parasitic widow birds synchronize their reproductive cycle with that of their host. One element in the cycle is that when courtship begins, the male widow bird displays a splendid and conspicuous plumage. The molt into breeding plumage occupies a period of four to six weeks. This means that the maturation of the widow bird's gonads, which initiates the molt, actually occurs sometime before the host's gonads mature.

To cite some examples, the paradise widow bird and the straw-tailed widow bird have hosts that breed during the rainy season. That requires the male

PARASITIC ADULTS, the female (a) and male (b) paradise widow bird, do not in any way mimic the appearance of the foster parents of their young, the female (c) and male (d) melba finch. The male paradise widow bird is seen in its bright breeding plumage.

widow birds to develop their breeding plumage before the end of the dry season. Similarly, two species of widow bird, the Togo paradise widow bird and the Kongo paradise widow bird, have hosts that breed during the dry season. The male parasites' molt must therefore take place toward the end of the rainy season. Under these circumstances it is clear that the onset of the parasites' reproductive activity cannot simply be triggered by the onset of the hosts'; it is evidently set in motion by some kind of seasonal rhythm. The specific rhythms remain unidentified but one is justified in assuming that, because the rainy and dry seasons follow each other with great regularity in most of Africa, the gonads of rainy-season breeders begin to be stimulated after the passage of a certain length of dry-season time, and vice versa.

Now, in any fine-tuned system of parasitic adaptation one of the greatest potentials for disruption is accidental hybridization between species of parasites that are closely related and coexist in the same area. As far as the widow birds are concerned, any hybrid offspring would display intermediate mouth markings rather than markings that match those of their foster siblings. That in turn means they would be refused food by the host species of both parents. This hazard, with its long-term potential for the extinction of all widow birds, has been avoided by the evolution of a kind of behavior that is unknown among other parasitic birds. In brief, when the male widow bird sings, it includes in its song certain unique finch-species phrases it has learned from its foster parents.

Widow birds are polygamous. At the start of the breeding season a male stakes out a large territory that it defends against potential rivals. Throughout the breeding season the male perches on certain selected trees and bushes in the territory and sings its unique melody. If one compares the songs of various widow-bird species, either by listening or by analyzing sound spectrograms, it soon becomes apparent that the songs of all species have a few phrases in common. For example, they all include harsh chattering sounds; the motif is evidently related to the vocalizations produced by the widow birds' closest relatives, the bishop birds of the subfamily Euplectinae. Since these chatterings appear to represent an ancient genetic heritage, we call them widow-bird phrases.

The major portion of any male widow bird's song consists of quite different vocalizations. Its motifs vary from one

YOUNG PARASITES, unlike adult parasites, closely resemble their host siblings in appearance. Profiles at left show a paradise widow-bird nestling (a) and a melba-finch nestling (b) 13 days after hatching; profiles at right show a straw-tailed widow-bird nestling (c) and a purple-grenadier nestling (d) 15 days old. The young are independent after five weeks.

species to another, and each species-specific repertory perfectly imitates the motifs characteristic of the widow bird's finch host. We therefore call such motifs host phrases. They include the finch's long-distance call, its contact call, its distress and anger calls, its greeting phrases and, of course, its routine song. The parasite's mimicry of this repertory is so exact that it sounds like the finch's to the human ear and looks like it in a spectrogram [see illustration on page 142]. Furthermore, the host finch itself cannot distinguish between the widow bird's imitation and the song of its own species.

Let us examine in detail the song of one widow-bird species. The straw-tailed widow bird of East Africa (Tetraenura fischeri) ranges over the scrub savanna of that region from southern Tanzania northward into Ethiopia and Somalia. The male's breeding plumage is black and yellow and its four conspicuously elongated central tail feathers look like yellowish blades of dry grass. During the breeding season the male perches high in small trees or thornbushes and twitters its song unceasingly from early morning until sunset. The first motif in its repertory is usually the contact call of its finch host, the purple grenadier,

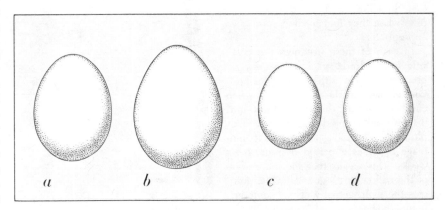

MIMICRY IN THE NEST extends to the size and shape of the egg laid by the parasite. At left is the egg of the melba finch (a) and of the paradise widow bird (b); the parasite's egg is only slightly larger and more rounded than the host's. At right is the egg of another host, the red-billed fire finch (c), and the egg of its parasite, the black-winged combassou (d). As centimeter grid shows, the host and parasite eggs are virtually identical in size and shape.

Uraeginthus ianthinogaster; this is a long trill. The mimicked phrase is repeated several times, increasing and decreasing in loudness. It is followed by other elements of the host's repertory. They include the brief "call" phrase used by the male to attract the female, the male's clacking song, a chase call, whispering nest calls and the shrill begging calls that fledglings use; in all, the mimic's repertory includes a total of eight finch motifs. Each mimicked vocalization is separated from the next by an interval, and each lasts the same length of time as the host's. The finch potpourri is interspersed with three different widow-bird phrases that, like the mimicked host melodies, occur in unpredictable sequences. The eight finch motifs give the male straw-tailed widow bird effective command of the entire vocabulary of the purple grenadier.

The same pattern of host-melody mimicry is apparent in almost all other species of widow birds. The shaft-tailed widow bird of South Africa (*Tetraenura regia*) mimics the repertory of its host, the violet-eared waxbill (*Uraeginthus granatinus*). Each of the five species of the paradise widow-bird genus (*Steganura*) copies the host melody of the particular species of the finch genus (*Pytilia*) it parasitizes. The same is true of the seven species in the genus of short-tailed black combassous (*Hypochera*) that mimic all seven species of the genus of fire finches (*Lagonosticta*). Indeed, in only two instances has it been impossible to demonstrate the presence of host phrases in a widow bird's song; these cases are the two species of the genus *Vidua,* the blue widow bird of East Africa and the pin-tailed widow bird, which is found in the sub-Saharan regions of the continent.

By determining which are the host phrases in a male widow bird's song, it is even possible to predict the identity of the host when it is not known. Such predictions have been made more than once in our laboratory after careful analysis of the songs of captive widow-bird males, and I was later able to prove their accuracy. In the field in East Africa we found the appropriate young parasites among the nestlings of exactly those species of estrildid finches that had been identified as probable hosts.

The use of host melodies in widow-bird singing serves a vital purpose: it is a barrier to hybridization. How this isolating mechanism is acquired is best shown by a review of the widow-bird life cycle. The young nestlings' first weeks of life, as they share the nest of

BREEDING PLUMAGE of parasitic widow birds develops during a four-to-six-week period. Seen here are a male straw-tailed widow bird (*a*) and its host, a purple grenadier (*b*), and below them a male purple combassou (*c*) and its host, a Jameson's fire finch (*d*). The parasite's breeding cycle must synchronize with its host's to ensure synchronous egg-laying; in order to grow breeding plumage, however, a parasite must enter its breeding cycle first.

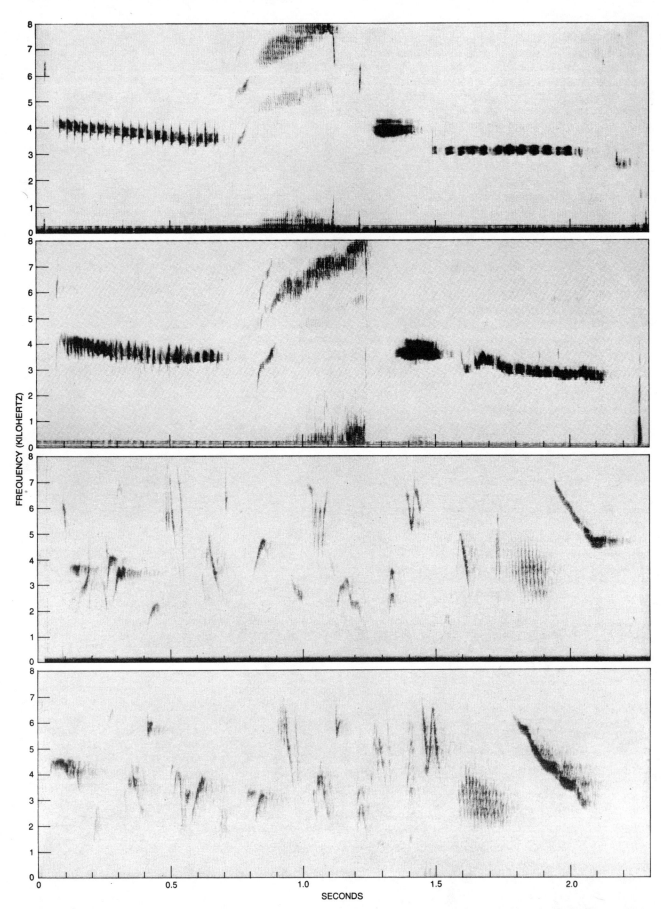

SONG MIMICRY, a meticulous imitation of the host male's call, learned by the parasite male while a juvenile, is demonstrated by these paired sonograms. The top sonogram shows the final seconds of a Damara melba finch's song; the sonogram directly below it is mimicry of this part of the finch's song by a male paradise widow bird. The third sonogram shows a two-second segment of the song of the violet-eared waxbill; the fourth is mimicry of this segment of the song by the waxbill's parasite, the shaft-tailed widow bird.

their host siblings, are ones of constant contact with their foster parents. The parasites become independent at the age of five weeks, but even after that they continue to live for some time among their host siblings. It is only later that they form a flock with other juveniles of their own species.

During these weeks of dependence and association the young widow birds are imprinted to the host species; this imprinting is a decisive factor in their later reproductive behavior. When, at the age of seven to eight weeks, the widow-bird juveniles first engage in premature sexual behavior as all songbirds do, there arises in the parasites a quite narrow and specific interest in the activities of the host species. Whereas other songbirds not only play at building nests during this phase but also feed and court one another, the attention of the juvenile widow bird is concentrated on two aspects of its hosts' adult behavior. The focus of interest depends on the sex of the widow bird.

As the female widow bird approaches sexual maturity her interest is concentrated on the hosts' reproductive cycle. The parasite checks frequently on the progress of the host pair's nest-building. The female also synchronizes her own ovulation with that of the host female; this ensures that her own egg is ready at about the time the host female produces her second or third egg. As a consequence the young parasite will hatch at about the same time that its host siblings do and will be their equal in the competition for food.

The male widow bird's concentration is differently oriented. During the most impressionable phase of its youth its interest is focused on the vocalizations of the male foster parent. The parasite evidently disregards the songs of any other species of bird; as a result, when it reaches maturity, it is able to reproduce its host's sound patterns exactly. Although the female parasite does not sing, she becomes similarly song-imprinted, and a memory of the host's repertory remains with her for life. The imprinting becomes evident when ovulation takes place and the female's sexual drive is aroused. Once the female widow bird is in search of a mate, she "knows" that the only appropriate male partner is one who can recite the same calls and notes she heard from her foster father while she was still a nestling. Only when the courting male corresponds to this prototype will the female widow bird allow mating.

Superficially it might seem that the male's signal is merely a kind of lure, for example "I grew up among the same host species that you did!" In my view the signal contains a message with a much deeper biological meaning; it might be paraphrased as "I have inherited the same adaptive characteristics that you have and so our offspring will have the same chances of survival."

It is clear that the adult widow birds' general adaptation to their hosts' song repertory, breeding rhythms and feeding habits, along with their nestlings' perfect mimicry of foster-sibling markings, movements and calls, can only have been the result of an extended evolutionary process. As we have come to know the identity of more and more host species of estrildid finches some of the mechanisms of the process responsible for these unusual adaptations have become apparent. It seems that in effect the potential for the evolution of a new species of widow bird has arisen in the past only when a new species of host finch has evolved.

Consider the relations between hosts and parasites from a taxonomic viewpoint. Two species of the widow-bird genus *Tetraenura* are known: the straw-tailed widow birds and the shaft-tailed. The estrildid finches parasitized by these two related widow-bird species are also two species of the same genus, *Uraeginthus*. In turn five species of paradise widow birds are the parasites of five species of another finch genus, and the seven species of combassous are linked to seven species of fire finch. Even though the specific hosts of the two widow birds of the genus *Vidua* remain unknown, we are certain that they are one or another of the species in the finch genus *Estrilda*.

Such species-to-species relations can be understood only if one assumes that a long time ago the widow birds began to arise from a single viduine prototype species that had started to parasitize a single species of estrildid finches. When this ancient estrildid species began to evolve into several species, the widow birds were forced to follow suit, because only by doing so could they evolve mouth markings sufficiently like those of the evolving new species of potential hosts. Of course, each widow-bird population that failed to achieve such an adaptive process was destined to become extinct, since its mismarked offspring would starve to death in the nests of hostile hosts. In this way the evolution of the adaptively successful widow-bird species both parallels and reflects the evolution of their chosen hosts among the numerous species of estrildid finches.

14

The Social System of Lions

by Brian C. R. Bertram
May 1975

*The organization of the social unit of lions, the pride,
and the behavior of the animals in it are a sensitive
reflection of the lion's adaptation to its environment*

One of the principal habitats of the lion (*Panthera leo*) is the Serengeti National Park in Tanzania, an area of 5,000 square miles where about 1,500 lions live with some two million other large mammals. Here other investigators and I, working at the Serengeti Research Institute, have been able to make detailed observations of the lion's habitat and way of life over a period of years. From these observations is beginning to emerge an overall picture of the integrated system consisting of the lion's environment, its prey, its hunting behavior, its social organization, its sexual behavior and its reproduction.

Lions are distinctive as hunters and as social animals. No other animal that hunts in groups on land approaches the lion in size. No other species of cat is social.

In the Serengeti National Park lions are the largest of the predators and, except for hyenas, the most numerous. Their prey includes a variety of large mammals; zebras and wildebeests are taken most often, but the lions also kill buffaloes, gazelles, hartebeests, impalas, warthogs and other animals. Lions are particularly adept at catching such prey in fairly open country, mainly because they tend to hunt cooperatively.

Even for lions prey animals are not easy to catch, since the prey animal can usually run faster than a lion. Therefore most hunting efforts by lions fail, even at night, when they do most of their hunting and when they most often succeed at it. When lions hunt in a group, they spread out and each one stalks toward the prey, which is therefore likely to be surrounded. A prey animal fleeing from a lion it has detected may run within range of one it has not seen and thus be caught. And when a prey animal is caught, all the lions in the hunting group feed on it.

Cooperative hunting tends to be more successful for the lions than solitary hunting, although solitary hunting is not unusual among them. The females do most of the hunting. One can speculate that they are more effective at it than the males because they do not have an unwieldy large mane and are lighter. (A female weighs about 120 kilograms, or 265 pounds, and a male about 180 kilograms, or 395 pounds.) In any case the males, being stronger than the females, can gain access to any kill made by the females.

The communal hunting and communal living of lions are clearly interrelated. The social unit of the lion—the pride—is a long-lasting entity. Its nucleus is a group of from three to 12 females of breeding age. They are accompanied by a smaller group of males of breeding age; most often a pride will have two such males, but the number can range from one to six. The pride also includes cubs of various ages.

The members of a pride are not likely to be together all the time or even most of the time. Individual lions sometimes go off on their own or in the company of one or two other lions. Some lions tend to be with certain individuals much more frequently than they are with others. All members of the pride encounter one another fairly often, however, and at such times they interact peaceably.

Each pride occupies a territory that is a few miles in diameter. Intruding lions are kept away, particularly by the males. The territories do not have sharply defined boundaries, so that between adjoining territories both overlapping and gaps are found. A territory may shift with the seasons, although if prey animals can be found in it throughout the year, such shifts are small. Most of the woodland area of the park can be visualized as being subdivided into a mosaic of separate, adjoining pride territories with little overlap.

In addition to the lions living in territorial prides one finds lions that are nomadic. They tend to follow the migratory herds of prey animals, and so they wander through the territory of prides and also through areas not permanently

PRIDE OF LIONS lives in the Serengeti National Park in Tanzania. This group, which is part of a pride, includes an adult

occupied by other lions. About 15 percent of the lion population live in this way. The nomads are not a different population but represent the surplus from the resident breeding population. Many of the nomads are males at a particular stage in their life cycle; some nomads are females that have been expelled from their pride. Since the nomads make up only a small part of the total population and are much less successful in reproducing, I shall not discuss them further and shall concentrate on the more typical resident lions, as exemplified by two adjacent Serengeti prides.

The data on the two prides, which occupy territories near Seronera in the middle of the park, were gathered over seven years of observation. George B. Schaller of the New York Zoological Society had started keeping records on the same prides in 1966. I continued the records from 1969 to 1973, and my successors at the Serengeti Research Institute are maintaining them now.

Since the prides are adjacent, the lions in the two different social groups are usually within four or five miles of each other. The climatic conditions in the areas are similar, and the densities of prey change seasonally in much the same way. Therefore one can reasonably assume that differences between the two prides are caused by factors operating within the prides rather than by gross seasonal factors common to both of them.

The prides are in an area frequented by tourists, and so the lions have become accustomed to vehicles. In an automobile one can get close to the animals without disturbing them. They are easy to observe at ranges of from 10 to 40 yards. Moreover, after watching a pride for a period of time one finds that the lions are recognizable individually by such markings as nicks in the ears (caused by squabbling at kills), scars, missing teeth and the arrangement of whisker spots. I made an identity card for each lion; it included photographs of the animal and notations on the animal's characteristic features.

Finding the lions was a bigger problem than recognizing them. In another area I had been locating the animals by tracking the radio signal emitted by a small transmitter in a collar fitted around each animal's neck. That, however, could not be done in the tourist area. Therefore I had to drive around in the territories of the prides in the hope of finding some of their members, and often I could find only a few. As a result my observations of individuals were necessarily intermittent and irregular. When I did find a group, I made a note of which individuals were there and also kept records of newborn cubs and lions that were pregnant, feeding, mating or sick.

Even though the observations were intermittent, the fact that they were conducted over a period of seven years made it possible to amass a great deal of information. (Observations must be made over a long period of time in any case because lions are long-lived and have a long generation time.) Where necessary I have drawn on data from other prides to supplement the records. Let us now examine what this information reveals

male eating a prey animal at right and an adult female standing at left. The other lions are cubs or subadults. A typical pride usually includes two or three adult males, from five to 10 adult females and a number of cubs. Lions are termed cubs up to the age of two years and subadults until they are four years old. In the background is a kopje, an outcropping of rock that is characteristic of the region.

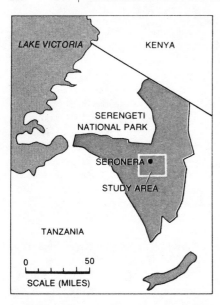

about the interaction of the social system of lions and their reproduction.

Considering the females first, it is significant that no strange female joined any of the prides observed. Therefore every female in a pride was born and reared in that pride. No prides were observed to die out, and no new pride was seen to form. (It is likely that both events occur but that they are uncommon.) If a pride lasts for some decades at least, it follows that all the females in it are likely to be related to one another. They are sisters, mothers, grandmothers, half-sisters, cousins and so on. The permanent nucleus of a pride is therefore a group of related females with a range of ages.

To be born in a pride, however, is not necessarily to remain in it. At the age of about three years a subadult female is either recruited into the pride or driven out of it. (Lions are termed cubs until they are two years old and subadults between the ages of two and four.) A lioness is more likely to be recruited if the pride has relatively few adult females than if it has a large number.

A lioness that is expelled becomes nomadic. She leaves the territory of the pride and roams over large distances in search of a living. Such females do not reproduce nearly as effectively as the resident females. Schaller has shown that they do not survive as long and that they produce smaller litters, whose members are less likely to survive.

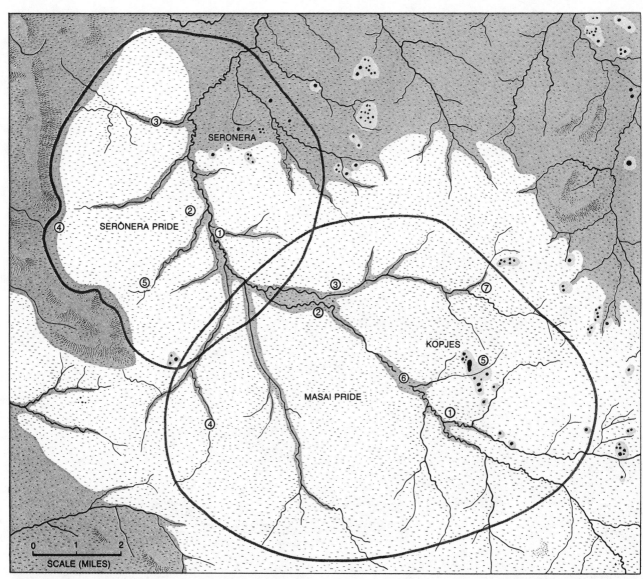

TERRITORIES OF TWO PRIDES in the Serengeti National Park show where the members might be on a typical day. In the Seronera pride's area are (1) a male, two females and five small cubs; (2) a male and female mating; (3) three females and seven large cubs; (4) a peripheral female, and (5) two subadult males, nomadic intruders. Masai pride has (1) a male alone; (2) two males with a kill; (3) four females and 11 cubs; (4) a female and four subadults; (5 and 6) females with new cubs, and (7) a female alone.

It would seem that if a group of nomadic lionesses were to come on a suitable area for a territory, they would settle there and establish a pride. It is likely, however, that any suitable area would already be occupied by a pride and that the members of it would not allow the newcomers to settle there or to join it.

A female that is recruited into a pride has an easier and more productive life. She will generally produce her first litter when she is about four years old and will continue bearing litters until within a couple of years before her death at the age of about 18. A recruited female therefore has a reproductive period of some 13 years.

For young males the ordinary course of events is quite different. At the age of about three they leave the pride in which they were born. They may leave voluntarily or they may be expelled. Males depart in small groups of up to half a dozen members each. Staying together as a group, they join the nomadic portion of the population.

At that age a young male has very little mane. He is sexually active, however, and will mate if he encounters an unattended estrous female. During the next two years or so the male gradually reaches adult size, and his mane develops from a scruffy fringe around the neck to a darkening cape from head to shoulders. The young males hunt for themselves and also scavenge what they can from the carcasses of animals that have died of natural causes or have been killed by other predators.

After about two years of this kind of life the group of males is likely to find a pride it can take over. It may be a pride whose males have left or died. If it is a pride with males still in residence, the new group may take over by driving them out.

The driving out of resident males is sometimes a slow process. The newcomers settle at the edge of the pride's territory and gradually expand their area of operation. In other cases the takeover is sudden and dramatic, being brought about by a fight that the resident males lose. In either case the new males come into possession of the pride and its territory.

The pride they have taken over is unlikely to be the pride they grew up in. It might be any of the prides in the park. As a result the males are unlikely to be genetically related to the females. On the other hand, the males are closely related to one another, because they are the collective offspring of a group of re-

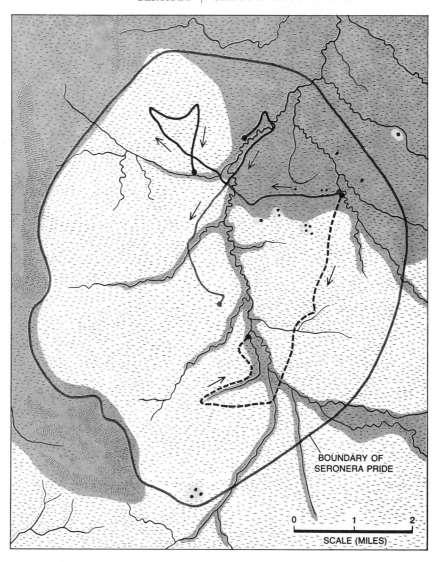

DAY'S MOVEMENTS by several members of the Seronera pride are plotted for a group of three adult females and seven cubs (black), an adult male (gray) and another adult male (broken line). Most activities, such as hunting, feeding, finding companions and patrolling the territory, take place at night. During most of the day the animals rest in the shade.

lated females and have stayed together since they left those females. This distinction accounts for the apparently unique social structure of lions: a permanent group of closely interrelated females and a smaller group of separately interrelated coequal males, which are with the females for a short period.

The length of time during which the males stay in possession of the pride varies. On the average it is two or three years before they in turn are expelled by a new group of younger, stronger or more numerous males. The lions that are expelled are unlikely to find another pride they can take over, because they are older now and also may have been injured. Moreover, they are probably less efficient at hunting than they were as young lions. In the pride they could depend on the females to do most of the hunting.

Meanwhile they have become bulkier and have acquired a mane. Therefore the conditions of life for males that have been expelled from a pride are harsh. For these reasons the effective reproductive period of the males is much shorter than that of the females, being only as long as the few years during which the males are in possession of a pride.

The brevity of the male reproductive period is a result of the strong competition between male groups. Males are likely to be expelled from a pride soon after they are past their prime or are reduced in number by injury, disease or death. A single male can almost never retain possession of a pride in the Serengeti because almost any pair of males, fighting in tandem as they do, can defeat him. Similarly, larger groups of males can defeat smaller groups, other things being equal, and so the larger groups have the longest tenure of prides. In addition large groups (of from three to six males) sometimes manage after a while to take over a second adjacent pride too and to keep possession of both prides concurrently. Thus it is to the selective advantage of a male to have male companions.

Lionesses do not appear to have a regular estrous cycle in the wild: they come into heat at variable intervals of between three weeks and a few months. A lioness usually remains in heat for a few days, and during that time she will mate on the average every 15 minutes. The first male to encounter a female coming into estrus temporarily gains dominance over other males. He mates with her and by his presence keeps other

males 20 yards or more away. Competition among males for an estrous female is rare. A female sometimes changes males while she is in heat, but she seldom does so more than once a day. Occasionally, if several females are in heat at the same time, a male may mate with more than one female.

The females of a pride are often in heat synchronously. The synchrony is unrelated in time to the synchrony in adjacent prides, so that it must be caused by some factor operating within the pride. Possibly the females are responding to signals from one another, such as pheromones in the urine, or to signals from the males. Another possibility is that they are responding to particular characteristics of their food intake, such as a series of kills of large prey animals. The data are scanty, and it is difficult

with lions in the wild to obtain information on the mechanisms involved.

Notwithstanding the energetic mating performance of lions, most mating periods do not result in the birth of cubs. It is difficult to tell why, again for want of adequate information. The male almost always ejaculates and probably produces viable sperm. Whether the low level of births has to do with ovulation, fertilization or abortion in the female is not known.

When a female does conceive, the gestation period is some 14 or 15 weeks, which is a remarkably short time for such a large animal. As a result the cubs are tiny at birth, weighing less than 1 percent of the adult weight. A litter usually consists of two or three cubs; the range is from one through five. I could not find

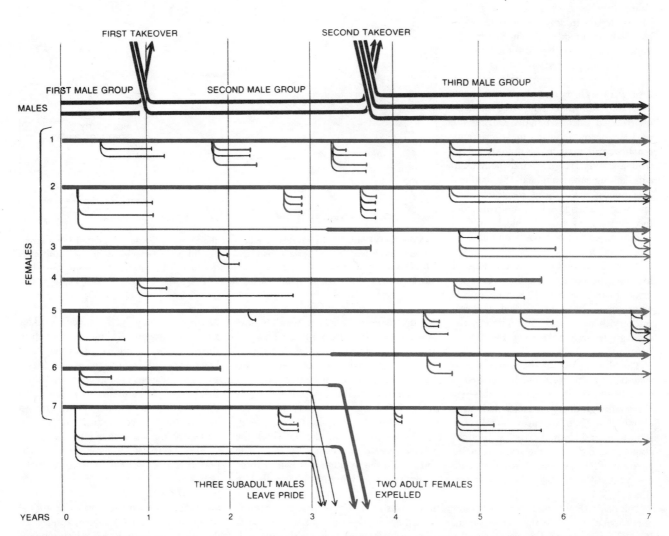

REPRODUCTIVE EVENTS in a pride of lions are portrayed for males (*black*), females (*color*) and cubs (*thin lines*). At the end of the first year the original group of males is driven out by a new group, which in turn is driven out some three years later. Below the events involving adult males one sees what happened to the pride's permanent group of interrelated females and their cubs. A horizontal line that stops means that the individual died. Mortality among cubs is high, often because males on taking over a pride will kill the cubs already there. One also sees, notably after the second takeover in this case, a tendency toward synchrony of births and enhanced survival of cubs at times when the pride has few or no older cubs. Cub survival also affects timing of a female's successive litters.

any factors that could be shown to influence the size of the litter.

Cubs can be born in any month of the year. The timing of births appears to be influenced by social factors rather than by the climate or the food supply. A lioness generally produces a litter when the cubs of her preceding litter are from 20 to 30 months old, provided that they have survived. If they have died, her next litter will be born some six to 12 months after the death of the last cub.

The females of a pride tend to produce litters at about the same time. The synchrony of estrus contributes to the synchrony of births, of course, but it cannot be the only factor because the synchrony of estrus is much more precise and because most periods of estrus do not result in cubs. The physiological causes, possibly pheromonal ones, for the synchrony of births remain to be determined, and it will not be easy to determine them with lions in the wild.

Evidence can be found for the influence of further social factors on the timing of births. For example, if one examines the timing of births in relation to the arrival of new adult males, one finds that for about six months after the new males take over few litters are born. Then the production of litters rises sharply.

Why should it be that males do not succeed in siring offspring until they have been in the pride for about three months? Possibly stress associated with their presence makes the females less likely to conceive. Their presence may also cause abortion among females that were pregnant when the new males arrived.

Lion cubs suckle as long as they can, which is from six to eight months, when the mother stops lactating. Long before that, at the age of two or three months, the cubs have started to eat meat, so that they undergo a period of gradual weaning. The cubs are totally dependent on the adults to get food for them until they are at least two years old.

Only a small proportion of cubs attain that age; the mortality among cubs is about 80 percent. Schaller has concluded that about a fourth of the deaths are violent, being brought about mainly by other lions but also by hyenas, buffaloes and accidents. Another fourth of the deaths are attributable to starvation, since cubs are the least able to get and keep even small pieces of meat at times when food is in short supply. Half of the deaths are of unknown origin, because

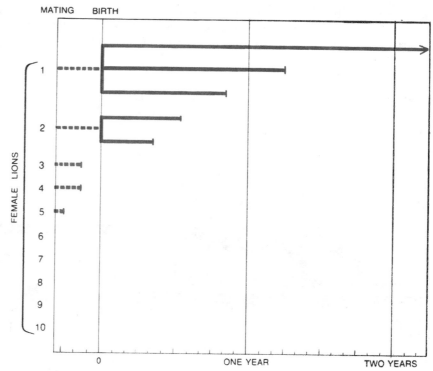

RESULTS OF MATING are indicated for 10 female lions. Usually a large number of matings is needed for a female to produce cubs. Because of this and the mortality rate of about 80 percent among cubs, number of matings per offspring reared to adulthood is about 3,000.

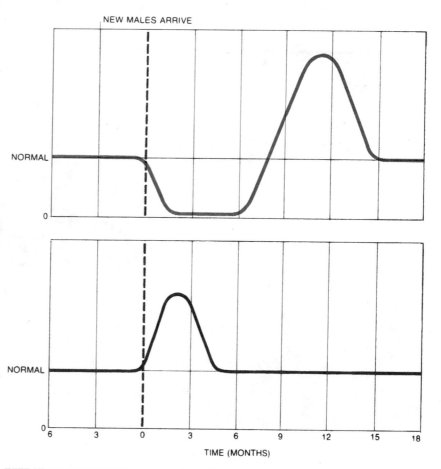

EFFECT OF TAKEOVER on the birth rate (color) and the death rate (black) of cubs is charted. At the point indicated by the broken line a pride is taken over by a new group of males. Mortality of cubs rises immediately, and few litters are born for several months.

the carcass is not found and the animal was not previously seen to be ailing.

I have looked for factors that might influence the survival of cubs. The mortality was higher at times when few prey animals were available. In the two prides near Seronera the level of mortality tended to be higher from October to March, but the times of food shortage differ greatly from year to year and from pride to pride, depending on the movements of the migratory prey species.

The time of year when cubs were born did not influence their chance of survival, but the time of birth in relation to the birth of other cubs in the pride was important. Litters tended to show poorer survival if appreciably older cubs were present in the pride. The reason is likely to be that at kills the stronger animals can plunder the share of the weaker ones.

Litters that were born synchronously survived better than litters that were not. Synchrony of births makes communal suckling and rearing possible, so that cubs have a more regular supply of milk

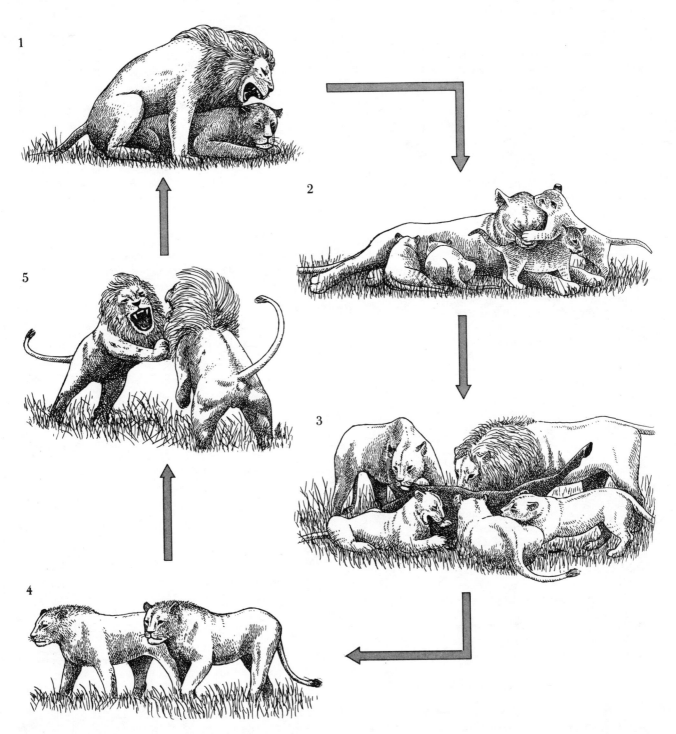

EVENTS IN LIFE CYCLE of a pride are portrayed, beginning with mating (*1*). Cubs suckle for several months (*2*) and until the age of two are dependent on adults to provide food by catching prey animals (*3*). Subadult males leave at about age three and become nomadic (*4*). Females usually remain with the pride but may be expelled at about age three if the pride has an excess of females. Expelled females become nomads. Young males from a pride tend to stay together. At about age five two or more or them will take over a pride by driving the resident males out (*5*). The pride is unlikely to be the pride in which the young males were reared.

and are less often left untended. Perhaps also more communal hunting by their mothers makes for more successful hunting and hence a larger food supply.

With survival as with births the take-over of a pride by new males appears to have an effect. On the basis of somewhat scanty data it seems that the mortality rate among cubs goes up for about three months after the new males arrive. An indirect cause might be stress effects on the females, making them produce less milk or hunt less efficiently. A direct cause is the killing of cubs by the new adult males. It is difficult to determine how prevalent this behavior is, but the available evidence suggests that males on taking over a pride are quite likely to kill cubs already there. The apparent avoidance of new males by females with cubs reinforces that evidence.

One is led to ask what evolutionary reason might explain the killing of cubs by new adult males. A plausible answer can be found in the short reproductive period of the males. It produces a strong selective pressure for the evolution of behavior that increases the reproductive output of the males during that period. A female whose cubs have died comes into estrus sooner and gives birth sooner to cubs fathered by a new male. Moreover, without the presence of older cubs to compete with new cubs for food the offspring of the new males will have a better chance of survival.

It is also interesting from an evolutionary point of view to ask why the matings of lions are so inefficient, that is, why the number of births is so low in relation to the number of matings. One reason appears to be that lions are under much less selective pressure than other animals to breed efficiently. They do not

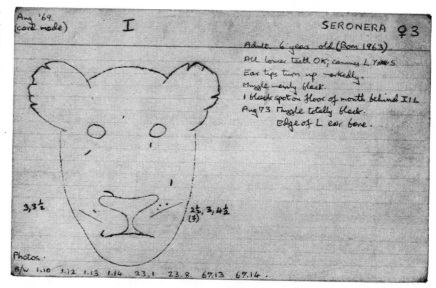

IDENTITY CARD was prepared by the author for each lion in the prides he observed. It included photographs of the animal and a drawing that showed the animal's distinctive physical features, such as scars, nicks in the ears and the positioning of its whisker spots.

have a breeding season that must not be missed, and they run no risk of predation that would increase the importance of mating quickly and efficiently.

The question might be modified to ask why the females have evolved in such a way that they mate at times when it is highly likely that they will not produce cubs as a result. An explanation could be that an effect of the system is to reduce competition among the males by reducing the genetic value to them of each copulation. Assuming that lions mate every 15 minutes for three days, that only one in five three-day mating periods results in cubs, that the mean size of litters is 2.5 cubs and that the mortality among cubs is 80 percent, then a male must mate on the average some 3,000 times

for each of his offspring reared to the next generation. When a single copulation is of such small potential consequence, the pressure on a male to fight another male for an opportunity to copulate is small. It is not to the evolutionary advantage of the females to have the males of a pride competing, because injury to the males causes them to be replaced more quickly by other males, which in turn results in increased mortality among the cubs of the females.

The same selective pressure may have a bearing on the synchrony of estrus among the females in a pride. It too tends to reduce competition among the males by ensuring that each male more often has an estrous female to be occupied with.

V

ORIENTATION
AND NAVIGATION

... It is a vast distance, if one calculates it, over which (the cranes) come from the eastern sea. They agree together when to start, and they fly high so as to see their route in front of them; they choose a leader to follow, and have some of their number stationed in turns at the end of the line to shout orders and keep the flock together with their cries.... Geese and swans also migrate on a similar principle.... They travel in a pointed formation like fast galleys, so cleaving the air more easily than if they drove at it with a straight front; while in the rear the flight stretches out in a gradually widening wedge, and presents a broad surface to the drive of a following breeze.

Pliny
NATURAL HISTORY, X, XXX, XXXII.

V ORIENTATION AND NAVIGATION

INTRODUCTION

In this Section we shall study some of the ways in which vertebrates orient themselves in their environments. Two highly specialized phenomena—electric location by fishes and infrared sensing by snakes—will be examined, as well as navigation, a complicated process dependent upon vision and perhaps detection of the earth's magnetic field. In fact, various vertebrates probably use combinations of senses, such as olfaction, hearing, and vision, as they carry out different behavioral patterns.

Olfaction is used in a wide variety of ways by fishes. For instance, a species of parasitic lamprey (*Petromyzon marinus*) detects prospective host fishes by smell. Even young lampreys fresh from metamorphosis respond to the odor of a potential host by oriented swimming behavior. Since these lampreys are not exposed to the odor of any other fish during larval life, this is a good example of a behavioral pattern that is built into the nervous system during development and maturation. It is innate behavior, which does not depend upon learning or conditioning. The substances that cause the reaction have been extracted from trout: they are amines. The same lamprey species that displays this orienting behavior uses a pulsating electric field to locate prey within 15 to 20 centimeters; the chemical sensing is used for long-distance orientation.

Other examples of odor-elicited behavior in fishes are the fright reaction and homing in order to reproduce. Particularly in a number of small schooling fishes, substances that cause an immediate fear reaction within the school are present in the skin cells. When the skin of one fish is injured, the chemical is released into the water, where the rest of the clustered fishes detect it and begin a rapid escape-swimming and dispersion. This substance is a warning device that helps others of the same species avoid imminent danger of predation.

Homing of salmon and eels is one of the most fascinating uses of olfaction by fish. Although it has not been proved, it seems reasonable to suppose that a salmon far out in the Pacific finds its way back to the general area of its river system by use of celestial navigation. This process, which involves orientation in relation to the sun or stars, is described in greater detail later in this Introduction. A number of experiments have established that fishes, like birds, show oriented swimming under clear skies but not under cloudy ones. Once the fishes are near the home river system, the senses of smell (and perhaps taste) can be assumed to lead the fish back to the stream where it grew up.

Consistent with this hypothesis is the finding that "home" water perfused into the nasal epithelium of a salmon elicits a burst of nerve impulse firing into the olfactory bulb. Water from nearby regions draining similar geological areas fails to alter the basic electroencephalogram firing pattern. Interestingly, the rate of the electrical activity in the olfactory bulb is high in young fish—at a time when they may be in the process of learning to recognize home water—but it is lower during adult life when the eyes are said to be the dominant sense. Finally, electrical activity increases again at the time of homing when smell apparently leads the fish to its spawning site.

How is detection of an odor translated into directional swimming? Some experiments suggest that, if the head is swung back and forth in the water (a normal occurrence during swimming of fish), it may be possible for first one nasal sac, and then the other, to sample for odorous molecules. Comparison of intensity in the central nervous system could then keep the fish "homed in" toward the spawning-site source of the odors. Moreover, branches of the trigeminal nerve have endings in the olfactory epithelium and connect centrally in such a way that the separate processing of information may be possible.

Young salmon may be trained in the laboratory with a given type of water, and they will return to it from the sea years later. Apparently they are imprinted with certain odorous substances of the home stream, so that these substances alone lead to a successful homing reaction. It is not known yet whether a single organic molecular type functions to mark the home territory, or whether groups of molecules combine to give a characteristic smell to each portion of each breeding stream.

Although fishes can employ smell for long-distance orientation, they can use electric locationing only for short distances. As pointed out by H. W. Lissmann in "Electric Location by Fishes," the sense is used under conditions of low visibility due either to darkness or to abundant suspended material in the water. The two families of fresh-water fishes in which the ability is found are quite distinct morphologically and only distantly related. Nevertheless, they have arrived at the same solution to problems of orientation in a difficult environment. In fact, we can assume that the only way such fishes can survive today is to be active at night, when nonelectric bony fishes are at rest in the water. This is because the mode of swimming of electric fishes—with a stiff body axis and propulsion by the fins—does not allow the quick movement needed for evasion of predators that is possible in typical sine-wave swimming.

Organs used to detect electric fields are probably modifications of the ordinary lateral line organs found on all vertebrate fishes. Each lateral line organ includes a structure that can be displaced by water currents; as a result of this displacement, sensory hairs are moved, and altered patterns of nerve impulses ensue. Thus, disturbances in the water pattern caused by objects near the fishes are detectable. In acting this way, the lateral line serves as an extension of the sense of touch.

Variations of the lateral line organ are found in sharks and their relatives and in electric fishes. In the electric fishes, the sensory epithelium is beneath the surface of the skin. In some electric receptors the canal leading from the sensory area to the surface is filled with jelly; this jelly has remarkable electrical conductivity (85 to 90 percent of that of sea water, compared to 45 percent in body fluids). In another type of electroreceptor, thin covering cells lie on top of the sensory cells. Neither these nor the cells under the jelly show any hint of ciliary derivation, a common feature of the lateral line neuromasts, as described above. The electric sensing cells are always found in close proximity to nerve endings (presumed to be sensory in nature). Even chemical coupling, like that in the main vertebrate nervous system, seems to exist between some of these sensory cells and nerve endings; for example, vesicles thought to contain a transmitter substance are found in the sensory cell cytoplasm, in the intercellular space, and in the nerve ending cytoplasm.

Electric organs can detect weak potential changes in the water—such as those caused by another fish or even by a bare metal wire—or they may be used to measure alterations in the electric field generated by the same organism. The remarkable feature of the electric organs is their great sensitivity; responses to currents as small as 2×10^{-5} microamperes per square centimeter have been observed. It seems clear that, in the course of the evolution of these various fishes, the basic mechanoreceptor properties of the lateral line have been modified to permit detection of electrical signals in the environment. The absence of any hint of cilia on the electric sensing cells emphasizes the change in sensory modality that has occurred in this evolutionary transition.

Certain snakes are able to detect still another portion of the electromagnetic spectrum in order to hunt prey or sense warm environments. As outlined by R. Igor Gamow and John F. Harris in "The Infrared Receptors of Snakes," these animals have evolved an energy-transducing system for heat which is more analogous to a green plant absorbing light rays in chlorophyll than it is to a vertebrate eye where light is used for "information" purposes rather than as an energy source. Unlike the plant, however, in which light energy is employed in synthetic processes, snakes use the energy-dependent nerve firing as a source of information about the environment.

Another mechanism is used by some marine vertebrates for orientation in the water as well as by terrestrial vertebrates for orientation in the air: it is echo-location. This capacity to use sound pulses and their echoes has evolved several times in the vertebrates. The oil birds found in caves in Venezuela, the dolphins and their relatives, the bats, and some insectivores all employ the mechanism. In each group the process is employed under conditions of low visibility, and, in each, rapid movement is made so that appropriate analysis of the environment by means of smell or taste is impossible. Cetaceans (dolphins, whales, and their relatives) and bats are reasonably close relatives of the insectivores, the ancestral group of modern mammals. Bats, of course, have retained the nocturnal habits of insectivores and many of them have adopted the echolocation method of orientation.

The last type of vertebrate orientation that we shall discuss—celestial navigation—is one of the most fascinating aspects of vertebrate biology. Birds, of course, provide the most widely quoted cases of long-range navigation, with flights of thousands of miles often occurring twice each year for many species. As pointed out by the late German behaviorist Gustav Kramer, true navigation requires that the organism have at least three pieces of information: (1) knowledge of its current position on the earth; (2) knowledge of the position (latitude and longitude) of the target of navigation; (3) knowledge of the direction from site one to site two. The basic question is whether birds or other animals are truly capable of such navigation, or whether compass direction alone is the parameter employed during oriented movement. Because it does seem likely that many animals use land marks extensively in initiating and terminating migration, compass direction might be the crucial information for them if navigation normally starts and ends only in known regions.

Sun navigation of birds has been investigated by many research workers since Kramer demonstrated "fluttering" orientation of starlings. Owing to altering day lengths, hormonal changes occur in a bird that bring it into a migratory condition. When caged, such birds become hyperactive and hop and flutter in the direction they would fly if free to migrate. This oriented behavior occurs *only when the sun is visible,* and the direction of fluttering shifts a predictable number of degrees when the image of the sun is displaced a known amount by mirrors (see Figure 1). This finding demonstrates that direction is determined in response to the sun; it does not prove true navigational ability, which demands, in addition, at least assessment of latitude. One of the best cases of measurement of the height of the sun (so that latitude can be determined) comes from fishes that give clear evidence of compensation in orientation when the sun appears to have an altered altitude in the sky through use of mirrors or by other means.

If the sun or the stars are to be used by an organism as guides in navigation, the organism must possess an internal biological "clock." As the sun moves across the sky, the bird is able to maintain a straight migratory course over the

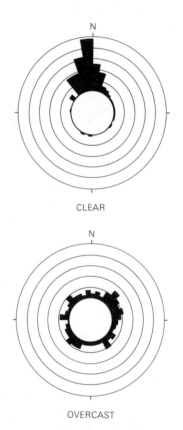

CLEAR

OVERCAST

Figure 1. **Flight directions taken by mallard ducks that had been displaced prior to release. Under clear skies, most flew north, but if the skies were cloudy, flight orientation was incorrect. Equivalent data were obtained when mallards were released on clear or overcast nights.**
[After F. C. Bellrose; in R. M. Storm, ed., *Animal Orientation and Navigation: Oregon State University Biology Colloquium, 1966,* Oregon State University Press, 1967.]

surface of the earth only if it can compensate for movement of the sun; thus, with the aid of its clock, the bird continually alters the angle between the sun and the axis of migration. The presence of the clock may be demonstrated by two types of experiments: In the first, a bird is displaced a long distance east or west. Despite the displacement in longitude the bird's clock still runs on the original "local" time, and, as shown in Figure 2, the bird orients as it normally would at home. A bird makes the same error in the second experiment, in which its clock has been "reset" by a gradual altering of the day-night cycle with artificial lighting. Afterward, even when it is exposed to sunlight in its home environment, it acts according to the new time and orients in the wrong direction.

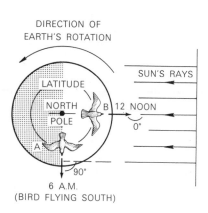

a. BEHAVIOR OF BIRD IN NATIVE LONGITUDE

b. BEHAVIOR OF BIRD WHEN DISPLACED 6 HOURS LONGITUDINALLY TO WEST OF NATIVE LONGITUDE, SAME LATITUDE

Figure 2. **Results of longitude displacement on bird orientation. Imagine that you are in space directly above the North Pole and the earth is rotating counterclockwise beneath you. In Figure *a* the bird flies directly away from the North Pole toward the south. If it begins its journey at 6:00 A.M., it displaces its angle of flight 90° with respect to the rays of the sun (position A). Six hours later, the earth has rotated on its axis, and now in order to fly "south," the bird has a 0° angle between flight path and sun's rays. In fact, all day long it compensates for the continually shifting sun** and so maintains a constant course on the earth's surface.

In *b* the bird is displaced to the west. As shown in the illustration, when its clock reads 6:00 A.M. the bird is perched on a branch in darkness. During the ensuing 6 hours, the earth rotates and the bird's clock advances. Then the sun appears. It is 6:00 A.M. local time, but the bird's clock reads 12 noon. It orients just as it did at position B of Figure *a* at noon; that is, it flies directly toward the sun. Thus, instead of flying south on the earth's surface, it flies eastward.

The pineal gland has been implicated in the cyclical, clocklike activity of birds (see "The Pineal Gland" by Richard J. Wurtman and Julius Axelrod, *Scientific American* Offprint 1015). In superb experiments conducted by M. Menaker and his colleagues, it was shown that the regular cyclical activity displayed by sparrows kept in constant darkness was lost when the pineal was removed. The sparrows regained this activity if the pineal was returned to some natural culture chamber, in this case, the anterior chamber of the eye. There the pineal liberated chemicals on a cyclical basis into the bird's blood vascular system and thereby reactivated the cyclical behavior pattern. Since no nerves reached such implanted pineals, this experiment emphasized the hormonal nature of the pineal's activity.

Next, Menaker took a pineal from a bird running on a biological "clock" set at one time and implanted the pineal into an arhythmic bird from which the pineal had been removed. The host immediately adopted the cyclical timing of the implant! This occurred even when pineals were taken from birds operating on widely different cycles, thus establishing that this remarkable gland is apparently the immediate source of clock-based behavior in sparrows.

The means by which birds use the stars or the earth's magnetic lines of force for navigation or homing are described in the articles by Stephen T. Emlen ("The Stellar Orientation System of a Migratory Bird") and by William T. Keeton ("The Mystery of Pigeon Homing"). One particularly important observation made recently by Walcott and his colleagues concerns the means by which birds detect the lines of force of the Earth's magnetic field. A small (1–2 millimeter) piece of tissue has been discovered between the pigeon's brain and skull. It contains tiny crystals of a naturally magnetic substance, almost certainly magnetite. Since the tissue has a rich supply of nerves, it is conceivable that the Earth's magnetic field could apply a torque to the crystals, and the resultant slight twisting could be detected by the nervous system. If so, there truly would be a sense organ for magnetism!

Experiments conducted in Germany by E. Gwinner provide an intriguing view of how the sun, stars, and magnetism may be used in real-life migration covering thousands of miles. Gwinner studied willow warblers and chiffchafs, birds that fly from Central Europe to North Africa or sub-Saharan Africa each winter. First, populations were studied that initially fly southwest over Spain and then turn southeast into Africa. If such a bird was allowed to show its endogenous migratory behavior while in a cage, it was found that during the early time of migration (September and October) it oriented southwest; then, during what would ordinarily be the later times of migration (October, November, and December) the bird turned in the cage and oriented southeast. In other words, direction and time are coupled. The direction the bird chooses to fly in response to orienting factors (such as stars, sun, magnetic field, and so on) is a variable under internal control.

Next, Gwinner observed a relationship between time and distance. Birds that, if free, would normally fly vastly different distances were all kept on a constant day-night cycle. Birds that would normally fly the longest distances consistently displayed longer migratory-activity behavior in their cages than did birds that would normally fly shorter distances. Since length of activity (flight) translates into distance over the earth, these results complement those for direction and imply that both distance and direction flown by these birds are in some sense innate properties of the organism. Distance, direction, ability to orient relative to the polar stars or earth, and drive to carry out migratory flight *per se* all seem to be part of an internal repertoire evoked by the bird's endocrine system.

Nevertheless there may be an important component of learning in bird migration, as experiments by A. C. Perdeck seem to show. Perdeck displaced a large number of marked adult and young migrating starlings from Holland several hundred kilometers southeast to Switzerland. The adults released after such displacement compensated, and they were recovered in their normal wintering area on the northwestern coast of France. They must have "navigated" to get there since they had to go in an abnormal *direction* to complete the trip. The young birds did not go northwest to the French coast but instead flew to southwestern France and even to Spain and Portugal independently of the old birds but on a route parallel to the normal migratory route from Holland. Furthermore, some of the young birds kept flying farther than they would have on the normal route; they did not stop at a specific longitude. The next spring, both young and old birds returned to the original nesting grounds in Holland. Then, the following autumn, the younger birds again went their separate way back to the first year's wintering ground in the southwest of France and in Spain.

Related observations indicated that "early" migrants, captured and displaced to a region in Spain suitable for wintering, continued to migrate beyond that place. But "late" migrants displaced to the same region soon searched for nesting sites there. These results paralleled Gwinner's results with willow warblers in indicating that time and distance are coupled and are, in a sense, innate. Perdeck's experiments demonstrate that untrained young birds have an innate migratory direction that they follow, and that they fly an approximate distance (or time). Obviously learning, too, occurs since the site of hatching and upbringing continued to be their home breeding ground, and the first-year wintering ground was equally permanent; and to go from one to the other and back again required this young population to fly in novel directions not taken by the parental birds. This appears to be true navigation.

15

Electric Location by Fishes

by H. W. Lissmann
March 1963

It is well known that some fishes generate strong electric fields to stun their prey or discourage predators. Gymnarchus niloticus produces a weak field for the purpose of sensing its environment

Study of the ingenious adaptations displayed in the anatomy, physiology and behavior of animals leads to the familiar conclusion that each has evolved to suit life in its particular corner of the world. It is well to bear in mind, however, that each animal also inhabits a private subjective world that is not accessible to direct observation. This world is made up of information communicated to the creature from the outside in the form of messages picked up by its sense organs. No adaptation is more crucial to survival; the environment changes from place to place and from moment to moment, and the animal must respond appropriately in every place and at every moment. The sense organs transform energy of various kinds—heat and light, mechanical energy and chemical energy—into nerve impulses. Because the human organism is sensitive to the same kinds of energy, man can to some extent visualize the world as it appears to other living things. It helps in considering the behavior of a dog, for example, to realize that it can see less well than a man but can hear and smell better. There are limits to this procedure; ultimately the dog's sensory messages are projected onto its brain and are there evaluated differently.

Some animals present more serious obstacles to understanding. As I sit writing at my desk I face a large aquarium that contains an elegant fish about 20 inches long. It has no popular name but is known to science as *Gymnarchus niloticus*. This same fish has been facing me for the past 12 years, ever since I brought it from Africa. By observation and experiment I have tried to understand its behavior in response to stimuli from its environment. I am now convinced that *Gymnarchus* lives in a world totally alien to man: its most important

sense is an electric one, different from any we possess.

From time to time over the past century investigators have examined and dissected this curious animal. The literature describes its locomotive apparatus, central nervous system, skin and electric organs, its habitat and its family relation to the "elephant-trunk fishes," or mormyrids, of Africa. But the parts have not been fitted together into a functional pattern, comprehending the design of the animal as a whole and the history of its development. In this line of biological research one must resist the temptation to be deflected by details, to follow the fashion of putting the pieces too early under the electron microscope. The magnitude of a scientific revelation is not always paralleled by the degree of magnification employed. It is easier to select the points on which attention should be concentrated once the plan is understood. In the case of *Gymnarchus*, I think, this can now be attempted.

A casual observer is at once impressed by the grace with which *Gymnarchus* swims. It does not lash its tail from side to side, as most other fishes do, but keeps its spine straight. A beautiful undulating fin along its back propels its body through the water—forward or backward with equal ease. *Gymnarchus* can maintain its rigid posture even when turning, with complex wave forms running hither and thither over different regions of the dorsal fin at one and the same time.

Closer observation leaves no doubt that the movements are executed with great precision. When *Gymnarchus* darts after the small fish on which it feeds, it never bumps into the walls of its tank, and it clearly takes evasive action at some distance from obstacles placed in

its aquarium. Such maneuvers are not surprising in a fish swimming forward, but *Gymnarchus* performs them equally well swimming backward. As a matter of fact it should be handicapped even when it is moving forward: its rather degenerate eyes seem to react only to excessively bright light.

Still another unusual aspect of this fish and, it turns out, the key to all the puzzles it poses, is its tail, a slender, pointed process bare of any fin ("gymnarchus" means "naked tail"). The tail was first dissected by Michael Pius Erdl of the University of Munich in 1847. He found tissue resembling a small electric organ, consisting of four thin spindles running up each side to somewhere beyond the middle of the body. Electric organs constructed rather differently, once thought to be "pseudoelectric," are also found at the hind end of the related mormyrids.

Such small electric organs have been an enigma for a long time. Like the powerful electric organs of electric eels and some other fishes, they are derived from muscle tissue. Apparently in the course of evolution the tissue lost its power to contract and became specialized in various ways to produce electric discharges [see "Electric Fishes," by Harry Grundfest; SCIENTIFIC AMERICAN, October, 1960]. In the strongly electric fishes this adaptation serves to deter predators and to paralyze prey. But the powerful electric organs must have evolved from weak ones. The original swimming muscles would therefore seem to have possessed or have acquired at some stage a subsidiary electric function that had survival value. Until recently no one had found a function for weak electric organs. This was one of the questions on my mind when I began to study *Gymnarchus*.

I noticed quite early, when I placed a

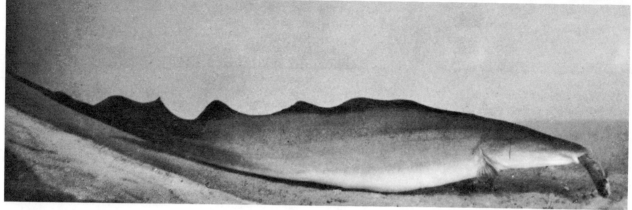

ELECTRIC FISH *Gymnarchus niloticus*, from Africa, generates weak discharges that enable it to detect objects. In this sequence the fish catches a smaller fish. *Gymnarchus* takes its name, which means "naked tail," from the fact that its pointed tail has no fin.

new object in the aquarium of a well-established *Gymnarchus*, that the fish would approach it with some caution, making what appeared to be exploratory movements with the tip of its tail. It occurred to me that the supposed electric organ in the tail might be a detecting mechanism. Accordingly I put into the water a pair of electrodes, connected to an amplifier and an oscilloscope. The result was a surprise. I had expected to find sporadic discharges co-ordinated with the swimming or exploratory motions of the animal. Instead the apparatus recorded a continuous stream of electric discharges at a constant frequency of about 300 per second, waxing and waning in amplitude as the fish changed position in relation to the stationary electrodes. Even when the fish was completely motionless, the electric activity remained unchanged.

This was the first electric fish found to behave in such a manner. After a brief search I discovered two other kinds that emit an uninterrupted stream of weak discharges. One is a mormyrid relative of *Gymnarchus;* the other is a gymnotid, a small, fresh-water South American relative of the electric eel, belonging to a group of fish rather far removed from *Gymnarchus* and the mormyrids.

It had been known for some time that the electric eel generates not only strong discharges but also irregular series of weaker discharges. Various functions had been ascribed to these weak dis-charges of the eel. Christopher W. Coates, director of the New York Aquarium, had suggested that they might serve in navigation, postulating that the eel somehow measured the time delay between the output of a pulse and its reflection from an object. This idea was untenable on physical as well as physiological grounds. The eel does not, in the first place, produce electromagnetic waves; if it did, they would travel too fast to be timed at the close range at which such a mechanism might be useful, and in any case they would hardly penetrate water. Electric current, which the eel does produce, is not reflected from objects in the surrounding environment.

Observation of *Gymnarchus* suggested another mechanism. During each discharge the tip of its tail becomes momentarily negative with respect to the head. The electric current may thus be pictured as spreading out into the surrounding water in the pattern of lines that describes a dipole field [*see illustration on the next page*]. The exact configuration of this electric field depends on the conductivity of the water and on the distortions introduced in the field by objects with electrical conductivity different from that of the water. In a large volume of water containing no objects the field is symmetrical. When objects are present, the lines of current will converge on those that have better conductivity and diverge from the poor conductors [*see top illustration on page 164*]. Such objects alter the distribution of electric potential over the surface of the fish. If the fish could register these changes, it would have a means of detecting the objects.

Calculations showed that *Gymnarchus* would have to be much more sensitive electrically than any fish was known to be if this mechanism were to work. I had observed, however, that *Gymnarchus* was sensitive to extremely small external electrical disturbances. It responded violently when a small magnet or an electrified insulator (such as a comb that had just been drawn through a person's hair) was moved near the aquarium. The electric fields produced in the water by such objects must be very small indeed, in the range of fractions of a millionth of one volt per centimeter. This crude observation was enough to justify a series of experiments under more stringent conditions.

In the most significant of these experiments Kenneth E. Machin and I trained the fish to distinguish between objects that could be recognized only by an electric sense. These were enclosed in porous ceramic pots or tubes with thick walls. When they were soaked in water, the ceramic material alone had little effect on the shape of the electric field. The pots excluded the possibility of discrimination by vision or, because each test lasted only a short time, by a chemical sense such as taste or smell.

The fish quickly learned to choose between two pots when one contained aquarium water or tap water and the other paraffin wax (a nonconductor). After training, the fish came regularly to pick a piece of food from a thread suspended behind a pot filled with aquarium or tap water and ignored the pot filled with wax [*see bottom illustration on page 164*]. Without further conditioning it also avoided pots filled with air, with distilled water, with a close-fitting glass tube or with another nonconductor. On the other hand, when the electrical conductivity of the distilled water was matched to that of tap or aquarium water by the addition of salts or acids, the fish would go to the pot for food.

A more prolonged series of trials showed that *Gymnarchus* could distinguish mixtures in different proportions of tap water and distilled water and perform other remarkable feats of discrimination. The limits of this performance can best be illustrated by the fact that the fish could detect the presence of a glass rod two millimeters in diameter and would fail to respond to a glass rod .8 millimeter in diameter, each hidden in a

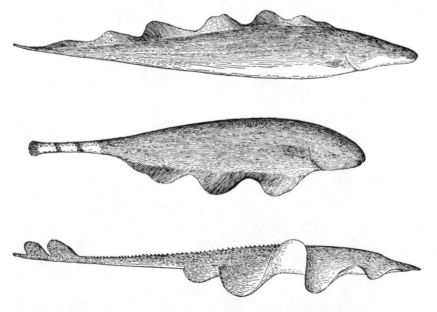

UNUSUAL FINS characterize *Gymnarchus* (*top*), a gymnotid from South America (*middle*) and sea-dwelling skate (*bottom*). All swim with spine rigid, probably in order to keep electric generating and detecting organs aligned. *Gymnarchus* is propelled by undulating dorsal fin, gymnotid by similar fin underneath and skate by lateral fins resembling wings.

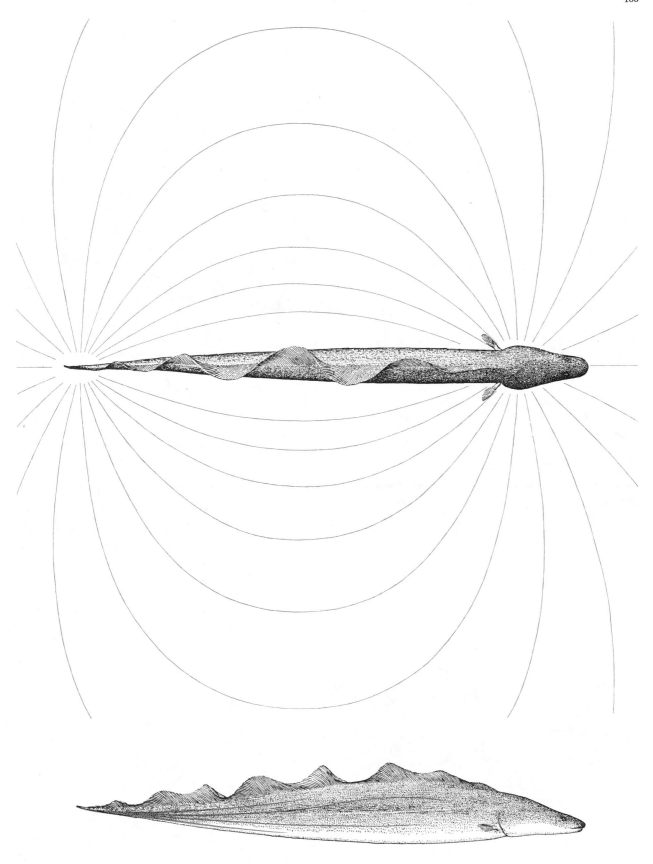

ELECTRIC FIELD of *Gymnarchus* and location of electric generating organs are diagramed. Each electric discharge from organs in rear portion of body (*color in side view*) makes tail negative with respect to head. Most of the electric sensory pores or organs are in head region. Undisturbed electric field resembles a dipole field, as shown, but is more complex. The fish responds to changes in the distribution of electric potential over the surface of its body. The conductivity of objects affects distribution of potential.

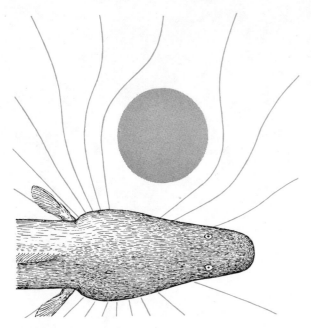

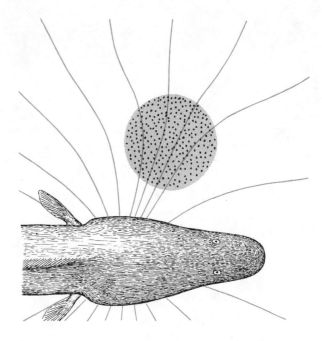

OBJECTS IN ELECTRIC FIELD of *Gymnarchus* distort the lines of current flow. The lines diverge from a poor conductor (*left*) and converge toward a good conductor (*right*). Sensory pores in the head region detect the effect and inform the fish about the object.

pot of the same dimensions. The threshold of its electric sense must lie somewhere between these two values.

These experiments seemed to establish beyond reasonable doubt that *Gymnarchus* detects objects by an electrical mechanism. The next step was to seek the possible channels through which the electrical information may reach the brain. It is generally accepted that the tissues and fluids of a fresh-water fish are relatively good electrical conductors enclosed in a skin that conducts poorly. The skin of *Gymnarchus* and of many mormyrids is exceptionally thick, with layers of platelike cells sometimes arrayed in a remarkable hexagonal pattern [*see top illustration on page 167*]. It can therefore be assumed that natural selection has provided these fishes with better-than-average exterior insulation.

In some places, particularly on and around the head, the skin is closely perforated. The pores lead into tubes often filled with a jelly-like substance or a loose aggregation of cells. If this jelly is a good electrical conductor, the arrangement would suggest that the lines of electric current from the water into the body of the fish are made to converge at these pores, as if focused by a lens. Each jelly-filled tube widens at the base into

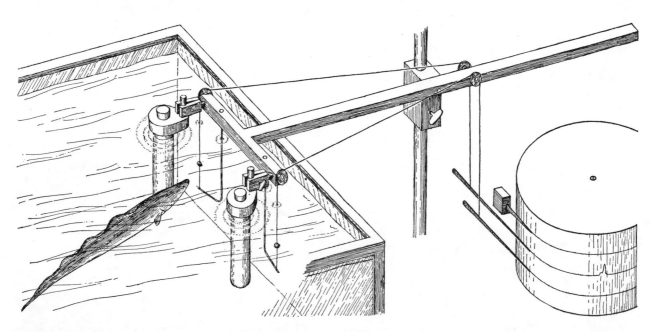

EXPERIMENTAL ARRANGEMENT for conditioned-reflex training of *Gymnarchus* includes two porous pots or tubes and recording mechanism. The fish learns to discriminate between objects of different electrical conductivity placed in the pots and to seek bait tied to string behind the pot holding the object that conducts best. *Gymnarchus* displays a remarkable ability to discriminate.

a small round capsule that contains a group of cells long known to histologists by such names as "multicellular glands," "mormyromasts" and "snout organs." These, I believe, are the electric sense organs.

The supporting evidence appears fairly strong: The structures in the capsule at the base of a tube receive sensory nerve fibers that unite to form the stoutest of all the nerves leading into the brain. Electrical recording of the impulse traffic in such nerves has shown that they lead away from organs highly sensitive to electric stimuli. The brain centers into which these nerves run are remarkably large and complex in *Gymnarchus,* and in some mormyrids they completely cover the remaining portions of the brain [*see illustration on next page*].

If this evidence for the plan as well as the existence of an electric sense does not seem sufficiently persuasive, corroboration is supplied by other weakly electric fishes. Except for the electric eel, all species of gymnotids investigated so far emit continuous electric pulses. They are also highly sensitive to electric fields. Dissection of these fishes reveals the expected histological counterparts of the structures found in the mormyrids: similar sense organs embedded in a similar skin, and the corresponding regions of the brain much enlarged.

Skates also have a weak electric organ in the tail. They are cartilaginous fishes, not bony fishes, or teleosts, as are the mormyrids and gymnotids. This means that they are far removed on the family line. Moreover, they live in the sea, which conducts electricity much better than fresh water does. It is almost too much to expect structural resemblances to the fresh-water bony fishes, or an electrical mechanism operating along similar lines. Yet skates possess sense organs, known as the ampullae of Lorenzini, that consist of long jelly-filled tubes opening to the water at one end and terminating in a sensory vesicle at the other. Recently Richard W. Murray of the University of Birmingham has found that these organs respond to very delicate electrical stimulation. Unfortunately, either skates are rather uncooperative animals or we have not mastered the trick of training them; we have been unable to repeat with them the experiments in discrimination in which *Gymnarchus* performs so well.

Gymnarchus, the gymnotids and skates all share one obvious feature: they swim in an unusual way. *Gymnarchus* swims with the aid of a fin on its back; the gymnotids have a similar fin on their

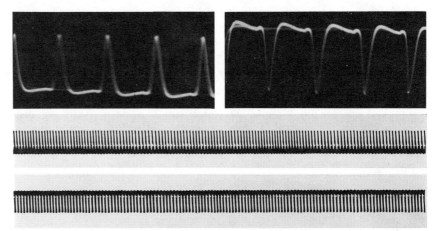

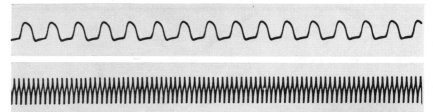

ELECTRIC DISCHARGES of *Gymnarchus* show reversal of polarity when detecting electrodes are rotated 180 degrees (*enlarged records at top*). The discharges, at rate of 300 per second, are remarkably regular even when fish is resting, as seen in lower records.

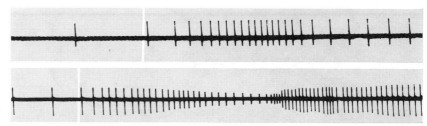

DISCHARGE RATES DIFFER in different species of gymnotids. *Sternopygus macrurus* (*upper record*) has rate of 55 per second; *Eigenmannia virescens* (*lower*), 300 per second.

VARIABLE DISCHARGE RATE is seen in some species. Tap on tank (*white line in upper record*) caused mormyrid to increase rate. Tap on fish (*lower record*) had greater effect.

underside; skates swim with pectoral fins stuck out sideways like wings [*see illustration on page 162*]. They all keep the spine rigid as they move. It would be rash to suggest that such deviations from the basic fish plan could be attributed to an accident of nature. In biology it always seems safer to assume that any redesign has arisen for some reason, even if the reason obstinately eludes the investigator. Since few fishes swim in this way or have electric organs, and since the fishes that combine these features are not related, a mere coincidence would appear most unlikely.

A good reason for the rigid swimming posture emerged when we built a model to simulate the discharge mecha-

nism and the sensory-perception system. We placed a pair of electrodes in a large tank of water; to represent the electric organ they were made to emit repetitive electric pulses. A second pair of electrodes, representing the electric sense organ, was placed some distance away to pick up the pulses. We rotated the second pair of electrodes until they were on a line of equipotential, where they ceased to record signals from the sending electrodes. With all the electrodes clamped in this position, we showed that the introduction of either a conductor or a nonconductor into the electric field could cause sufficient distortion of the field for the signals to reappear in the detectors.

In a prolonged series of readings the

slightest displacement of either pair of electrodes would produce great variations in the received signal. These could be smoothed to some extent by recording not the change of potential but the change in the potential gradient over the "surface" of our model fish. It is probable that the real fish uses this principle, but to make it work the electrode system must be kept more or less constantly aligned. Even though a few cubic centimeters of fish brain may in some respects put many electronic computers in the shade, the fish brain might be unable to obtain any sensible information if the fish's electrodes were to be misaligned by the tail-thrashing that propels an ordinary fish. A mode of swimming that keeps the electric field symmetrical with respect to the body most of the time would therefore offer obvious advantages. It seems logical to assume that *Gymnarchus*, or its ancestors, acquired the rigid mode of swimming along with the electric sensory apparatus and subsequently lost the broad, oarlike tail fin.

Our experiments with models also showed that objects could be detected only at a relatively short distance, in spite of high amplification in the receiving system. As an object was moved farther and farther away, a point was soon reached where the signals arriving at the oscilloscope became submerged in the general "noise" inherent in every detector system. Now, it is known that minute amounts of energy can stimulate a sense organ: one quantum of light registers on a visual sense cell; vibrations of subatomic dimensions excite the ear; a single molecule in a chemical sense organ can produce a sensation, and so on. Just how such small external signals can be picked out from the general noise in and around a metabolizing cell represents one of the central questions of sensory physiology. Considered in connection with the electric sense of fishes, this question is complicated further by the high frequency of the discharges from the electric organ that excite the sensory apparatus.

In general, a stimulus from the environment acting on a sense organ produces a sequence of repetitive impulses in the sensory nerve. A decrease in the strength of the stimulus causes a lower frequency of impulses in the nerve. Conversely, as the stimulus grows stronger, the frequency of impulses rises, up to a certain limit. This limit may vary from one sense organ to another, but 500 impulses per second is a common upper limit, although 1,000 per second have been recorded over brief intervals.

In the case of the electric sense organ of a fish the stimulus energy is provided by the discharges of the animal's electric organ. *Gymnarchus* discharges at the rate of 300 pulses per second. A change in the amplitude—not the rate—of these pulses, caused by the presence of an object in the field, constitutes the effective stimulus at the sense organ. Assuming that the reception of a single discharge of small amplitude excites one impulse in a sensory nerve, a discharge of larger amplitude that excited two impulses would probably reach and exceed the upper limit at which the nerve can generate impulses, since the nerve would now be firing 600 times a second (twice the rate of discharge of the electric organ). This would leave no room

BRAIN AND NERVE ADAPTATIONS of electric fish are readily apparent. Brain of typical nonelectric fish (*top*) has prominent cerebellum (*gray*). Regions associated with electric sense (*color*) are quite large in *Gymnarchus* (*middle*) and even larger in the mormyrid (*bottom*). Lateral-line nerves of electric fishes are larger, nerves of nose and eyes smaller.

to convey information about gradual changes in the amplitude of incoming stimuli. Moreover, the electric organs of some gymnotids discharge at a much higher rate; 1,600 impulses per second have been recorded. It therefore appears unlikely that each individual discharge is communicated to the sense organs as a discrete stimulus.

We also hit on the alternative idea that the frequency of impulses from the sensory nerve might be determined by the mean value of electric current transmitted to the sense organ over a unit of time; in other words, that the significant messages from the environment are averaged out and so discriminated from the background of noise. We tested this idea on *Gymnarchus* by applying trains of rectangular electric pulses of varying voltage, duration and frequency across the aquarium. Again using the conditioned-reflex technique, we determined the threshold of perception for the different pulse trains. We found that the fish is in fact as sensitive to high-frequency pulses of short duration as it is to low-frequency pulses of identical voltage but correspondingly longer duration. For any given pulse train, reduction in voltage could be compensated either by an increase in frequency of stimulus or an increase in the duration of the pulse. Conversely, reduction in the frequency required an increase in the voltage or in the duration of the pulse to reach the threshold. The threshold would therefore appear to be determined by the product of voltage times duration times frequency.

Since the frequency and the duration of discharges are fixed by the output of the electric organ, the critical variable at the sensory organ is voltage. Threshold determinations of the fish's response to single pulses, compared with quantitative data on its response to trains of pulses, made it possible to calculate the time over which the fish averages out the necessarily blurred information carried within a single discharge of its own. This time proved to be 25 milliseconds, sufficient for the electric organ to emit seven or eight discharges.

The averaging out of information in this manner is a familiar technique for improving the signal-to-noise ratio; it has been found useful in various branches of technology for dealing with barely perceptible signals. In view of the very low signal energy that *Gymnarchus* can detect, such refinements in information processing, including the ability to average out information picked up by a large number of separate sense organs,

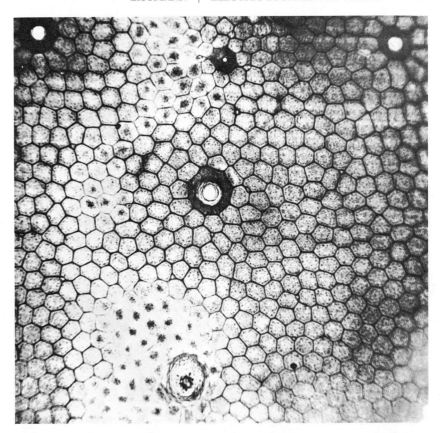

SKIN OF MORMYRID is made up of many layers of platelike cells having remarkable hexagonal structure. The pores contain tubes leading to electric sense organs. This photomicrograph by the author shows a horizontal section through the skin, enlarged 100 diameters.

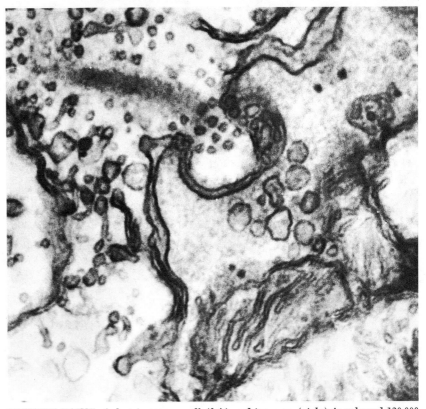

MEETING POINT of electric sensory cell (*left*) and its nerve (*right*) is enlarged 120,000 diameters in this electron micrograph by the author and Ann M. Mullinger. Bulge of sensory cell into nerve ending displays the characteristic dense streak surrounded by vesicles.

appear to be essential. We have found that *Gymnarchus* can respond to a continuous direct-current electric stimulus of about .15 microvolt per centimeter, a value that agrees reasonably well with the calculated sensitivity required to recognize a glass rod two millimeters in diameter. This means that an individual sense organ should be able to convey information about a current change as

small as .003 micromicroampere. Extended over the integration time of 25 milliseconds, this tiny current corresponds to a movement of some 1,000 univalent, or singly charged, ions.

The intimate mechanism of the single sensory cell of these organs is still a complete mystery. In structure the sense organs differ somewhat from species to species and different types are also found

in an individual fish. The fine structure of the sensory cells, their nerves and associated elements, which Ann M. Mullinger and I have studied with both the light microscope and the electron microscope, shows many interesting details. Along specialized areas of the boundary between the sensory cell and the nerve fiber there are sites of intimate contact where the sensory cell bulges into the fiber. A dense streak extends from the cell into this bulge, and the vesicles alongside it seem to penetrate the intercellular space. The integrating system of the sensory cell may be here.

These findings, however, apply only to *Gymnarchus* and to about half of the species of gymnotids investigated to date. The electric organs of these fishes emit pulses of constant frequency. In the other gymnotids and all the mormyrids the discharge frequency changes with the state of excitation of the fish. There is therefore no constant mean value of current transmitted in a unit of time; the integration of information in these species may perhaps be carried out in the brain. Nevertheless, it is interesting that both types of sensory system should have evolved independently in the two different families, one in Africa and one in South America.

The experiments with *Gymnarchus*, which indicate that no information is carried by the pulse nature of the discharges, leave us with a still unsolved problem. If the pulses are "smoothed out," it is difficult to see how any one fish can receive information in its own frequency range without interference from its neighbors. In this connection Akira Watanabe and Kimihisa Takeda at the University of Tokyo have made the potentially significant finding that the gymnotids respond to electric oscillations close in frequency to their own by shifting their frequency away from the applied frequency. Two fish might thus react to each other's presence.

For reasons that are perhaps associated with the evolutionary origin of their electric sense, the electric fishes are elusive subjects for study in the field. I have visited Africa and South America in order to observe them in their natural habitat. Although some respectable specimens were caught, it was only on rare occasions that I actually saw a *Gymnarchus*, a mormyrid or a gymnotid in the turbid waters in which they live. While such waters must have favored the evolution of an electric sense, it could not have been the only factor. The same waters contain a large number of

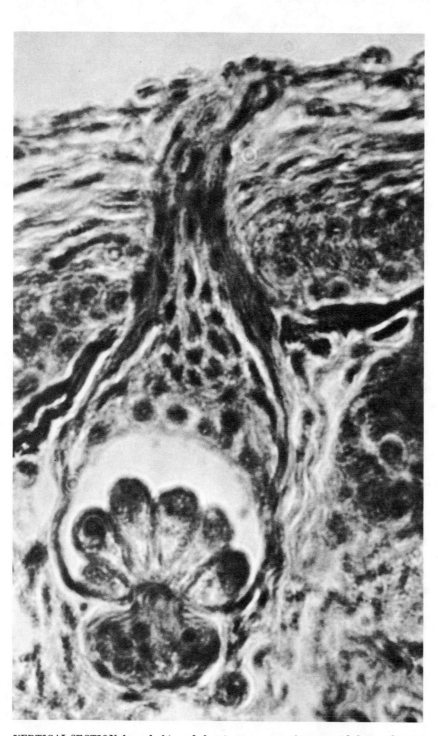

VERTICAL SECTION through skin and electric sense organ of a gymnotid shows tube containing jelly-like substance widening at base into a capsule, known as multicellular gland, that holds a group of special cells. Enlargement of this photomicrograph is 1,000 diameters.

other fishes that apparently have no electric organs.

Although electric fishes cannot be seen in their natural habitat, it is still possible to detect and follow them by picking up their discharges from the water. In South America I have found that the gymnotids are all active during the night. Darkness and the turbidity of the water offer good protection to these fishes, which rely on their eyes only for the knowledge that it is day or night. At night most of the predatory fishes, which have well-developed eyes, sleep on the bottom of rivers, ponds and lakes. Early in the morning, before the predators wake up, the gymnotids return from their nightly excursions and occupy inaccessible hiding places, where they often collect in vast numbers. In the rocks and vegetation along the shore the ticking, rattling, humming and whistling can be heard in bewildering profusion when the electrodes are connected to a loudspeaker. With a little practice one can begin to distinguish the various species by these sounds.

When one observes life in this highly competitive environment, it becomes clear what advantages the electric sense confers on these fishes and why they have evolved their curiously specialized sense organs, skin, brain, electric organs and peculiar mode of swimming. Such well-established specialists must have originated, however, from ordinary fishes in which the characteristics of the specialists are found in their primitive state: the electric organs as locomotive muscles and the sense organs as mechanoreceptors along the lateral line of the body that signal displacement of water. Somewhere there must be intermediate forms in which the contraction of a muscle, with its accompanying change in electric potential, interacts with these sense organs. For survival it may be important to be able to distinguish water movements caused by animate or inanimate objects. This may have started the evolutionary trend toward an electric sense.

Already we know some supposedly nonelectric fishes from which, nevertheless, we can pick up signals having many characteristics of the discharges of electric fishes. We know of sense organs that appear to be structurally intermediate between ordinary lateral-line receptors and electroreceptors. Furthermore, fishes that have both of these characteristics are also electrically very sensitive. We may hope one day to piece the whole evolutionary line together and express, at least in physical terms, what it is like to live in an electric world.

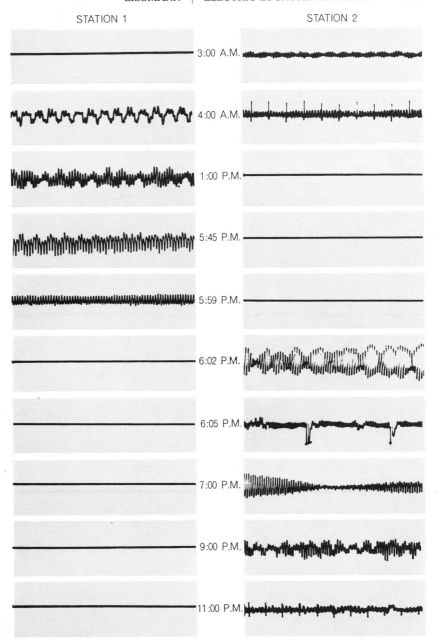

STATION 1   STATION 2

3:00 A.M.
4:00 A.M.
1:00 P.M.
5:45 P.M.
5:59 P.M.
6:02 P.M.
6:05 P.M.
7:00 P.M.
9:00 P.M.
11:00 P.M.

TRACKING ELECTRIC FISH in nature involves placing electrodes in water they inhabit. Records at left were made in South American stream near daytime hiding place of gymnotids, those at right out in main channel of stream, where they seek food at night.

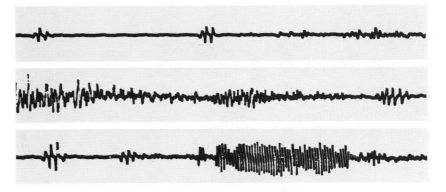

AFRICAN CATFISH, supposedly nonelectric, produced the discharges shown here. Normal action potentials of muscles are seen, along with odd regular blips and still other oscillations of higher frequency. Such fish may be evolving an electric sense or may already have one.

The Infrared Receptors of Snakes

by R. Igor Gamow and John F. Harris
May 1973

The snakes of two large families have sensitive organs that can detect the heat radiation emitted by their prey. The performance of these detectors is investigated with the aid of an infrared laser

A boa constrictor will respond in 35 milliseconds to diffuse infrared radiation from a carbon dioxide laser. A sensitive man-made instrument requires nearly a minute to make what is essentially the same measurement. Reflecting on this comparison, one wonders what feat of bioengineering nature has performed to make the snake's sensor so efficient. One also wonders if a better understanding of the animal's heat-sensing apparatus would provide a basis for improving the man-made ones. It was the pursuit of these questions that gave rise to the somewhat unusual situation in which a group of workers in an aerospace engineering laboratory (our laboratory at the University of Colorado) was investigating snakes.

Snakes belong to one of the four large orders that comprise the living members of the class Reptilia. The order Testudinata contains such members as the turtles, the tortoises and the terrapins. The order Crocodilia contains the crocodiles and the alligators. The third order, the Rhynchocephalia, has only one member: the tuatara of New Zealand. In the fourth order, the Squamata, are the lizards and the snakes.

On the basis of outward appearance one might suppose that the lizards are more closely related to the crocodiles and the alligators than to the snake. Evolutionary evidence, however, clearly indicates that the snakes arose from the lizard line. Although the lizard is therefore the snake's closest relative, the two animals have developed pronounced differences during the course of evolution. Most lizards have limbs and no snakes have limbs, although vestigial ones are found in certain snakes. Most lizards have two functional lungs, whereas most snakes have only one. Again a few snakes have a small second lung, which

is another indication of the direction of evolution from the lizard to the snake.

Today most herpetologists would agree that the first step in the evolution of the snake occurred when the animal's ancestral form became a blind subterranean burrower. In evolving from their lizard-like form the ancestral snakes lost their limbs, their eyesight and their hearing as well as their ability to change coloration. Later, when the animals reappeared on the surface, they reevolved an entire new visual system but never regained their limbs or their sense of hearing.

Today the snakes constitute one of the most successful of living groups, being found in almost every conceivable habitat except polar regions and certain islands. They live in deep forests and in watery swamps. Some are nocturnal, others diurnal. Some occupy freshwater habitats, others marine habitats. Certain snakes are arboreal and survive by snatching bats from the air, others live in the inhospitable environment of the desert. Their success is indicated by the fact that their species, distributed among 14 families, number more than 2,700.

Two of the 14 families are distinguished by the fact that all their members have heat sensors that respond to minute changes of temperature in the snake's environment. The snake employs these sensors mainly to seek out and capture warm-bodied prey in the dark. It seems probable that the snake also uses the sensors to find places where it can maintain itself comfortably. Although snakes, like all reptiles, are cold-blooded, they are adept at regulating their body temperature by moving from place to place. Indeed, a snake functions well only within a rather narrow range of temperatures and must actively seek environments of the proper temperature.

A case in point is the common sidewinder, which maintains its body temperature in the range between 31 and 32 degrees Celsius (87.8 and 89.6 degrees Fahrenheit). One advantage of a heat sensor is that it enables the snake to scan the temperature of the terrain around it to find the proper environment.

One of the families with heat receptors is the Crotalidae: the pit vipers, including such well-known snakes as the rattlesnake, the water moccasin and the copperhead. The other family is the Boidae, which includes such snakes as the boa constrictor, the python and the anaconda. Although all members of both families have these heat receptors, the anatomy of the receptors differs so much between the families as to make it seem likely that the two types evolved independently.

In the pit vipers the sensor is housed in the pit organ, for which these snakes are named. There are two pits; they are located between the eye and the nostril and are always facing forward. In a grown snake the pit is about five millimeters deep and several millimeters in diameter. The inner cavity of the pit is larger than the external opening.

The inner cavity itself is divided into an inner chamber and an outer one, separated by a thin membrane. A duct between the inner chamber and the skin of the snake may prevent differential changes in pressure from arising between the two chambers. Within the membrane separating the chambers two large branches of the trigeminal nerve (one of the cranial nerves) terminate. In both snake families this nerve is primarily responsible for the input from the heat sensor to the brain. Near the terminus the nerve fibers lose their sheath of myelin and fan out into a broad, flat,

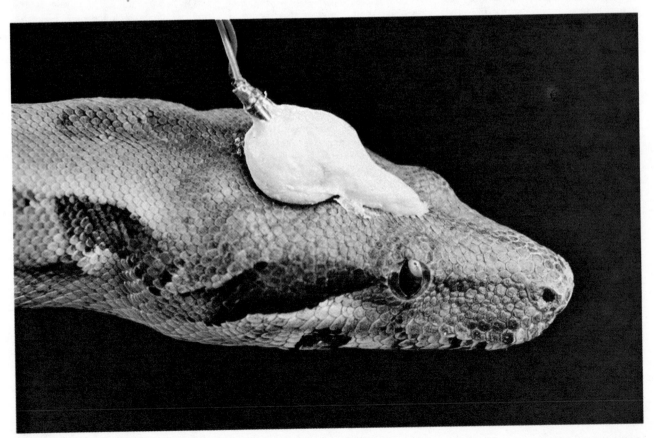

GREEN PYTHON of New Guinea (*Chondropython viridis*) is a member of the family Boidae that has visible pits housing its infrared detectors. The pits extend along the jaws. Photograph was made by Richard G. Zweifel of the American Museum of Natural History.

BOA CONSTRICTOR is a boid snake with infrared detectors that are not visible externally, although they are in the same location as the green python's. This boa wears an apparatus with which the authors recorded responses of the brain to infrared stimuli.

palmate structure. In this structure the nerve endings are packed full of the small intracellular bodies known as mitochondria. Evidence obtained recently by Richard M. Meszler of the University of Maryland with the electron microscope strongly suggests that the mitochondria change morphologically just after receiving a heat stimulus. This finding has led to the suggestion that the mitochondria themselves may constitute the primary heat receptor.

In the family Boidae there are no pit organs of this type, although somewhat different pits are often found along the snakes' upper and lower lips. Indeed, it was once thought that only the boid snakes with labial pits had heat sensors. An extensive study by Theodore H. Bullock and Robert Barrett at the University of California at San Diego has shown, however, that boid snakes without labial pits nonetheless have sensitive heat receptors. One such snake is the boa constrictor.

For experimental purposes the boid snakes are preferable to the pit vipers because the viper is certain to bite sooner or later, and the bite can be deadly. The boids, in contrast, can be described as friendly, and they get along well in a laboratory. When our laboratory became interested several years ago in the possibility of using an infrared laser as a tool to help unravel the secrets of the mode of operation of the snake's heat sensor, we chose to work with boid snakes.

Bullock and his collaborators have done most of the pioneering work on the heat receptors of snakes. In their original experiments, using the rattlesnake, they first anesthetized the animal and then dissected out the bundle of large nerves that constitute the main branches of the trigeminal nerve. It is these branches that receive the sensory information from the receptor.

Bullock and his colleagues found by means of electrical recording that the frequency of nerve impulses increased as the receptor was warmed up and decreased as it was cooled. The changes were independent of the snake's body temperature; they were related only to changes of temperature in the environment. The Bullock group also determined that the operation of the sensor is phasic, meaning that the receptor gives a maximum response when the stimulus is initiated and that the response quickly subsides even if the stimulus is continued. (Many human receptors, such as the ones that sense pressure on the skin, are phasic; if they were not, one would be constantly conscious

of such things as a wristwatch or a shirt.)

Our work was built on the foundations laid by Bullock and his associates. In addition we had in mind certain considerations about electromagnetic receptors in general. Biological systems utilize electromagnetic radiation both as a source of information and as a primary source of energy. Vision is an example of electromagnetic radiation as a source of information, and photosynthesis is a process that relies on electromagnetic radiation for energy.

All green plants utilize light as the source of the energy with which they build molecules of carbohydrate from carbon dioxide and water. To collect this energy the plants have a series of pigments (the various species of chlorophyll molecules) that absorb certain frequencies of electromagnetic radiation. Indeed, green plants are green because they absorb the red part of the spectrum and reflect the green part. Because the chlorophyll molecule absorbs only a rather narrow spectral frequency, it can

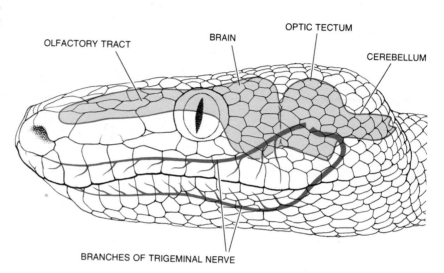

ANATOMY OF RECEPTOR in a boa is indicated. The scales along the upper and lower jaws have behind them an elaborate network of nerves, which lead into the two branches of the trigeminal nerve shown here. When the system detects an infrared stimulus, the trigeminal nerve carries a signal to the brain. A response can be recorded from the brain within 35 milliseconds after a boid snake receives a brief pulse of infrared radiation.

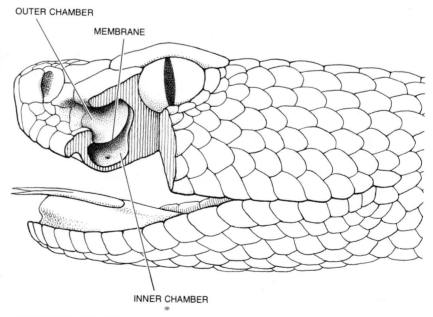

STRUCTURE OF PIT in a pit viper, the rattlesnake *Crotalus viridis*, differs from the anatomy of the infrared receptor in boid snakes. A pit viper has two pits, located between the eye and the nostril and facing forward. Each pit is about five millimeters deep, with the opening narrower than the interior. The elaborate branching of the trigeminal nerve is in the thin membrane that separates the inner and outer chambers of the pit organ.

DIRECTIONALITY OF SENSOR in a pit viper is indicated by the location of the two pits on the snake's head and by the geometry of the pits. It appears certain that in stalking its prey, which include birds and small mammals, such a snake can establish the direction in which the prey lies by shifting its head as it does in using eyesight. Rattlesnake and copperhead are among the pit vipers.

be called a frequency (or wavelength) detector.

The eye is also a frequency detector, but it does not use radiation as an energy source. The incoming radiation triggers the release of energy that has been stored in the nerves previously, having been produced by normal metabolism. The eye, like other frequency detectors, operates within a narrow band of the electromagnetic spectrum, namely at wavelengths from about 300 to about 1,000 nanometers (billionths of a meter). One can see how narrow the band is by recalling that man-made instruments can detect electromagnetic wavelengths from 10^{-20} meter to 10^5 meters, a full 25 orders of magnitude, whereas the range of the human eye is from $10^{-6.4}$ to about $10^{-6.1}$ meter. Within this range the eye can resolve thousands of different combinations of wavelengths, which are the number of shades of color one can recognize. Although the eye is a good frequency detector, it is a poor energy detector: a dim bulb appears as bright as a bright one to the dark-adapted eye, which is to say that the eye adjusts its sensitivity according to the conditions to which it has become adapted.

Why has nature chosen this frequency range for its photobiology? From an evolutionary point of view the answers seem clear. One reason is that 83 percent of the sunlight that reaches the surface of the earth is in that frequency range. Moreover, it is difficult to imagine a biological sensor that would detect X rays or hard ultraviolet radiation, because the energy of the photons would be higher than the bonding energy of the receiving molecules. The photons would destroy or at least badly disrupt the structure of the sensor. Low-fre-

quency radiation presents just the opposite difficulty. The energy of long-wavelength infrared radiation and of microwaves is so low that the photons cannot bring about specific changes in a molecule of pigment. Hence the sensor must operate in a frequency range that provides enough energy to reliably change biological pigment molecules from one state to another (from a "ground" state to a transitional state) but not so much energy as to destroy the sensor.

Early workers on the heat detectors of snakes had determined that the receptor responded to energy sources in the near-infrared region of the spectrum. The work left unanswered the question of whether the sensor contained a pigment molecule that trapped this long-wave radiation, thus acting as a kind of eye, or whether the sensor merely trapped energy in proportion to the ability of the tissue to absorb a given frequency and was thus acting as an energy detector. We therefore directed our experiments toward trying to resolve this issue.

To make sure that the response we obtained was maximal, we wanted to work with snakes that were functioning as close to their normal physiological level as possible. First we studied the normal feeding behavior of boa constrictors that were healthy and appeared to be well adjusted. The work entailed seeing how the snake sensed, stalked and captured prey animals such as mice and birds. Since the snake can capture prey in complete darkness as well as in light, it is clear that the heat receptors play a crucial role.

Barrett, while he was a graduate student working with Bullock, went further with this type of behavioral study. He found that the snake would strike at a

warm sandbag but not at a cold, dead mouse. On the other hand, the snake would swallow the cold mouse (after a great deal of tongue-flicking and examination) if the mouse was put near the snake's mouth, but it never tried to swallow a sandbag. Barrett concluded that the snake has a strike reflex that is triggered by the firing of the heat receptors, whereas another set of sensory inputs determines whether or not the snake will swallow the object.

In searching for a reliable index that would tell us whether or not the heat receptor was responding to an infrared stimulus, we first tried measuring with an electrocardiograph the change in heartbeat after the snake received a stimulus. This venture ran afoul of the difficulty of finding the heart in such a long animal. (It is about a third of the way along the body from the head.) A more serious difficulty was our discovery that a number of outside influences would change the rate of the heartbeat, so that it was hard to establish a definite stimulus-response relation.

We next turned to a method that proved to be much more successful. It entailed monitoring the electrical activity of the snake's brain with an electroencephalograph. A consistent change in the pattern of an electroencephalogram after a stimulus has been received by the peripheral nervous system is called the evoked potential. When a neural signal from a sensory receptor arrives at the cortex of the brain, there is a small perturbation in the brain's electrical activity. When the signal is small, as is usually the case, it must be extracted from the electrical background noise. The process is best accomplished by averaging a substantial number of evoked potentials. This procedure results in a highly sensi-

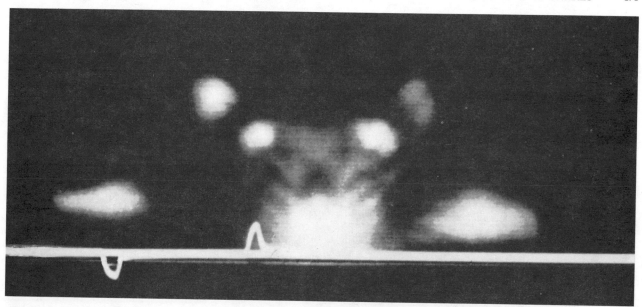

INFRARED VIEW OF RAT suggests what a snake "sees" through its infrared detectors when it is stalking prey. Snakes with such detectors prey on birds and small mammals. This view was obtained with a Barnes thermograph, which detects infrared radiation. In a thermogram the coolest areas have the darkest appearance and the warmest areas, such as the nose of the rat, appear as white spots.

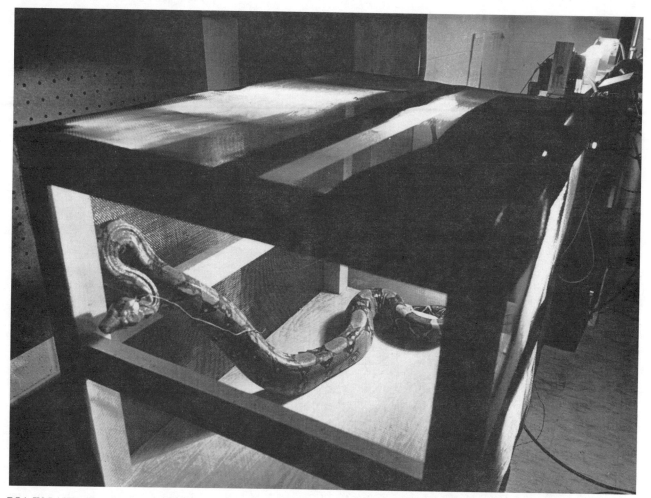

BOA IN LASER BEAM was tested in the authors' laboratory at the University of Colorado for responses to infrared radiation. The carbon dioxide laser, emitting in the infrared at 10.6 microns, appears as a glowing area in the upper right background. Its beam is spread by a lens so that the snake, even when moving about in the cage, is doused in infrared radiation delivered in occasional pulses lasting eight milliseconds each. A brain signal recorded by the electrode assembly on the snake's head goes to a preamplifier and then to an oscilloscope and to a signal-averager. Electroencephalograms recorded in this way appear in the illustration on page 177.

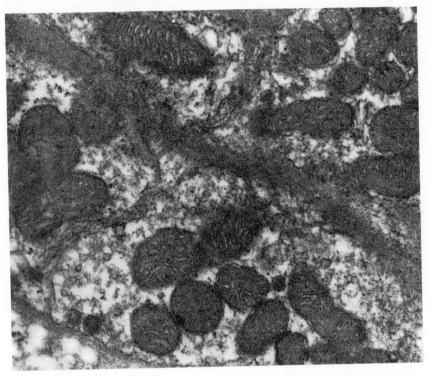

APPEARANCE OF MITOCHONDRIA in the nerve endings of the infrared receptor of a cottonmouth moccasin (*Agkistrodon piscivorus piscivorus*) after exposure of the receptor to an infrared stimulus is shown in this electron micrograph made by Richard M. Meszler of the University of Maryland. The enlargement is 34,000 diameters. In contrast to the mitochondria in the micrograph below, which was made when the receptor was exposed to a cold body, these mitochondria are condensed, as shown by the dense matrix and the organization of the inner membrane. Change in morphology of the mitochondria after a heat stimulus has led to the suggestion that they constitute the primary receptors in the detector.

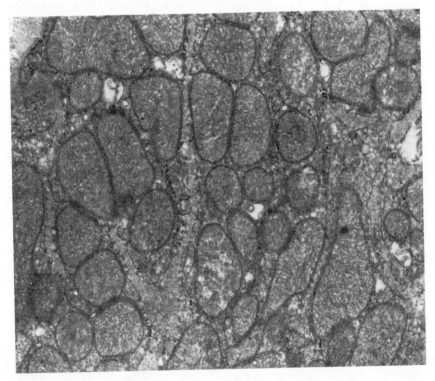

CONTRASTING APPEARANCE of the mitochondria in the infrared detector of a cottonmouth moccasin when the receptor was exposed to a cold body is evident in this electron micrograph made by Meszler. The enlargement is 27,000 diameters. A cold body, in contrast to a warm one, is known to reduce the firing of discharges by the heat-sensitive receptor.

tive measure of a physiological response.

The boa constrictors used in our study ranged from 75 to 145 centimeters in length and from 320 to 1,200 grams in weight. For several weeks before we involved them in experiments they lived under normal conditions in our laboratory. To prepare a snake for the experiments we anesthetized it with pentobarbital and then installed an electrode assembly on its head. After a postoperative recovery period the animal appeared to behave in the same way as snakes that had not been operated on.

A brain signal recorded by this apparatus went to a preamplifier and then to an oscilloscope and to a signal-averager. The signal-averager, which is in essence a small computer, is the workhorse of our system. By averaging the electroencephalogram just before and just after a stimulus it extracts the evoked potential, which would otherwise be buried in the background noise of the brain. In general we average the evoked potentials from about 20 consecutive stimuli.

The birds and mammals that the boa constrictor hunts emit infrared radiation most strongly at wavelengths around 10 microns. A carbon dioxide laser is ideal for our experiments because it produces a monochromatic output at a wavelength of 10.6 microns. We pulse the laser by means of a calibrated camera shutter so that it will deliver a stimulus lasting for eight milliseconds. The opening of the shutter also triggers the signal-averager, thus establishing precisely the time when the stimulus is delivered.

After the beam passes through the shutter it is spread by a special infrared-transmitting lens, so that the entire snake is doused in the radiation. The intensity of the radiation is measured by a sensitive colorimeter placed near the snake's head. This is the instrument we mentioned at the outset that takes nearly a minute to measure the power, whereas the snake gives a maximal response within 35 milliseconds after a single eight-millisecond pulse. Another indication of the sensitivity of the snake's receptor can be obtained by putting one's hand in the diffused laser beam; one feels no heat, even over a considerable period of time.

In order to verify that the responses of the snake resulted directly from stimulation of the heat receptor, we repeated the entire procedure with a common garter snake, which has no heat sensor. Even at laser powers far exceeding the stimulus given to the boas, we found no response in recordings from the garter snake. On the other hand, both species showed clear responses to visible light.

Our data strongly suggested an answer to the question of whether the receptor is a photochemical frequency detector like an eye or is an energy detector. The answer is that the receptor is an energy detector. One argument supporting this conclusion is that the stimulus is so far out in the low-energy infrared region of the spectrum (10.6 microns) that it would not provide enough power to activate an eyelike frequency detector, and yet the snake shows a full response. Another argument has to do with the 35-millisecond interval between the stimulus and the response. Photochemical reactions are quite fast, occur-

ring in periods of less than one millisecond. Although the time a nerve impulse from the eye takes to reach the cortex is about the same (35 milliseconds) as the time the nerve impulse from the heat sensor takes, the neural geometry of the two systems is quite different. The visual pathway incorporates a large number of synapses (connections between neurons), which account for most of the delay. In the trigeminal pathway no synapses are encountered until the signal reaches the brain. We therefore believe the delay found in the heat-receptor response is largely a result of the time required to heat the sensor to its threshold.

We also tested the snake's receptor in the microwave region of the spectrum, where the signals have longer

wavelength, lower frequency and lower energy than in the infrared. The reason was that in view of the many problems that have arisen in contemporary society about exposure to radiation we wanted to see whether an organism experienced physiological or psychological effects after being exposed for various periods of time to low-energy, long-wavelength radiation. There is no question that high-intensity microwave radiation can be detected not only by snakes but also probably by all animals; after all, a microwave oven can cook a hamburger in a matter of seconds. Our concern was with the kind of exposure arising from leaky microwave appliances such as ovens and from the increasing use of radar.

Testing the snakes with microwave radiation as we did with infrared, we obtained a clear-cut response [*see illustration on this page*]. Our result provides what we believe is the first unambiguous physiological demonstration that a biological system can indeed be influenced by such low-energy microwave radiation. Our conviction that the snake's heat receptor functions entirely as an energy detector is therefore reinforced.

The question of how much energy is required to activate the detector can be answered with certain reservations: it is approximately .00002 (2×10^{-5}) calorie per square centimeter. The reason for the reservations is that it is difficult to obtain an absolute threshold of sensitivity for any biological phenomenon. For one thing, a biological system shows considerable variability at or near its threshold of response. Moreover, there is always a certain amount of variation in the amount of energy put out by our sources of energy. With these reservations we have determined that the snake can easily and reliably detect power densities from the carbon dioxide laser ranging from .0019 to .0034 calorie per square centimeter per second. Since this density is administered in a short time period (eight milliseconds), the total energy that the snake is responding to is about .00002 calorie per square centimeter. The density of microwave power that is needed for a reliable response from the snake is about the same as the amount of laser power.

Our studies have shown that the heat-sensing snakes have evolved an extremely sensitive energy-detecting device giving responses that are proportional to the absorbed energy. It will be interesting to see whether the growing understanding of the snake's heat sensor will point the way toward an improvement in man-made sensors.

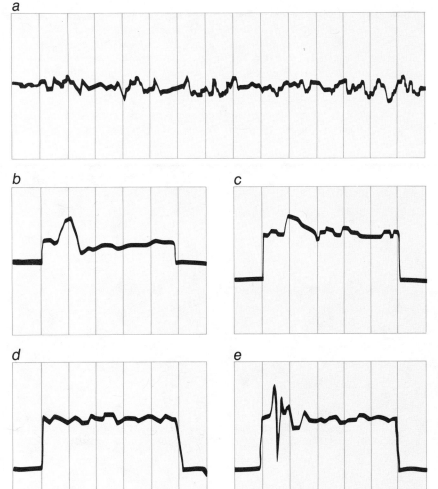

a

b

c

d

e

ELECTROENCEPHALOGRAMS of a boa constrictor were recorded under various conditions. The normal activity of a boa's brain (a) was traced directly on a strip recorder; the interval of time between each pair of colored vertical lines was 100 milliseconds. The remaining electroencephalograms were recorded through a signal-averager that reflected both the stimulus and the response. In each case the first rise shows the time of the stimulus and the next rise, if any, shows the response. The time interval is 100 milliseconds in every tracing but b, where it is 50 milliseconds. Traces show the averaged evoked response after an infrared stimulus (b) and a microwave stimulus (c), in the control situation in which the snake was shielded from the stimulus (d) and after a series of visible-light flashes (e).

17

The Stellar-Orientation System of a Migratory Bird

by Stephen T. Emlen
August 1975

When the indigo bunting is put in a planetarium, it exhibits an ability to orient itself by the stars. This, however, can be only one of the cues it uses for long-distance navigation

The blackpoll warbler is a small, inconspicuous songbird that breeds during the summer in the stunted conifer forests of Alaska and northern Canada. When fall approaches, these birds embark on a remarkable migratory journey. They leave the northern forests and fly east-southeast across the North American continent to the Atlantic coast. During this stage of the journey they stop to feed and build up stores of subcutaneous fat that will serve as a vital energy reserve for the next stage. The blackpolls concentrate near the coast of New England and the Maritime Provinces of Canada, waiting for the right weather conditions. Then, as the next high-pressure cell moves in from the west, bringing with it winds from the north or northwest, the blackpolls depart again, this time over the ocean on a nonstop flight that will take them three to five days. They fly over Bermuda, the Antilles and Puerto Rico, stopping only when they make landfall on the northeastern coast of South America. It is a tremendous feat: a nonstop flight of

EXPERIMENTAL ARRANGEMENT in a planetarium for testing the ability of the indigo bunting to orient by the stars is shown. The photograph was made in the Southern Cayuga Atmospherium-Planetarium in Poplar Ridge, N.Y. Projectors are not visible to birds in the cages. Stars at any latitude and longitude, as well as celestial motion, can be projected onto the planetarium dome.

more than 2,400 miles over water by a bird weighing less than 20 grams! Any error in navigation would obviously lead to disastrous results. Any misinterpretation of the weather would also lead to disaster, since a blackpoll that encounters stiff headwinds or a storm has no chance of landing to rest or to find shelter.

The blackpoll warbler is an extreme but not atypical case of birds that migrate south. Fully two-thirds of the species of songbirds that breed in the northern U.S. travel south in the winter. The distances of migration typically range from 600 to 1,800 miles, but each fall some songbirds make one-way trips of up to 4,000 miles. The following spring the birds fly back to their breeding grounds. Year after year the adult birds return with amazing precision to the same several square miles of territory at both their breeding and their wintering grounds.

How do the migrating birds select the appropriate flight directions? Can they determine when they have been blown off course, and can they correct appropriately? How do they know when they have arrived at the latitude of their destination? What is the fate of young birds flying alone on their first migratory trip?

Biologists have long been intrigued by the phenomenon of bird migration, but it is only during the past two decades that significant progress has been made toward answering the fundamental questions about it. Today scientists around the world are focusing their attention on questions of animal navigation. Hundreds of thousands of migrant birds are being individually marked with a leg band so that field investigators can determine their migratory paths by plotting their recapture locations. In the laboratory other workers are testing the ability of birds to detect different potential directional cues and are examining how such cues are used. Ornithologists are tracking "unseen" migrating birds with radar and following individual migrants by attaching small radio transmitters to them.

The results of these various studies have shown that bird navigation is not a simple affair; it is not entirely dependent on any single cue or sensory system. It seems that migrating birds make use of a variety of cues to determine their direction and maintain it in flight. A recent article by my colleague William T. Keeton describes some of the types of cues available and the interplay among them [see "The Mystery of Pigeon Homing," by William T. Keeton, *beginning on page 188*]. Here I shall concentrate on

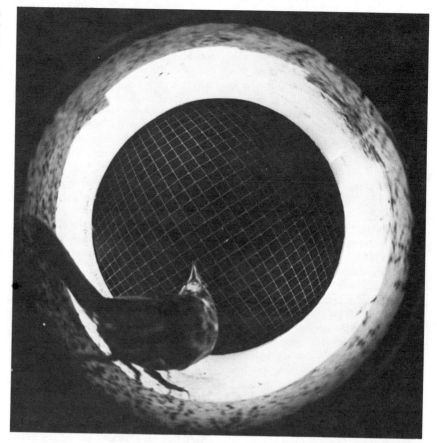

INDIGO BUNTING is shown hopping onto the side of a test cage. The photograph was made by placing a camera with a "fish-eye" lens at the bottom of the cage. Black marks on the white side of the cage are footprints made by the bird. The screen is the top of the cage.

one cue system that has been studied intensively and that appears to be of major importance to night-migrating birds: orientation by the stars. The reader should bear in mind that I am covering only one aspect of how birds navigate. It is only by dissecting out the various aspects and studying them one at a time that an understanding of the full story will ultimately be achieved.

The scientific study of the directional orientation of migratory birds had its breakthrough with the pioneering work of Gustav Kramer, an ornithologist at the Max Planck Institute for Marine Biology at Wilhelmshaven. It had long been known that when migrant songbirds were kept in captivity, they displayed intense activity at night during the periods of their normal spring and fall migration. When Kramer placed songbirds in circular cages during their migratory period, he discovered that they would spontaneously orient their activity in a particular direction. By manipulation of the cues available to the caged birds the determinants of direction finding could be studied. Through this technique migration could be "brought into the laboratory."

In the late 1950's another German ornithologist, E. G. F. Sauer of the University of Freiburg, carried out a long series of experiments with European warblers. It was he who first hypothesized that these warblers determine their migratory direction from the stars in the night sky. Since then much has been learned about star orientation by birds. Numerous species have been examined, and it appears that the ability to orient by the stars is widespread among birds that migrate at night.

For some years I have been studying one night-flying migratory bird, the North American indigo bunting (*Passerina cyanea*). Enough information is now available to piece together a fairly complete picture of how the stellar-orientation system in this species operates. The species breeds throughout the eastern U.S., where the brilliant blue male is a well-known songster. During the fall migration the buntings fly up to 2,000 miles to winter in the Bahamas and in southern Mexico and Central America south to Panama.

When indigo buntings are kept in captivity and exposed to the same pattern of day lengths they would experience in nature, they exhibit intense nocturnal activity in April and May and again in September and October, the times of normal migration. When this restlessness appeared, I tested the birds by placing them individually in circular cages. The cage was constructed of a piece of white blotting paper, rolled and stapled to form a funnel, mounted on a base consisting of an ink pad and topped with either a clear plastic sheet or a hardware cloth screen. A bird inside the cage can see only the sky overhead, since all ground objects are blocked from view.

A bunting in migratory condition stands in one place or turns slowly in a circle, its bill tilted upward and its wings partly spread and quivering rapidly. At frequent intervals the bird hops onto the sloping paper funnel, only to slide back and continue its pointing and quivering. Each hop from the ink pad leaves a black print on the paper. The accumulation of inked footprints provides a simple record of the bird's activity; they can later be counted and analyzed statistically.

In the first stage of my studies I placed the buntings in their funnel cages outdoors on clear, moonless nights. In September and October most of the birds exhibited a distinct preference for jumping toward the southern sector of the cage, the same direction in which they would have migrated if they could. When the birds are tested in late April and May, however, the preferred direction is to the northeast, the appropriate direction for spring migration.

Since the wall of the test cage completely screens the horizon from view, it seems likely that the buntings are able to determine their migration direction when the only visual cues are those provided by the night sky. This hypothesis was reinforced by the changes observed when the birds were placed outdoors at night under cloudy conditions. As the stars disappeared behind the clouds, the orientation of the birds deteriorated considerably [*see lower illustration on opposite page*].

In order to test the stellar hypothesis under more rigorous conditions, I took the buntings into a planetarium, an approach also used by Sauer in some of his experiments with European warblers. In September and October, when the birds were exhibiting nocturnal restlessness, I projected the normal fall stars onto the planetarium dome. When the buntings were tested in the funnel cages, they oriented to the south. Birds tested in April and May under a spring sky in the planetarium consistently oriented to the north and northeast. When the North

Star in the artificial sky was shifted to the east or west, the buntings changed their orientation to the new "south" or the new "north," depending on the season of the year. The change in orientation behavior was consistent and predictable. In control experiments I turned off the star projector in the planetarium and exposed the birds to a diffusely illuminated dome. Their behavior paralleled their response to overcast conditions outdoors: the accuracy of their orientation deteriorated considerably.

Since indigo buntings are willing to exhibit meaningful orientation in spite of the confinement of captivity, I was able to further modify the visual cues in the night sky and thus dissect out the detailed workings of the birds' stellar-orientation system. The experiments were designed to answer the following questions: Which stellar cues are important? How are such cues employed? What kind of information does the bird obtain? How accurately does it obtain it?

In theory there are two ways birds could determine direction by the stars. One way would be for the bird to locate a critical star or a group of stars and then guide itself by flying at a particular angle with respect to the star or the group of stars. The absolute position of a star, however, is not constant throughout the night: stars shift from east to west as the result of the rotation of the earth. In order to maintain a given compass direction the bird would have to alter its angle of flight with respect to the selected star in such a way as to compensate for the apparent motion of the star [*see top illustration on page 183*]. Such a mechanism would be analogous to the sun-compass orientation in which a daytime bird migrant, making use of an internal time sense, correctly compensates for the daily movement of the sun across the sky.

The requirements for a stellar-navigation system are much more demanding than those for a system that depends on the sun for determining compass direction by day. There is only one sun, and it moves at a regular rate, but there are thousands of stars and different stars are visible above the horizon at different times of the night and at different seasons. A nocturnal migrant presumably would need to be able to consistently locate a specific star or a specific group of stars, and that would require it to possess some form of pattern recognition.

In addition the rate of compensation for apparent motion will differ, depending on the star or stars selected. Celestial motion is an apparent motion produced

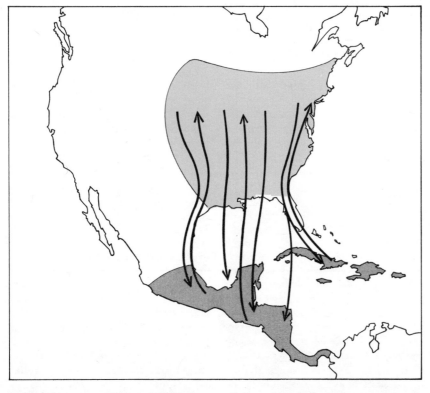

MIGRATION OF INDIGO BUNTINGS proceeds along a broad front. Buntings migrate from their wintering grounds in the Bahamas, southern Mexico and Central America (*gray areas*) in late April and arrive at their breeding grounds in the eastern U.S. (*colored area*) throughout the month of May. They depart for wintering grounds in September and October.

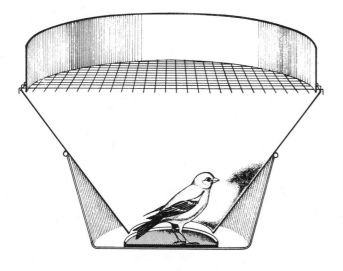

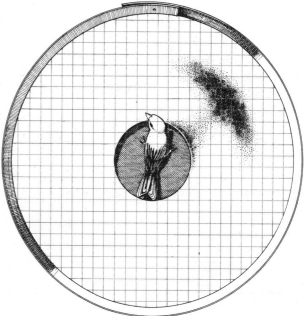

CIRCULAR TEST CAGE for determining the directional preference of an indigo bunting is shown in cross section and in a top view. Funnel portion of the cage is made of white blotting paper.

The bunting stands on an ink pad, and each time it hops onto the sloping funnel wall it leaves black footprints. The bird's view is limited to a 140-degree overhead sector of sky when it hops up.

by the earth's rotation once every 24 hours. All stars move with an angular velocity of 15 degrees per hour, as the sun does. The linear velocity of a star will vary, however, depending on how close it is to the North Star. Stars near the North Star move through a small arc, whereas stars near the celestial equator move through a large one. If a bird were to use star groups in different parts of the sky, it would have to compensate at a different rate for each group. Finally, the direction of compensation depends on whether the guiding stars are in the north or in the south: northern stars would require clockwise compensation, southern stars counterclockwise.

According to the second theoretical

model, the bird would use patterns of stars to determine directional reference points. Human beings easily recognize the Big Dipper by the characteristic arrangement of its stars. And by visually extending the line joining the two pointer stars of the Big Dipper, they can readily locate the North Star and hence geographic north. Star patterns such as the Big Dipper also move across the sky, but the shape of each pattern remains constant and each preserves a distinct relation to the North Star.

Since each star bears a fixed geometric relation to other stars, it would be theoretically possible for a bird to determine a directional reference point from any number of star patterns [*see middle illustration on page 183*]. The major difference between this model and the first one is that a compass direction can be determined from the geometric patterns of the stars independently of an internal time sense, or "biological clock."

I tested these alternative hypotheses in a planetarium by creating an artificial situation in which the astronomical time would be out of phase with a bird's internal time sense. If a time sense is involved in star navigation, then presenting the stars in positions that are advanced or retarded from local time should cause the bird to orient in the wrong direction. On the other hand, there should be no change if the bird relies only on the geometric patterns of the

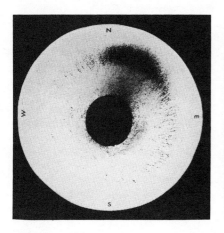

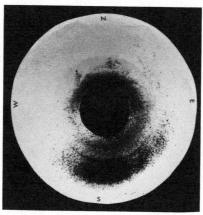

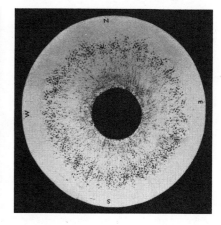

FOOTPRINT RECORDS of male indigo buntings tested in a circular cage placed outdoors under the stars on moonless nights are shown. In the spring the bird typically orients its hopping to the north (*left*). In the fall its hopping is oriented to the south (*middle*). When the stars are obscured by clouds, the bunting's activity remains high but orientation of the hopping is random (*right*).

stars. I tested buntings when the planetarium sky was three, six and 12 hours ahead of local time and when it was three, six and 12 hours behind local time. The results were clear: the birds generally maintained their normal migratory orientation under all these conditions. Apparently indigo buntings do not make use of an internal time sense to orient by the stars but obtain directional information from star patterns, much as human beings do.

I then turned to the question of which star groupings are of particular importance to buntings. Once again the planetarium was an invaluable tool, since one can block from view selected stars, constellations or entire areas of the sky. In a series of experiments I systematically removed and later reinserted portions of the artificial sky. I found that most buntings rely for direction finding on the northern area of the sky that lies within

about 35 degrees of the North Star. The major constellations in this area are the Big Dipper, the Little Dipper, Draco, Cepheus and Cassiopeia. The birds relied on the northern circumpolar stars not only during the spring migrating season when they normally fly north but also during the fall migrating season when they fly south.

An important corollary finding was that there is considerable redundancy in the buntings' recognition of star patterns. The birds are familiar with several star groups, and the removal of one group of stars, say the Big Dipper, often merely forces them to rely on some alternate constellation. Since birds frequently migrate on nights when there is variable cloud cover, such redundancy is obviously adaptive.

Navigation can be regarded as involving a two-step process. Consider a man equipped with a map and a compass. In order to determine how to reach a par-

ticular geographic destination he must first calculate his position on the map with respect to his goal and then use the compass to select the appropriate direction. The navigation problems of a migrating bird can be viewed in the same way, that is, as a "map and compass" process.

Theoretically an accurate knowledge of the absolute positions of the stars coupled with a stable and highly accurate internal time sense could provide enough map-and-compass information for a bird to determine its absolute geographic position. If the bird retains a precise memory of the temporal position of the stars at its destination point, it could in principle select the appropriate direction to the goal from the displacement of the stars in the sky overhead.

The finding that the indigo bunting does not integrate a temporal component with its use of stellar cues implies, however, that the bird does not detect or correct for longitudinal displacement, at least not by celestial cues. Hence star orientation in the indigo bunting appears to be a compass sense that enables the bird to select and maintain a particular direction but does not provide the information that makes it orient to a particular goal.

What, then, does determine how the star compass will be used? The bunting may be able to locate the Big Dipper or other constellations, but why does it use them to orient north or south rather than east or west? And how does the bunting select north in the spring and south in the fall?

The sidereal, or astronomical, day is four minutes shorter than the solar day. Because of this inequality the temporal positions of the stars change with the seasons, with the result that the stellar information available from the fall night sky is quite different from that available from the spring night sky. Does the indigo bunting have a specific northerly directional response to the stellar stimuli in the spring sky and a southerly response to the stellar stimuli in the fall sky?

To test this possibility I captured 15 adult male buntings in their summer breeding territories near Ithaca, N.Y., and divided them into two groups. The weight, fat level and molt status of each bird was recorded weekly until the testing period the following spring.

The control group of eight birds lived in a flight room where the length of the day simulated what they would have normally encountered in nature. An astronomical time clock maintained a day length equivalent to the day length at

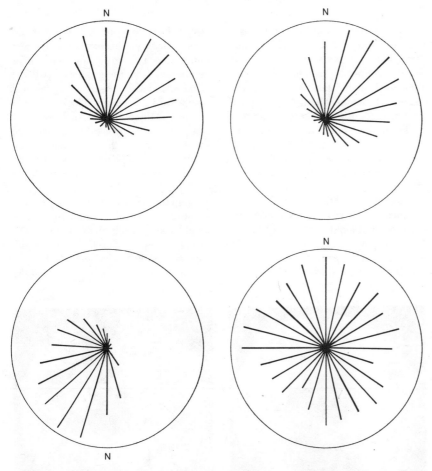

VECTOR DIAGRAMS show the similarity of the orientation of a bunting outdoors under the stars in the spring (*top left*) and under a simulated spring night sky in a planetarium (*top right*). When the planetarium stars were shifted so that the North Star was at the true south, the bird reversed its orientation (*bottom left*). When the stars were turned off and the planetarium was diffusely illuminated, the bunting's orientation became random (*bottom right*). The radius of each circle is equal to the largest amount of activity in any one 15-degree sector, and the vectors for the other 15-degree sectors are proportional to the radius.

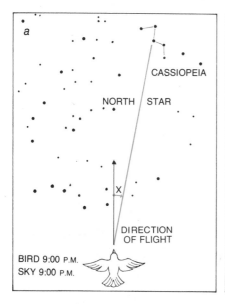
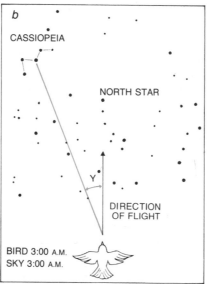
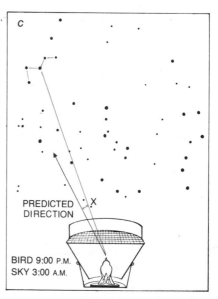

STELLAR-ORIENTATION HYPOTHESIS I proposes that the indigo bunting guides itself by flying at an angle to a particular star or group of stars. Since the positions of the stars change throughout the night, the bird would have to use an internal time sense to compensate for the motion of the stars. For example, a bunting going north at 9:00 P.M. would fly at angle X with respect to a critical star (a). At 3:00 A.M. the bird compensates for the rotation of the stars by flying at angle Y to the critical star (b). According to the hypothesis, when a bunting whose physiological time is at 9:00 P.M. is presented with a 3:00 A.M. star pattern in a planetarium, it should compensate in the wrong direction, that is, it should orient at angle X with respect to the critical star instead of at angle Y (c).

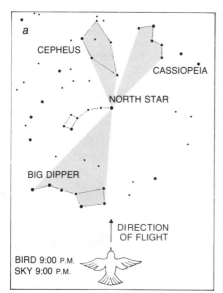
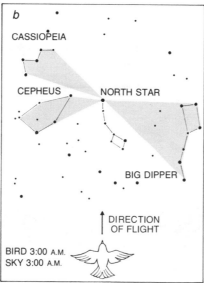
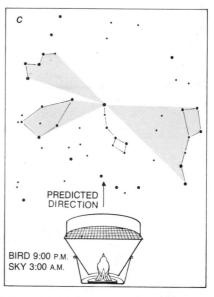

STELLAR-ORIENTATION HYPOTHESIS II states that the bunting obtains directional information from the configuration of the stars. The bird can determine a reference direction such as north from fixed geometric relation of the stars regardless of the time of night (a, b). When the bunting is exposed to a time-shifted sky in a planetarium, there should be no change in its orientation (c).

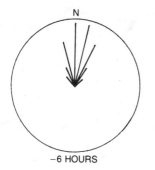
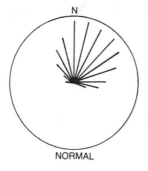
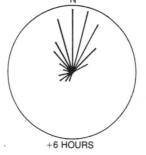
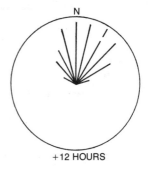

RESULTS OF PLANETARIUM TEST of the two star-navigation hypotheses show that buntings continue to orient correctly regardless of whether planetarium stars are shifted ahead of or behind the bird's normal physiological time. This indicates that the bunting does not incorporate its biological clock in the star-orientation process and obtains only directional information from star patterns.

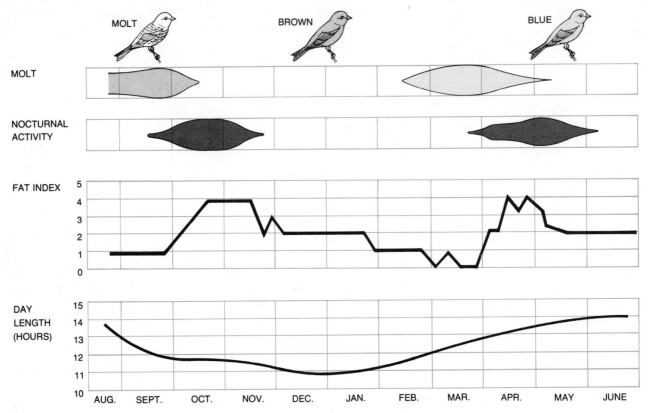

CONTROL GROUP OF ADULT MALE BUNTINGS lived in a flight room where an astronomical time clock maintained the day length equivalent to length at their wintering grounds. The birds molted normally in the fall and again in the spring. After each molt they built up fat reserves and became active at night. In May directional preference of the birds was tested in a planetarium.

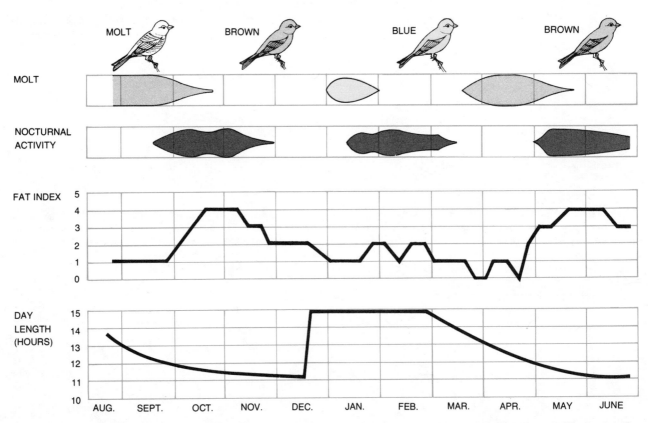

EXPERIMENTAL GROUP OF ADULT BUNTINGS was given an accelerated photoperiod regimen. Beginning in mid-December birds were exposed to the longer day lengths typical of spring. They molted into their blue spring plumage in January. The day lengths were shortened in March, and the birds molted into their brown winter plumage. Their directional orientation was tested in May.

their wintering ground in Guatemala. The male buntings in this group molted normally and acquired their bright blue prenuptial plumage between February and early April. After molting they built up substantial reserves of subcutaneous fat. They became nocturnally restless in April and May.

The experimental group of seven buntings was subjected to the same day-length regime as the control group until mid-December. The birds were then exposed to a spring day length of 15 hours, which caused them to molt into their blue spring plumage in January. Beginning on March 1 the day length was progressively shortened to simulate the day lengths of fall in the buntings' summer breeding territory, and the birds molted out of their blue plumage and into their brown winter coloration. After the molt the buntings built up fat reserves, and nocturnal activity began in May.

The directional preferences of the control group and the experimental group were tested in the planetarium under identical spring night skies. There was a marked dichotomy in the orientation behavior of the two groups. The control group of blue buntings, which were ready for their normal spring migration, oriented to the north and northeast, whereas the experimental group of brown buntings, which were ready for their normal fall migration, oriented to the south.

The results indicate that the physiological state of the indigo bunting affects its migratory orientation. We have already seen that the same northern circumpolar stars are used as the chief stellar reference in both spring and fall. It now appears that the polarity of the migratory orientation—whether it is toward or away from the northern circumpolar stars—is under hormonal control. Recent studies by Albert H. Meier and D. D. Martin of Louisiana State University support this hypothesis. They report being able to reverse the orientation behavior of another nocturnal migrant, the white-throated sparrow, by altering its physiological state with the administration of two hormones, prolactin and corticosterone, which appear to have a synergistic effect in stimulating the birds' migratory activity.

The finding that the physiological state of the indigo bunting affects the direction in which it orients does not, however, fully answer the question of how it chooses a specific direction for its migration. The young of many species of birds migrate independently of the

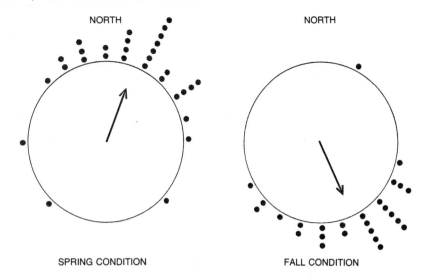

NORTH

NORTH

SPRING CONDITION

FALL CONDITION

DIFFERENCE IN ORIENTATION of indigo buntings in the control group (*top illustration on opposite page*) and the experimental group (*bottom illustration*) shows that the physiological state of the bunting affects its migratory direction. The control group, which was physiologically in the normal spring migratory condition, oriented to the north-northeast. The experimental group, which was in an induced fall migratory condition, oriented to the south-southeast. The mean vector of the birds in each group is shown by the arrow. The length of the vector represents degree of agreement among the birds in selecting a direction.

adults, setting out on a course they have never traveled before and without the benefit of experienced companions. What causes young, inexperienced birds to select a southerly direction for their first migration? To many investigators the fact that they do so implied that early experience was not important in the development of normal orientation abilities. Some workers even proposed that birds possess a genetically inherited "star map."

Field studies have produced evidence that makes the inherited-star-map hypothesis unlikely. These investigations reveal that there are differences between the navigational abilities of young birds and those of adult birds. When birds of some species, for example the European chaffinch, are captured and displaced from their normal fall migration routes, the adults may correct for the displacement and fly to their regular winter grounds, but young birds on their first migration do not. Prior experience obviously improves the navigational ability of birds.

In my studies adult indigo buntings have always been more accurate and consistent in their orientation than young birds. If very young buntings are prevented from the time of their capture from viewing the normal night sky, aberrant orientation behavior develops. One summer I located numerous nests of indigo buntings near Ithaca, and I carefully removed the young birds when they were between four and 10 days old. The

nestlings were hand-reared in the laboratory, where their visual experience with celestial cues could be controlled. One group of young buntings never saw a point source of light. They lived in a windowless room with diffuse fluorescent lighting until they began depositing migratory fat and exhibiting intense nocturnal activity in September and October. When I tested these birds in the planetarium under a night sky that matched the normal one for the season, they were unable to select a migratory direction. The young birds were highly motivated, jumping with great frequency in the test cages, but the orientation of the jumps was random. They were unable to obtain directional information from the stars.

A second group of young birds was prevented from seeing the sun, but the birds were allowed to view the night sky in a planetarium every other night during August and September. The star projector rotated at a speed of one revolution every 24 hours, thus duplicating the normal pattern of celestial rotation. When these birds began displaying nocturnal activity, they were tested in the planetarium under the same sky as the group of birds that had had no visual experience with the stars. Unlike the inexperienced birds, the buntings that had already been exposed to the night sky were able to orient to the south. In some way the exposure to the stars was of extreme importance for the normal maturation of star-orientation abilities. Finally, a third group of young buntings was exposed to the night sky in the planetarium but with one important difference. I modified the star projector by constructing a special arm that allowed the projector to be rotated around any axis of my choosing. I selected Betelgeuse, a bright star in Orion, as the new polestar around which all other stars rotated. The star patterns and constellations remained unchanged; only their positions and movements with respect to the new axis of rotation were altered.

The young buntings of the third group were exposed to the sky with the Betelgeuse polestar every other night for two months. When they became nocturnally active, they were tested under a normal sky with the North Star at the pole position. The buntings consistently oriented their activity 180 degrees away from Betelgeuse—the appropriate "southerly" direction as defined with respect to the polestar of their early experience [see illustration below].

On the basis of experiments of this kind I have hypothesized that young buntings respond to the apparent rotational motion of stars in the night sky. The stars near the North Star move through much smaller arcs than the stars near the celestial equator, and this enables the young birds to determine a

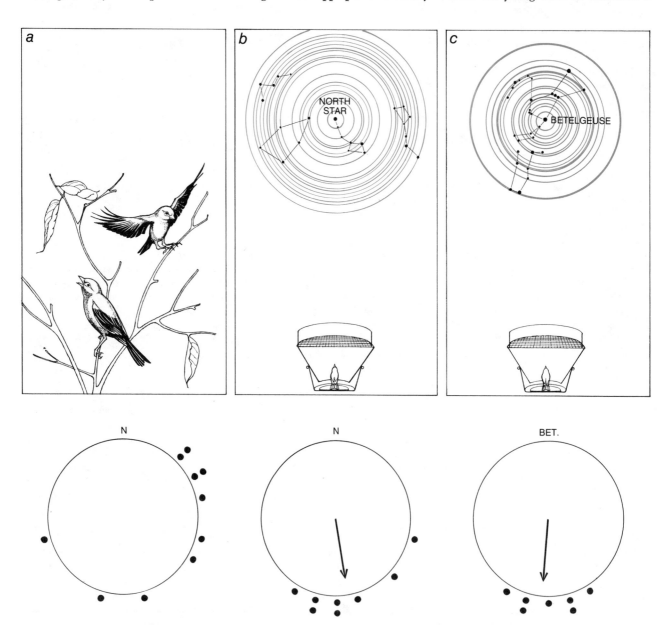

EARLY VISUAL EXPERIENCE of young indigo buntings was found to play an important role in the development of their celestial-orientation abilities. Three groups of nestlings were captured and hand-reared in the laboratory. The first group (a) lived in a windowless room with diffused lighting and never saw a point source of light. In the fall the birds began to display intense nocturnal activity. When they were tested under a stationary night sky in a planetarium, they were unable to select a migratory direction. The second group (b) never saw the sun and was exposed to a night sky in a planetarium every other night for two months. Normal celestial rotation was simulated. When the birds were tested in a planetarium under a normal sky during their fall migratory period, they oriented to the south. The third group (c) also never saw the sun and was exposed to a modified night sky in a planetarium every other night for two months. Betelgeuse, a star in Orion, became the new polestar around which all other stars rotated. When the birds were tested in the fall under a normal night sky, they continued to regard Betelgeuse as the polestar and oriented their activity away from it. The experiment shows that young buntings initially learn the north-south axis from the rotation of stars and that star patterns by themselves are not useful cues to a naïve bunting. Star patterns take on directional meaning only after they have become part of the bird's general orientational framework, the formation of which is influenced, at least in part, by observing the rotation of stars.

north-south directional axis. Individual stars and patterns of stars are of no value for direction finding until their positions with respect to some reference framework have been learned. The axis of the rotation of the stars appears to function as one reference system. Once the stellar information and the rotational have been coupled, however, the bunting can locate the rotational axis from star patterns alone. This is suggested by the finding that adult indigo buntings orient accurately even under stationary planetarium skies. Celestial motion thus becomes a secondary or redundant cue for adult birds.

One cannot help but speculate about the possible selective advantage of a maturation process that makes use of celestial rotation for a directional reference system. One possible explanation lies in the long-term unreliability of the stellar cues themselves. The rotation of the earth can be viewed as being analogous to the spinning of a top. And like most spinning tops, the spinning earth wobbles. This slight wobble, usually described as the precession of the equinoxes, causes the direction of the earth's spin axis to shift. Over a period of 26,000 years the precession of the equinoxes causes the earth's spin axis to trace on the celestial sphere a full circle with a radius of 23.5 degrees. This motion gives rise to marked seasonal and latitudinal changes in the apparent position of stars. The spring stars of the present become the fall stars in 13,000 years, and vice versa. The values of declination also change: as the polar axis moves through its circle Vega becomes the new polestar, and the present North Star shifts to 43 degrees north [*see illustration on this page*]. Similar changes occur for all stars.

The possible implications of these changes for the star-navigation system of birds are obvious. If birds were to rely on a genetically fixed star map, the rate of genetic change would have to be extremely rapid to allow for the change in position of the stars. A maturation process in birds that involves finding the north-south axis by the rotation of stars, however, minimizes the problem. Of course, several reference cues may play a role, but the axis of celestial rotation is well suited to function as one such reference because that axis is aligned with geographic north-south regardless of which particular stars and patterns of stars are located near the celestial pole.

Experiments in the planetarium have enabled us to learn a great deal about the orientation of night-migrating birds. Young birds develop a north-south reference axis as a result of early exposure

to celestial rotation. Then they learn the patterns of stars around the northern celestial pole, which they use in a configurational manner to determine a direction of migration. The precise direction that is selected depends on the hormonal and physiological state of the bird and not on seasonal differences in the position of the stars. The star-orientation process of the indigo bunting is basically one of pattern recognition that does not involve an internal time sense.

The experiments described here are equally important for what they do not explain about migratory orientation. Although the star-orientation process enables a bird to maintain a given course during its migratory trip, as an explanation for the orientation abilities of birds the process by itself is not entirely adequate and not absolutely essential. I say not entirely adequate because the direction-finding system described for the indigo bunting is basically only a star-compass system that enables the bird to select and maintain a given direction. The system does not provide any information about actual geographic location.

The star compass does not tell the bird it has been blown off course to the east or to the west, nor does it tell the bird when it has reached the latitude of its destination. That is because the star-orientation capability I have described still lacks the map component of the map-and-compass hypothesis. It is quite possible that most migrating birds are not at all goal-oriented during the major portion of their migratory flight. Their process of orientation may be fundamentally different from that of homing pigeons and they may only revert to a homing type of process during the very final stages of the migratory flight.

I say not absolutely essential because we now know that migratory birds have numerous directional cues available to them. In addition to the use of the sun and the stars, experiments have shown that songbird migrants can make use of the position of the sunset, the directionality of the winds aloft, the direction of the earth's magnetic field, the presence of topographic landmarks and the activity and the call notes of other birds of the same species as sources of information, enabling them either to select or at least to maintain a given migratory direction.

Birds thus have access to many sources of directional information, and natural selection has favored the development of abilities to make use of them all. Some cues may give more accurate information than others; some may be available throughout the flight, whereas others may be useful only at specific geo-

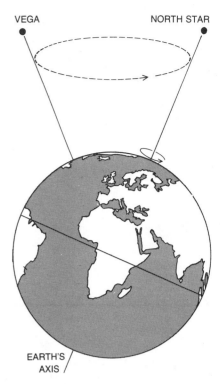

PRECESSION OF EARTH'S AXIS produces change in the apparent position of the stars. In 13,000 years, as the polar axis moves through half of its circle, the star Vega will become the new polestar and the present North Star will shift to 43 degrees north.

graphic locations; some may be available regardless of flight conditions, whereas others may be functional only under optimal meteorological situations.

The realization that birds have multiple cues at their disposal is in itself a finding of major importance. Although I have concentrated here on the star compass, I do not mean to imply that the other cue systems are not important. The discovery of a hierarchy of redundant directional cues makes the search for a single mechanism of migratory orientation obsolete.

The nocturnally migrating bird should be viewed as an animal whose behavior has been shaped by aeons of intensive selection pressure. It is the combination of the bird's skill as a meteorologist and as a navigator that accounts for its successful traversing of thousands of miles of environmentally inappropriate terrain each fall and spring. Although our understanding of the migratory navigation of birds has come a long way since Kramer began the experimental approach some 20 years ago, the total of our knowledge is still not enough to fully explain how an individual bird finds its way between its breeding territory and its wintering grounds.

18

The Mystery of Pigeon Homing

by William T. Keeton
December 1974

Recent findings have upset previous explanations of how pigeons find their way home from distant locations. It appears that they have more than one compass system for determining direction

How does a homing pigeon find its way back to its home loft from hundreds of miles away? The answer does not lie in visible landmarks; pigeons taken in covered cages to areas they have never seen before have little trouble finding their way home. Nor does it lie entirely in the bird's ability to determine compass directions from the sun or the earth's magnetic field. Even when a pigeon can determine compass directions, how can it know which way home is? Although the homing prowess of the pigeon has long engaged the curiosity of man, the full story of how the bird navigates still remains a mystery. Nonetheless much has been learned about the pigeon's navigational abilities in the past two decades, particularly in the past six years.

The modern homing pigeon, a de-

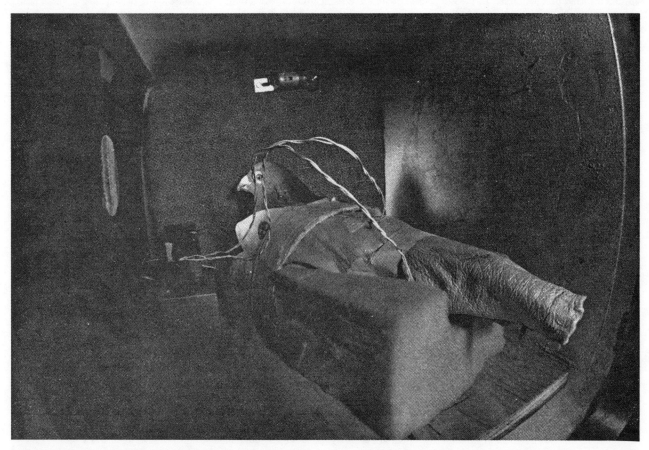

PIGEON IN ISOLATION CHAMBER has been prepared for tests of its unusual sensory capabilities. Two wires go to electrodes on the pigeon that give it a mild electric shock during the test, and two other wires are connected to electrodes that pick up the bird's heartbeat. The pigeon is restrained by a harness to keep it from moving. In a typical experiment the bird receives a shock following a specific stimulus, which might be a change in the strength of an induced magnetic field, a change of air pressure or a change in the plane of polarized light falling on the pigeon's eye. If the bird is able to sense the changes in the stimulus, it begins to anticipate the shock and its heart rate increases at the beginning of the stimulus, which is given at random time intervals. Experiments in the author's laboratory indicate that the pigeon is capable of sensing tiny fluctuations in atmospheric pressure. In addition pigeons, like honeybees, can detect changes in the plane of polarized light. The photograph was made with a camera placed inside the chamber.

scendant of several earlier breeds of pigeons, was developed in Belgium in the middle of the 19th century. Today, in addition to serving as message carriers, homing pigeons are raised for competitive racing. This sport is widespread and very popular in Europe, and has become firmly established in many parts of the U.S. as well. Often several thousand pigeons are entered in a single race. The birds are shipped to a designated point, usually between 100 and 600 miles away, and are then released simultaneously. After the owners of the birds have recorded the arrival times at home, using special devices designed for the purpose, the speed of the individual birds is calculated to determine the winners. Speeds of 50 miles per hour are common; the best pigeons can make it home from 600 miles away in a single day.

The remarkable ability of pigeons to find their way home has been known for at least as long as there has been written history. The armies of the ancient Persians, Assyrians, Egyptians and Phoenicians all sent messages by pigeon from the field. It is known that regular communication via pigeon existed in the days of Julius Caesar. During the siege of Paris in 1870 more than a million messages reached Parisians by means of pigeons that had been smuggled out of the city in balloons. Pigeons did such valuable service in both world wars that monuments in their honor were erected in Brussels and in the French city of Lille. In the U.S. some famous pigeon "heroes" were stuffed and mounted after their death; they are on display at the Army Signal Corps Museum and the National Museum.

In 1949 Gustav Kramer and his students at the Max Planck Institute for Marine Biology at Wilhelmshaven in Germany demonstrated that a pigeon in a circular cage with identical food cups at regular intervals around its periphery could easily be trained always to go to a food cup located in a particular direction, for example the northwest, even though the cage was rotated and the visual landscape around it was changed. They found that the pigeon's ability to determine a direction depended on the bird's being able to see the sun. Under a heavy overcast the bird's choice of food cups became random. If the sun's apparent position was altered by mirrors, the pigeon's choice of food cups was correspondingly altered.

It is obvious that if birds can use the sun as a compass to determine directions, they must be able to compensate for the change in the sun's apparent position during the day. In the Northern Hemi-

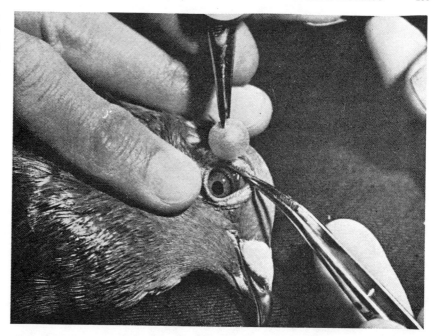

FROSTED CONTACT LENS is placed on a pigeon's eye before a test release. When both of the pigeon's eyes are covered by the lenses, the pigeon is unable to see objects that are more than a few yards away. Control pigeons have clear lenses put on their eyes and are released at the same time from the site. Experiments by Klaus Schmidt-Koenig and H. J. Schlichte of the University of Göttingen, who developed the technique, have demonstrated that pigeons wearing the frosted lenses are able to orient their flight in a homeward direction when they are released at a distant site, and that some pigeons are able to fly back to their home loft. Lenses currently in use are made of a gelatin that dissolves in a few hours.

NIGHT NAVIGATION of pigeons is being studied by Cornell workers. Pigeons with radio transmitters on their back are released and tracked by a radio receiver in the truck. In this time exposure three successive light flashes were used to illuminate the flying pigeon.

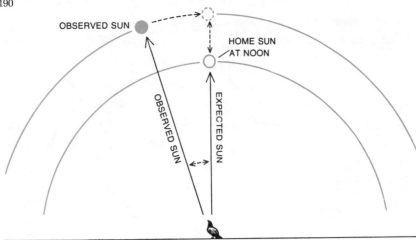

SUN-ARC HYPOTHESIS was proposed by G. V. T. Matthews of the University of Cambridge in the 1950's to explain how pigeons could obtain from the sun alone all the information required to determine their north-south and east-west displacement from home. For example, if a pigeon were released at noon at an unfamiliar site that is southwest of its home loft, the bird would observe the sun's motion and quickly extrapolate along the sun's arc across the sky to the noon position. It would then compare the sun's noon altitude with the remembered altitude at noon at home. Since the bird is south of home, the sun would be higher at the release site, and the pigeon would know that it has to fly north to make the sun appear lower. In order to determine its east-west displacement the pigeon would compare the position of the sun at the release site with the position the sun should have according to the bird's internal clock. In this instance the bird's clock would inform it that the time at home is noon. The sun at the release site, however, is at an altitude lower than that at noon, so that the bird knows it must fly east. By combining the two displacements, the bird would know that it should start flying to the northeast to get to its home loft.

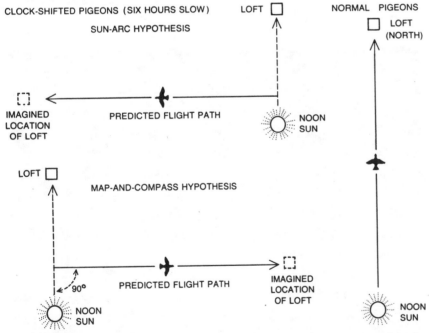

TESTS OF TWO SUN NAVIGATION HYPOTHESES were made with pigeons whose internal clocks had been shifted six hours slow by altering their day-and-night time periods in the laboratory. According to the sun-arc hypothesis, when the clock-shifted pigeons are taken south of their home loft and released at noon, their internal clock tells them that it is 6:00 A.M. at home. They observe that the sun at the release site is too far along its arc for 6:00 A.M. and that they therefore should fly west (*top left*). The alternative map-and-compass hypothesis suggests that the pigeons know where they are relative to home from some kind of map, and that they use the sun only to get compass direction. Their internal clock says it is 6:00 A.M., and they therefore assume that the sun is in the east. Since the sun is really in the south, they should begin flying east, thinking that this direction is north (*bottom left*). When the experiment was carried out, the clock-shifted pigeons flew east, thereby supporting the map-and-compass hypothesis and contradicting the sun-arc hypothesis. Normal pigeons released at the same site departed in the correct homeward direction (*right*).

sphere the sun rises in the east, moves through south at noon and sets in the west. If a pigeon is to determine a particular direction, it cannot simply select a constant angle relative to the sun. It must change the relative angle by about 15 degrees per hour, which is the average rate of change of the sun's position throughout the day. In short, the bird must have an accurate sense of time, an internal clock, and that clock must somehow be coupled with the position of the sun in the sky if an accurate determination of direction from the sun is to be possible.

In a simple but elegant fashion Kramer demonstrated that birds do indeed compensate for time when they are using the sun as a compass. He trained some birds, in this case starlings, to use the sun to go in a particular direction to get to a food cup. He then substituted a stationary light for the sun. The starlings responded to the light as though it were moving at 15 degrees per hour. Since the light was in fact stationary, the bearing taken by the birds shifted approximately 15 degrees per hour.

Klaus Hoffmann, one of Kramer's students, went an important step further in demonstrating the role of the internal clock in sun-compass orientation. He kept starlings for several days in closed rooms where the artificial lights were turned on six hours after sunrise and turned off six hours after sunset. It is known that the internal clocks of most organisms can be shifted to a new rhythm in this manner; the process is very similar to what is experienced by a human being who flies from the U.S. to Europe in a few hours and then takes several days to adjust to European time. When the starlings whose internal clocks had been shifted six hours slow were tested in a circular cage under the real sun, they selected a bearing 90 degrees to the right of the original training direction. Since their internal clocks were a quarter of a day out of phase with sun time, they made a quarter-circle error in their selection of food cups.

Although Kramer and his colleagues had clearly demonstrated that some birds, including pigeons, can use the sun as a compass, their discovery by itself cannot explain how pigeons home. As I have indicated, homing requires more than a compass. If you were taken hundreds of miles away into unfamiliar territory, given only a magnetic compass and told to start walking toward home, you would not be able to get there. Even though you could determine where north was, you would not know where you were with respect to home, hence such compass information would be nearly useless.

In 1953 G. V. T. Matthews, who was then working at the University of Cambridge, suggested that pigeons get far more information than compass bearings from the sun. He hypothesized that the sun gives them all the information they need to carry out true bicoordinate navigation. Stated briefly, on release at a distant site a pigeon would determine its north-south displacement from home by observing the sun's motion along the arc of its path across the sky, extrapolating to the sun's noon position on that arc, measuring the sun's noon altitude and comparing it with the sun's noon altitude at home (as the bird remembered it). If the sun's noon altitude at the release site was lower than it was at home, the bird would know that it was north of home; if the sun was higher than it was at home, the bird would know that it was south of home. To calculate the east-west displacement the bird would determine local sun time by observation of the sun's position on its arc at the release site and compare the local time with home time as indicated by its internal clock. A local time ahead of home time would indicated the bird was east of home; a local time behind home time would indicated the bird was west of home. Thus, according to Matthews, the bird would determine its north-south displacement from the sun's altitude and its east-west displacement by the time difference; combining these data would indicate the homeward direction [*see top illustration on page 190*].

Matthews' sun-arc hypothesis was a major stimulus to further research on pigeon homing, and it formed the basis for many of the experiments conducted in the following decade. Unfortunately, however, nearly all the results of these experiments contradicted the hypothesis, and investigators actively engaged in research on pigeon homing today no longer regard it as being probable. The evidence against the hypothesis is so extensive that most of it cannot be discussed here. For the moment I shall mention only one kind of experiment to help the reader understand some of the more recent research.

Klaus Schmidt-Koenig, another of Kramer's students, showed in 1958 that when pigeons whose clocks have been artificially shifted are released at a distant site, their initial choice of direction is shifted. Their vanishing bearings (the bearings at which they vanish from the

MAGNETIC-FIELD HYPOTHESIS, proposed more than a century ago, had been rejected until recently because earlier experiments failed to show that putting a magnet on a pigeon disorients its homing. Recent tests show, however, that pigeons with bar magnets attached to them are disoriented when they are released at an unfamiliar site under a total overcast but are not disoriented when the sun is visible. Control pigeons with brass bars attached to them show little difference in their mean vanishing bearing under the sun or an overcast. The vanishing bearings of individual pigeons, as determined by an observer with binoculars, are shown by the solid circles. The broken line indicates the true home bearing. The mean vector, or directional tendency, of all the birds in a test group is shown by the arrow. The length of the mean vector is a statistical representation of the degree of agreement among the birds in selecting a direction. Perfect agreement would give a vector length equal to the circle's radius; the more scattered the departing directions, the shorter the vector.

view of an observer using high-power binoculars) deviate from those of normal pigeons by 15 degrees for each hour the birds have been clock-shifted.

Let us examine a test involving clock-shifted pigeons to see whether or not the results agree with what would be predicted by the sun-arc hypothesis. Suppose we shift the birds' internal clock so that it is six hours slow and then release them at noon 100 miles south of their home loft. According to the sun-arc hypothesis, the birds would observe that it is noon at the release site, but their internal clock would tell them it is only 6:00 A.M. at home. They should therefore react as though they were thousands of miles east of home, and they should start flying almost due west. When such an experiment is actually performed,

however, the birds vanish nearly due east, exactly opposite what the sun-arc hypothesis predicts [*see bottom illustration on page 190*].

Is there any way we can make sense of these results? The answer is yes, but to do so we must turn from Matthews' sun-arc hypothesis to an alternative model proposed by Kramer. Kramer emphasized that all the evidence supports the conclusion that pigeons get only compass information from the sun and nothing else. They appear to behave in a manner analogous to a man who uses both a map and a compass, as though they first determine from some kind of map where they are relative to home and in which direction they must fly to get home and then use the sun compass to locate that direction.

Since Kramer could never explain

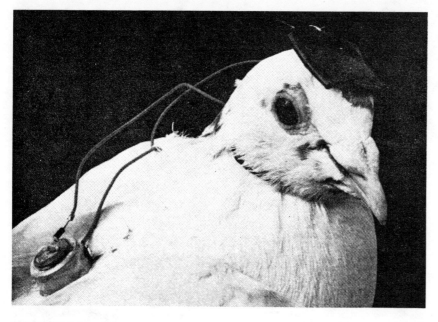

HELMHOLTZ COILS above the pigeon's head and around its neck induce a relatively uniform magnetic field through its head. The coils are powered by a small mercury battery on the bird's back. Direction of the induced field can be reversed simply by reversing the connections of the battery. The strength of the magnetic field can be varied by controlling the amount of current passing through the coils. Battery is exhausted in two or three hours.

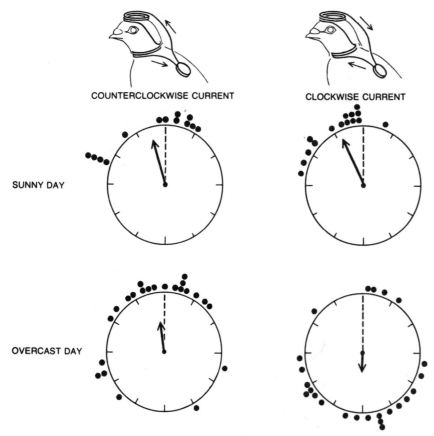

PIGEONS WITH HELMHOLTZ COILS in which the current flows counterclockwise (south-seeking pole of a compass in the induced magnetic field points up) fly almost directly homeward on both sunny and overcast days. When the current in the coils is made to flow clockwise (north-seeking pole of a compass in the induced magnetic field points up), the pigeons still fly homeward on sunny days, but on overcast days they fly almost 180 degrees away from home. These results were obtained in several experiments conducted by Charles Walcott and Robert Green of the State University of New York at Stony Brook.

what the source of the map information might be, let us for the sake of our example pretend that before we release each pigeon we whisper in its ear, "Home is due north." Now the bird must use its sun compass to locate north. Its internal clock says it is 6:00 A.M., when the sun should be in the east; hence north should be approximately 90 degrees counterclockwise from the sun. Remember, however, that the bird's clock is six hours slow; it is actually noon, when the sun is in the south. Hence the bird's choice of a bearing 90 degrees counterclockwise from the sun sends it east, not north. We can summarize by saying that no matter what combination of directions and clock-shift we use in actual experiments, the results come out consistent with the predictions of Kramer's map-and-compass model and not with Matthews' sun-arc hypothesis.

Because the pigeon's use of the sun compass in orientation was the one thing that was firmly established, there was a tendency in the 1960's for many investigators to assume that the sun is essential for homeward orientation at an unfamiliar release site. Several discrepancies, however, led me and my colleagues at Cornell University to doubt it. First, I knew of numerous instances of fast pigeon races under heavy overcast. Second, our pigeons seemed to perform well under overcast if they had first been made to fly in the vicinity of the home loft in rainy weather. Third, the published evidence that pigeons were disoriented under heavy overcast was not entirely consistent, and fourth, we and others had been able to get pigeons to home at night.

We set out to reexamine the importance of the sun in pigeon navigation. In our most important experiments we too used clock-shifted pigeons. As we expected, when pigeons whose internal clocks had been shifted six hours fast or slow were released under sunny conditions, their vanishing bearings were roughly 90 degrees to the right or left of the vanishing bearings of control pigeons whose internal clocks had not been altered. When the pigeons were released in total overcast, however, the results were quite different: both the clock-shifted birds and the control birds vanished toward home and there was no significant difference in their bearings. This was true even when the release site was completely unfamiliar to the pigeons.

These results led us to several conclusions: (1) Pigeons accustomed to flying in inclement weather are able to orient homeward under total overcast.

Since there is no difference between the bearings of control and clock-shifted birds under such conditions, it is clear they are not able to see the sun through the clouds and hence are no longer using the sun compass. (2) There must be redundancy in the pigeons' navigation system. They use the sun compass when it is available, but they can substitute information from other sources when it is not. (3) The alternative information used in lieu of the sun compass does not require time compensation. (4) The alternative system cannot be pilotage by familiar landmarks, because pigeons can correctly orient themselves homeward under overcast even in distant, unfamiliar territory.

Recognition of the fact that pigeons are able to use alternative cues, depending on the circumstances involved, meant that the results of many older experiments could no longer be accepted at face value. For example, if an experimenter altered cue A while keeping other conditions optimum, and if the pigeons continued to orient well, they may simply have used cue B as an alternative to A. Similarly, if B was altered while everything else was kept at an optimum and the birds oriented well, they may have used A as an alternative to B. In short, we would have been wrong if we had concluded from these experiments that cues A and B are not elements in the pigeons' navigation system. In fact, such experiments show only that neither A nor B alone is essential for proper orientation under the particular test conditions.

This kind of reasoning led us to conduct experiments in which we varied several possible orientational cues simultaneously, on the assumption that if we could interfere with enough of them at the same time we could hope to learn which cues are more important and how they interact with one another.

We chose first to look again at the old idea that birds might obtain directional information from the earth's magnetic field. Although this hypothesis had been known for more than a century, there was no evidence for it and much experimental evidence against it. Nonetheless, it seemed worth reexamining. And it was! When we repeated older experiments of putting bar magnets on pigeons to distort the magnetic field around them, we found, as had others before us, that the birds had no difficulty orienting on sunny days. When the test releases were conducted on totally overcast days, however, the birds carrying

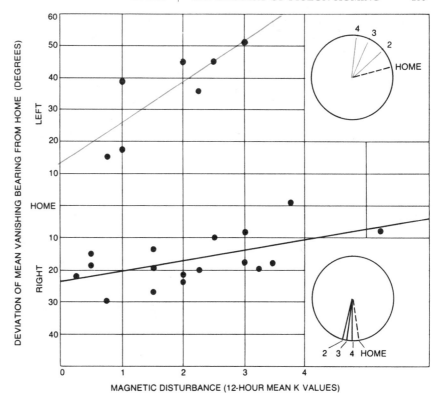

DISTURBANCES OF THE EARTH'S MAGNETIC FIELD caused by solar activity appear to affect a pigeon's initial choice of bearing when it is released at a distant site under sunny conditions. The K-index scale is used to indicate the degree of magnetic activity, ranging from quiet (less than 2) to a major magnetic storm (6 or more). In 1972 a series of releases of Cornell University pigeons from a site 45 miles north of the home lofts revealed that, as the degree of magnetic disturbance increases, the vanishing bearing of the birds steadily shifts to the left as seen by an observer facing homeward (*black curve*). At this release site the shift to the left brought the birds' vanishing bearings closer to the true home bearing, but success in homing was not improved. In another series of tests, pigeons from a different loft were released from a site west of their home. A similar leftward shift of vanishing bearings with increasing magnetic disturbance was found (*colored curve*). In this instance the shift to the left caused the vanishing bearings to recede away from the true home bearing.

magnets usually vanished randomly whereas control birds carrying brass bars of the same size and weight vanished toward home. Several other workers have since repeated these experiments, with the same results.

More recently Charles Walcott of the State University of New York at Stony Brook and his student Robert Green have gone one step further. Instead of working with bar magnets, they put a small Helmholtz coil on the pigeon's head like a cap and another coil around its neck like a collar. Power is supplied from a battery on the bird's back. This device makes it possible to induce a relatively uniform magnetic field through the bird's head. The direction of the induced magnetic field can be made to point up through the bird's head or to point down simply by hooking up the battery to make the current in the coils flow clockwise or counterclockwise. Under sunny conditions Walcott and Green found that the direction of the induced magnetic

field did not affect the pigeon's ability to orient homeward. Under total overcast, however, the direction of the induced magnetic field had a dramatic effect: when the north-seeking pole of a compass in the induced field pointed up, the pigeons flew almost directly away from home, whereas when the south-seeking pole of a compass in the induced field pointed up, the pigeons oriented toward home.

Our results, together with Walcott's, suggest that magnetic information may play a role in the pigeon navigation system. This is consistent with the recent discovery by Freidrich Merkel and Wolfgang Wiltschko of the University of Frankfurt that European robins in circular cages can use magnetic cues to orient themselves in a particular direction. William Southern of Northern Illinois University also has reported that the orientation of ring-billed gulls is influenced by magnetic activity.

Recently Martin Lindauer and Herman Martin of the University of Frankfurt have demonstrated that honeybees give orientational responses to magnetic cues several thousand times weaker than the earth's field. Only a few years ago biologists were debating whether or not any organism could detect a magnetic field as weak as the earth's (approximately half a gauss). The responses of honeybees to magnetic cues now makes us wonder if one gamma (10^{-5} gauss) will not prove to be the lower limit. Indeed, a study that my colleagues and I have recently conducted suggests that the magnetic-detection sensitivity of pigeons may rival that of honeybees. In four long series of tests over a period of three years we have found that fluctuations of less than 100 gamma (and probably less than 40 gamma) in the earth's magnetic field, caused by solar flares and sunspots, appear to have a small but significant effect on the pigeons' choice of an initial bearing at the release site.

The question of how organisms detect magnetic stimuli is unanswered. We have very little idea what a magnetic sense organ should look like, or even where in the body we should expect to find it. Since magnetic flux can pass freely through living tissue, magnetic detectors might be anywhere inside the body. The search for these detectors has already begun in our laboratory and in others throughout the world. It promises to be a challenging undertaking.

Exciting as the discovery that magnetism plays a part in avian navigation systems may be, we are in a sense back where we started. The weight of the evidence at present suggests that magnetism simply provides a second compass, not the long-sought map. Hence we must continue our search. What other sources of information might the birds have?

One possibility that comes readily to mind in this age of long-range rocketry is that the birds might be capable of inertial guidance, that they might somehow detect and record all the angular accelerations of the outward journey to the release site, then double-integrate them to determine the direction home. Intriguing as this possibility is, all the evidence is against it. Pigeons have been carried to release sites while riding on turntables or in rotating drums, yet this input of additional inertial "noise" has no effect; the birds orient homeward as accurately as control birds not so treated. Other pigeons that were carried to the release site while they were under deep anesthesia were able to determine the direction home with no difficulty. Pigeons with a variety of surgical lesions of the semicircular canals—the principal detectors of acceleration in vertebrates—orient themselves accurately, whether they are tested under sunny conditions or overcast ones.

The hypothesis that pigeons may be able to use olfactory information in navigating has been advocated by Floriano Papi and his colleagues at the University of Pisa. The probability that this is the case seems low in view of the relatively poor development of the pigeon's olfactory system. Nonetheless Papi has some interesting experimental results, and it is too early to make a judgment on his proposal. We are currently conducting experiments to test his ideas.

By now the reader may well be wondering why so little has been said about what might seem the most obvious possible cue for homing pigeons: familiar landmarks. The reason is that there is abundant evidence that landmarks play a very small role in the homing process. In the course of tracking pigeons by airplane Walcott and his colleague Martin Michener have repeatedly noted that when pigeons flying on an incorrect course encounter an area over which they have recently flown, they seldom give any indication of recognizing the familiar territory. Several other investigators, including members of our group, have found that pigeons clock-shifted six hours and released less than a mile from

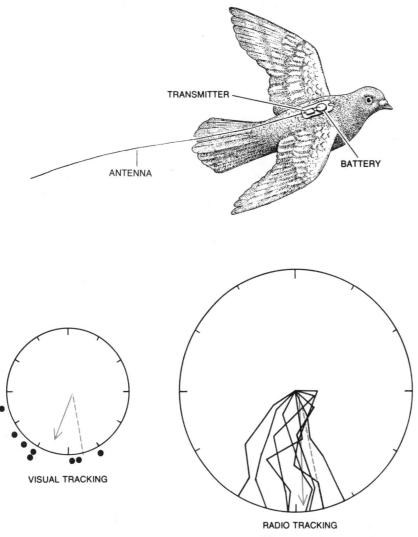

RADIO TRACKING OF PIGEONS is carried out from a receiver on the ground at the release site or sometimes by a receiver on an aircraft. An FM transmitter and a battery are glued on the pigeon's back (*top*). The 19-inch antenna trails behind the bird as it flies. Data from radio tracking reveal that pigeons do not continue to fly in a straight line after they leave a release site but frequently alter their course. The vanishing bearings for eight pigeons, as determined by observers with binoculars, and the bearings determined by simultaneous radio tracking from the site are compared (*bottom*). The scale of the two circles is arbitrary. Visual tracking extends to one or two miles, depending on the flying height of the bird. Radio tracking extends to eight miles or more. Broken line indicates home direction.

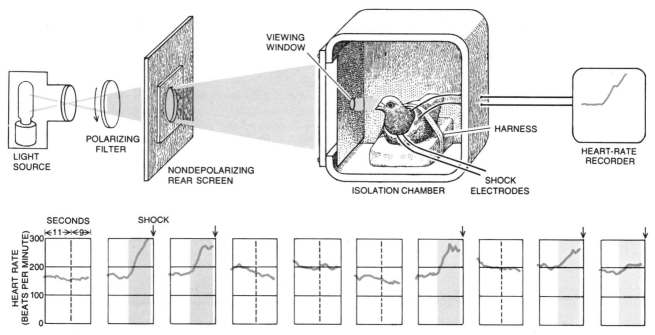

PIGEON PERCEPTION OF POLARIZED LIGHT is being tested in the author's laboratory by Melvin L. Kreithen. After electrodes for administering electric shock and for detecting the heart rate are attached, the pigeon is put in a harness and then is placed in a sealed soundproof chamber. Light is projected through a polarizing filter on a rotating mount and then through a nondepolarizing rear-projection screen. Part of the light enters the isolation chamber through a small window and falls on the pigeon's eye. The light comes on at random intervals. In some trials, which are also determined at random, the polarizing filter starts to rotate after 11 seconds; in others it does not. When the filter rotates, the pigeon receives a shock at the end of the light signal. When the filter does not rotate, no shock is given. After a number of trials the pigeon's heart rate begins to rise rapidly at the beginning of the rotation of the polarizing filter, indicating that the bird is able to sense the change in the plane of polarized light and is anticipating the shock that is to come. Recordings from a series of tests of a pigeon are shown (*bottom*). The colored block indicates the interval during which the polarizing filter rotates. In control runs during the corresponding interval (*to right of broken line*) no rotation occurs.

home, in territory over which they have flown daily during their exercise period, often vanish 90 degrees away from the home direction. Only a direct view of the loft building itself takes precedence over what their navigation system is telling them; nearby buildings or trees apparently do not serve as reference points under these conditions. In fact, even a view of the loft is not always effective, particularly at distances of a mile or more.

Perhaps most convincing of all are experiments conducted by Schmidt-Koenig and H. J. Schlichte of the University of Göttingen. They put frosted contact lenses over the eyes of pigeons, thus making it impossible for the birds to see any object that is more than a few meters away. Not only do these pigeons orient homeward when they are released as far as 80 miles away but also a surprising number of them actually get home. Schmidt-Koenig conducted some of his experiments at our Cornell lofts, and thus I had the opportunity of observing them at first hand. It was a remarkable experience. The birds arrived very high overhead and fluttered down to a landing in the fields around the loft.

Being unable to see the loft, they waited for us to pick them up and carry them the last few feet. These results suggest that the pigeon navigation system is often accurate enough to pinpoint the home location almost exactly without reliance on familiar landmarks; vision is necessary only for the final approach, frequently at a distance of less than 200 yards.

It will be apparent from all I have said that the task of uncovering the pigeon's navigation system is going to be a difficult one. The old idea that birds use a single method to determine the home direction has given way to the realization that there are probably multiple components in the system and that these components may be combined in a variety of ways, depending on such factors as weather conditions, the age of the bird and the bird's experience.

One approach that holds much promise for helping us tease apart the many elements in the system is the study of the ontogeny of navigational behavior. For example, we have found that bar magnets disrupt the initial orientation of very young pigeons released away from home for the first time in their life, even when the sun is visible. Moreover, normal first-

flight youngsters cannot orient under total overcast, even if they have previously been released for exercise in inclement weather. It seems, then, that inexperienced homing pigeons need both sun information and magnetic information. Various other manipulations that have little effect on experienced birds also disorient first-flight pigeons.

Perhaps with experience a pigeon learns to orient accurately with less information. Or perhaps experience is necessary to enable the pigeons to settle on a weighting scheme that allows them to decide what to do when they get conflicting information from different sources. Early results from some current experiments indicate that by training very young pigeons under conditions in which we severely restrict, or eliminate altogether, certain normally important environmental cues such as the sun, we may be able to induce the birds to settle on weighting schemes for evaluating directional cues that are quite different from normal. The availability of such birds would greatly facilitate the carrying out of experiments designed to clarify the roles of cues that are normally difficult to alter.

Another approach we are actively pursuing is an attempt to learn more about the sensory capabilities of pigeons. The more we learn, the more we become convinced that birds live in a sensory world very different from our own. For example, my student Melvin L. Kreithen has recently demonstrated that pigeons are remarkably sensitive to tiny changes in barometric pressure. Such a sensitivity might enable pigeons to get navigationally useful information from pressure patterns in the atmosphere. Kreithen and I have also recently obtained experimental evidence that pigeons can detect the plane of polarized light, which might

mean that pigeons, like honeybees, can continue using the sun as a compass on partially overcast days, when the sun's disk is hidden from sight but some blue sky remains visible.

It has long been known that the bearings chosen by pigeons at distant release sites, although roughly in the homeward direction, are almost never oriented directly toward home. Moreover, the mean bearings of repeated releases at any given site usually show a consistent deviation from home; there is, in effect, a relatively stable "release-site bias" that is characteristic of each location. At some

sites the bias is apparent only in the vanishing bearings obtained by visual tracking. The bias often becomes less marked when the final-contact bearings obtained by radio tracking are used, but at some sites the bias is still manifest when the birds move out of radio range (between six and 10 miles from the release site). In the hope that these biases might prove to be a key to local geophysical factors that could provide at least part of the map information for pigeons, we chose for intensive study several release sites where the biases were unusually large.

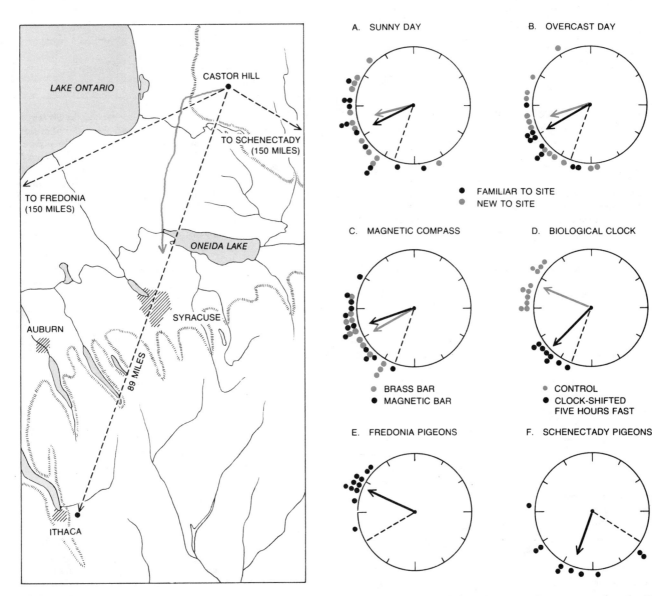

CASTOR HILL is approximately 89 miles northeast of the Cornell University pigeon lofts in Ithaca, N.Y. At Castor Hill the Cornell pigeons regularly choose an initial bearing that deviates clockwise from the true home direction. This occurs both on sunny days and on overcast days (A and B); thus the characteristic bias must not depend on the sun compass. The bias probably is not due primarily to the magnetic compass either, since under sunny conditions birds with bar magnets have about the same vanishing bearing as pigeons with brass bars (C). Pigeons whose internal

clocks have been shifted five hours fast choose a more homeward direction than normal pigeons, but the clock-shifted pigeons are less successful at getting home (D). Pigeons from another loft at Fredonia, N.Y., 150 miles west of Ithaca (E), and pigeons from Schenectady, N.Y., 150 miles east of Ithaca (F), choose bearings that also deviate clockwise from the true home bearing. It appears that the bias is a function of the site and not of the pigeons. The actual flight path of normal Cornell pigeons released at Castor Hill and tracked by an airplane is shown by the colored line on the map.

One such site is the Castor Hill Fire Tower, located 89 miles north northeast of our Cornell lofts. Here our pigeons regularly depart with a mean bearing that deviates roughly 60 degrees clockwise from the direction of home [*see illustration on opposite page*]. In a long series of experiments we have found that this clockwise bias is evident not only with experienced pigeons new to the site but also with pigeons that have been released at the site before. The same bias is found in very young pigeons on their first homing flight. It is found on both sunny and overcast days, so that it apparently has nothing to do with the sun compass, and it is found in pigeons wearing magnets, so that it probably has nothing to do with the magnetic compass either. It is even found when the pigeons are wearing frosted contact lenses; hence it must not depend on anything the pigeons see.

Wild bank swallows captured near Cornell and released at Castor Hill showed the same clockwise bias, indicating that the biasing factor, whatever it may be, affects other bird species in the same way. Pigeons borrowed from lofts 150 miles east and west of Cornell and released at Castor Hill show a similar clockwise departure bias relative to their home. Finally, pigeons clock-shifted five hours fast depart nearly straight toward home from Castor Hill but nonetheless have poorer homing success than control pigeons that depart with the usual 60-degree bias. In a joint experiment with Walcott, normal pigeons with radio transmitters attached to them were tracked by an airplane after their release from Castor Hill. We found that the birds turn onto a more homeward course when they are approximately 14 to 18 miles west of Castor Hill. It may be that the clock-shifted birds that have

poor homing success make a corresponding turn when they are a similar distance from the release site and thus become directed away from home. We hope soon to learn if this is so.

We conclude, then, that the bias in the birds' initial bearings is not a biological error, that it is due not to some peculiarity of the birds but to a peculiarity of the location. The birds are probably reading the map cues correctly but the map itself is twisted clockwise at Castor Hill. Perhaps if we can learn what geophysical factors are responsible for this distortion of the map we will finally be on the way to understanding the ancient mystery of how pigeons home.

VI

VERTEBRATES IN THEIR HABITATS

The crocodile belongs to the Nile; it is a curse on four legs, and equally pernicious on land and in the river. It is the only land animal not furnished with a tongue and the only one that bites by pressing down the mobile upper jaw, and it is also formidable because of its row of teeth set close together like a comb. In size it usually exceeds 18 ells. It lays as many eggs as a goose, and by a kind of prophetic instinct incubates them always outside the line to which the Nile in that year is going to rise at full flood. . . . This creature, when sated with a meal of fish and sunk deep in sleep on the shore with its mouth always full of food, is tempted by a small bird to open its mouth wide to enable the bird to feed; and first it hops in and cleans out the mouth, and then the teeth and inner throat also, which yawns open wide as possible for the pleasure of this scratching.

Pliny
NATURAL HISTORY, VIII, xxxvii

VERTEBRATES IN THEIR HABITATS VI

INTRODUCTION

In previous Sections of this book our emphasis has been upon particular organ systems, modes of behavior, or adaptations. Here, we shall consider five types of animals from many points of view. The emphasis is upon integration—how a variety of morphological, physiological, and behavioral attributes fit together to permit a given life style.

An important lesson comes from actually seeing the way in which an organism such as a crocodile behaves with respect to its eggs and young. This subject is dealt with in "The Nile Crocodile," by Anthony C. Pooley and Carl Gans. Carrying the hatchlings in the mouth is not predictable from what we see in terms of bodily structure or physiology. Similarly, the cooperative transportation of large food would not be guessed. These sorts of observations and conclusions are of particular interest in view of the fact that crocodiles are one of the few animals that enable us to obtain any real hint about how living dinosaurs might have functioned. We must not forget, however, that it is only *hints* we obtain. For example, it has been observed that monotremes such as echidnas only generate extra heat by muscular means, not by increasing the rate of nonshivering thermogenesis. Actual measurements of living animals had to be made before this significant conclusion could be reached. (This is described in the Introduction to Section IV of *Vertebrates: Physiology*, the companion volume to this book.) These considerations should make us cautious indeed, therefore, in attempting to assign physiological properties or particular behaviors to extinct creatures such as dinosaurs. The same sort of cautions should be applied to the sweeping generalizations offered by some paleontologists about behavior, hunting abilities, and intelligence of the early relatives of humans (*Australopithecus*, *Homo habilis*, and so on) whenever chipped stone implements are found in the same geological strata as the fossils.

The same lessons can be applied to kangaroos (see "Kangaroos," by T. J. Dawson). Who would guess, for instance, that the peculiar, hopping gait of such animals is more efficient at high speeds than is the running or galloping of other quadrupedal mammals? The marsupial reproductive scheme is an intriguing variation upon that of placental mammals. Although implantation of the embryo in the wall of the uterus does occur, and a true placenta forms, the time spent *in utero* is relatively short; the offspring are born precociously, and then must continue a long developmental period while being supported by the mammary glands in the mother's pouch. In contrast, Eutherian placental embryos reside in the uterus for much longer relative times, are typically more mature at birth, and may not be dependent upon the mother's milk for so long. The reason for the precocious birth of the marsupial's young is not known, though the possibility has been suggested that the embryo's barriers to maternal immune attack are ineffective.

In "The Archer Fish," by K. H. Lüling, we see a specialized mode of feeding that depends upon morphological and behavioral adaptations. The mouth parts of such fish are modified as a kind of "water pistol," and the corneas of the eyes permit vision in both air and water. A number of fish have developed "split" visual systems of this sort. The differing refractive indexes of air and water require a means of compensation in the eye's focusing system if an image is to be focused on the neural retina of the eye. Human divers employ equivalent compensating devices, as is explained in "The Diving Women of Korea and Japan," by Suk Ki Hong and Hermann Rahn, in *Vertebrates: Physiology,* edited by Norman K. Wessells, W. H. Freeman and Company, 1980, *Scientific American* Offprint 1072. The women divers use special goggles when underwater; these ensure that air in the goggle is next to the cornea so that the human cornea and lens may function normally. When a fish views objects in water, the fish uses its spherical lenses for focusing. Fish have evolved two strategies for viewing objects in air; these are flattened lenses or flattened corneas. In flying fish and some tropical surf fish that spend some time in the air (fish of the genus *Mnierpes,* for example), flattened segments of the cornea reduce bending of light rays, so that the moveable lens can compensate sufficiently to allow focus to be achieved. In mudskippers, or "four-eyed fish," on the other hand, the lens itself is no longer spherical but is flattened to varying degrees, so that its movement permits effective vision in the air. The genetic and developmental plasticity of the vertebrates has allowed this basic physical problem to be solved independently on a number of occasions.

"The Weddell Seal," by Gerald L. Kooyman also presents evidence of anatomical and behavioral adaptations that permit life in a habitat seemingly inhospitable for a mammal, the dark reaches beneath ice sheets in the Antarctic. The most anterior teeth of this seal—the incisors and canines—are set at an angle that permits them to be used for cutting ice; coupled to an extraordinarily wide gape of the jaws, this permits the seal to reopen frozen airholes in the ice sheet from below, and so avoid drowning. What is still not clear is the means by which Weddell seals find their air holes, inasmuch as a dive beneath the ice sheet is made during the Antarctic winter darkness. Very little light from stars and moon can penetrate the ice sheet, yet the seals are able to relocate their holes. Since a variety of sounds (including "clicks" somewhat like those of dolphins) have been heard or recorded in the vicinity of Weddell seals swimming beneath the ice, it seems reasonable to speculate that these creatures may have a capability for echo-location, and that this may be their means of finding an air hole in the Antarctic darkness.

The final article of this Section deals primarily with social and behavioral aspects of dolphins, and secondarily with the physiological bases of those phenomena. Since Bernd Würsig, the author of "Dolphins," does not discuss mechanisms of sound generation and perception by dolphins, a few words may be useful as background.

Dolphins emit two types of sound: high-frequency pulses that sound like "clicks" to us, and whistling noises reminiscent of canary "chirps." Clicks are apparently generated by passing air from one nasal sac to the other through specially constructed valves. Whistles (and low frequency "moans" made by whales) are believed to emanate from the larynx, the regular mammalian voice box. As mentioned in "The Head of the Sperm Whale" in Section III, some sounds may be processed, altered, and focused by special structures such as the spermaceti organ. It is noteworthy that no air is lost from the bodies of dolphins and whales when they make their various noises. Clicks are echo-locating pulses, and just as a bat increases the pulse repetition rate when a flying insect is detected, so a dolphin increases its rate of pulsing when a fish is thrown into the water in front of it. Here the rate may rise from five to several hundred pulses per second. The frequency of sound (200 hertz to 200 kilohertz) emitted with each pulse is complex, and it seems likely that low frequencies (sometimes coupled with high intensity and long duration) are used for long-range general assessment of the environment, whereas brief, high-frequency ultrasonic pulses are of use in short-range discrimination. The low frequency clicks are also emitted during communication between dolphins.

Dolphin whistles are thought to be used both in communication and in echo-location. Each emission of a whistle lasts for longer periods (for as much as half a second, versus milliseconds for a click), and the frequency of sound rises from 4 to 20 kilohertz as the whistle continues. Winthrop N. Kellogg has proposed that such sounds may be used as a form of frequency-modulated echo-location, in contrast to the pulse-modulation of clicking. The constantly changing emission of the whistle would produce echoes of constantly changing frequency, so that any one echo would be of a frequency different from that of the sound being emitted. Thus the echoes could be heard even though sound was still being generated. It is, however, for communication that whistles are most common. The complexity of whistles between dolphins increases in a variety of situations where we, as observers, would expect communication to occur. Communication whistles may be a simpler form of the diversity of sounds made by whales or dolphins (sounds described as mewing, mooing, squeaking, roaring, trilling, barking, and howling).

Directionality in echo-location has not yet been studied extensively, but two interesting mechanisms are known. Some dolphins can emit highly directional clicks, with most of the sound energy being directed in a narrow cone directly in front of the animal. The other form of behavior is seen in the bottlenose dolphin as it swims toward a small target in the water: the animal's head moves back and forth in the water once every 2 to 3 seconds (the head moves about 5 degrees on each side of the axis of swimming movement). These head oscillations are obviously a form of auditory scanning, with the ears being moved back and forth to establish the source of the echo. As the echoes reach one ear and then the other, phase differences, intensity differences, and time differences can be detected by moving the head in this manner. Thus, the bottlenose dolphin has arrived at the same basic solution to locating its target in space as have certain bats.

The ears of cetaceans are truly remarkable (see Figure 1). Because the ear is necessarily streamlined into the head, there is no external ear. Studies by Bullock and Ridway have shown that the lower jaw region of the porpoise body is the most sensitive to sound. By comparison, the external ear region itself is quite insensitive. It is likely that echoes strike the mandible and are transmitted through a fat-filled canal to the middle ear where the normal hearing process can go on. The fats present in the outer ear are unique triacylglycerols not found in the blubber elsewhere in the dolphin's body. Because they have a very low density, sound travels more slowly through them, so that the canal containing these lipids acts like a sound lens to focus echoes upon the internal ear structures.

In order for such an arrangement to be useful in directional hearing, it is essential that sound-induced vibrations do not reach the middle ear by other

Figure 1. **A diagrammatic section through the ear region of a dolphin. The internal ear parts are located in the tympanic cavity enclosed by the ear bones. "Foamy" material insulates the bones from the skull.** [After Reysenbach deHaan, *Acta Oto-Laryngologica*, Supplementum 134, 1957.]

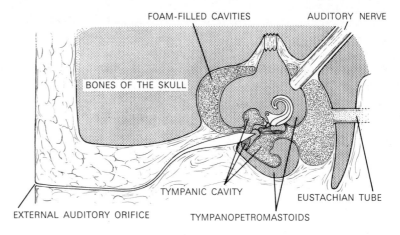

FOAM-FILLED CAVITIES   AUDITORY NERVE

BONES OF THE SKULL

TYMPANIC CAVITY   EUSTACHIAN TUBE

EXTERNAL AUDITORY ORIFICE   TYMPANOPETROMASTOIDS

routes. The secret may lie in the fact that the middle and inner ears are encased in special massive bones called the tympanopetromastoids. In contrast to the structures in most terrestrial mammals, these bones are not attached directly to the skull or to other body skeletal parts. Consequently, they do not vibrate when the rest of the skeleton does. In addition, the bones are so massive that they probably do not vibrate in response to sound waves over 150 hertz. Finally, a gelatinous "foam" insulates the internal ear parts from the ear bones themselves; this air-filled foam probably reflects and absorbs most of the sound energy that succeeds in passing through the ear bones. The result of all of these structural adaptations is to insulate the ear effectively from vibrations in all directions but one—from each side of the lower jaw via the fat-lined canals. This means that sound is detected from a circumscribed region in front of and below the animal and that echo-location of objects in the water is feasible.

In cetaceans, the auditory nerve leading from the cochlea of the ear to the brain is very large, and the lateral lobes of the cerebral cortex are huge. This is a major reason why the large complex brain of a dolphin has a width greater than its length. In humans, for whom sight is primary, or in a dog, for whom smell is so important, expansion of the frontal or olfactory centers causes the brain to be lengthened in the anterior-posterior direction. The brain of the bottlenose dolphin is an example of the opposite extreme, for the olfactory lobes and tracts are missing altogether, and the brain is short and wide. The degree of specialization of different portions of the vertebrate brain is emphasized by Bullock, who points out that the inferior colliculus alone of a porpoise (the area where "click" echoes are processed) is larger than the *whole* brain of a bat and is almost as large itself as some small bats. Why so many nerve cells are present in this area of a porpoise brain is a mystery of evolution and physiology.

How else, aside from catching food, do cetaceans use their echolocation abilities? One example is observed in whales during annual migrations. As gray whales migrate from the Arctic Ocean to Baja California each year, they tend to travel fairly close to the shore. Many other whales also remain above continental shelves for portions of their long migrations (see Figure 2). Typical echo-location signals have been recorded for most of them, and it is supposed that in regions of shallow water they use their sound systems as a fathometer, or depth-gauging device. In regions of deep oceanic trenches, it is common to see whales surfacing repeatedly, coming high out of the water, almost to the level of their fins. The biologist Kenneth Norris has proposed that they use eyesight at such times to guide them across the deep water to areas where depth detection is once more possible. Since in the course of most whale migrations extended periods of deep-water cruising take place, it seems likely that celestial navigation or some other undefined orienting device may be used. In effect, we can assume that the sensory system best suited for particular circumstances will be employed, and that echo-location is utilized in shallow-water swimming.

As Würsig outlines in his article, dolphins use sound for communication. Perhaps the most unexpected observations about whale sound were made by Roger Paine, who recorded the "songs" of humpback whales in the vicinity of Bermuda. What is so very astonishing is the stereotyped, repetitious nature of the songs. An individual song may last from 7 to 30 minutes and include a fantastic array of frequencies, modulations, and intervals between notes and "themes." However, when the song is done, it is begun over again and is repeated, sound by sound, theme by theme. Thus, despite the undeniable complexity of the sounds, and resultant capacity for great information content, the stereotyped character of the songs raises deep and completely unanswered questions about the significance of whistle-type communication in these marine mammals.

Figure 2. The main migratory routes of humpback whales in the southern hemisphere. Note the prevalence of coastal routes for long portions of the journeys. Many whales spend winter in the tropics where they breed. They migrate to the Antarctic and its abundant food supply for the summer and then return to the tropics where the young are born (often in shallow coastal waters). The young are weaned the summer following the return journey to the Antarctic. [After E. J. Slijper, *Whales*, Basic Books, 1962.]

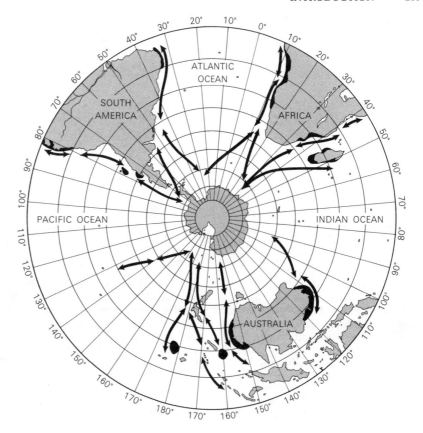

Whales, dolphins, and bats are not the only vertebrates that can use echoes or orientation. It has been known for many years that blind human beings have "facial vision," that is, an ability to detect objects in their vicinity, presumably by sensing differences in air pressure on the face. In fact, this ability is dependent upon the sense of hearing, so that plugging the ears causes disorientation of the subject. When blind people were allowed to select noises that they found most effective at producing echoes (for an "awareness" of objects in the environment), all chose forms of "hissing" or tongue "clicking." The skill in detecting objects in the environment varies from person to person; a typical subject identified a 7.1-centimeter target, located 61 centimeters from his face, 100 percent of the time, and a 30.5-centimeter target, 274 centimeters away, 95 percent of the time. The frequency of correct identification diminishes with decreasing target size, no matter what the distance is.

The sounds emitted by these subjects varied in frequency from 170 hertz to 16 kilohertz and even the clicks were of relatively long duration (0.025 seconds). This meant that echoes were returning before the sound production and emission were completed. Clearly, this learned adaptation of atypical human beings does not equal the highly evolved system of a bat.

Do these human subjects have good directional sensing ability? The experiments showed that some people could emit clicks and localize targets with the use of only one ear. With both ears uncovered and sound directed from the mouth straight in front of the face, these subjects could detect targets as much as 90 degrees to the right or the left and point them out with good accuracy. All of these results attest to the adaptability of the human organism in employing substitute sensory mechanisms when normal primary orientation by vision is not possible.

19

The Nile Crocodile

by Anthony C. Pooley and Carl Gans
April 1976

Found both in the Nile and in most other African rivers and lakes, it has a remarkably large repertory of social behavior that includes parental protection of the young

The largest of the living reptiles belong to the order Crocodilia. There are some 20 members of the order; for most people the best-known are the alligator of the New World and the Nile crocodile of the Old World. The Nile crocodile (*Crocodylus niloticus*) is now scarce in the lower Nile, but it is common in the upper waters of the river and in the other rivers and lakes of tropical and southern Africa, from the Sudan and Somalia southward well into South Africa and up the west coast of Africa as far as Senegal. Its tolerance of salt water has allowed it to occupy the coastal waters, and it has colonized not only the continent's offshore islands but also the great island of Madagascar, 250 miles out in the Indian Ocean. The same tolerance once extended the Nile crocodile's range along the Mediterranean coast as far west as Tunis and as far north as Syria.

The extent of this range, which overlaps the ranges of three smaller African crocodiles (the sharp-nosed *Crocodylus cataphractus* and two species of a dwarf genus, *Osteolaemus*), is an indication of the outstanding biological success of the Nile crocodile. Until a decade or so ago, however, this success was difficult to explain on the basis of the animal's recorded natural history. The consensus of incidental observations made by hunters, missionaries and an occasional naturalist was that the Nile crocodile was a lethargic hulk that spent most of its life basking in the sun, now and again rousing itself long enough to attack an unwary animal at the water's edge. Today, thanks to the systematic observation of individual crocodiles and of crocodile populations over periods of months and even years, we know that the animal is a remarkably efficient predator with a life cycle that includes substantial social components. The Nile crocodile often hunts cooperatively and has been seen to divide its prey. The adults are at least seasonally monogamous, and they protect their young by means of a complex repertory of behavior. The animals' adaptations to their water's-edge environment suggest why it is that crocodilians are found in all tropical and subtropical regions and how it is that the Nile crocodile has come to occupy such a large territory.

The adaptations that account for the success of the Nile crocodile are structural, physiological and behavioral. To deal with structure and physiology first, *C. niloticus* is among the largest of the crocodilians. Adults weighing 1,000 kilograms have been reported. The maximum recorded length, from the end of the snout to the tip of the tail, is five meters. Among reptiles only a few sea turtles and the estuarine crocodile, *C. porosus*, are heavier than the Nile crocodile, and only a few snakes and the estuarine crocodile are longer.

The most impressive part of the crocodile is its head. Between half and two-thirds of its length consists of the animal's elongated jaws. They bear a single row of stout, conical and slightly curved teeth that do not meet but interdigitate when the jaws are closed. The teeth of the upper jaw generally lie slightly to the outside of the teeth of the lower jaw, so that they are visible when the jaws close; the lower teeth fit into sockets in the upper jaw. In crocodiles, but not in alligators or caimans, an enlarged fourth lower tooth fits into a notch in the upper jaw rather than into a socket; it therefore remains visible when the jaws are closed and is a distinctive feature of the crocodile's appearance.

The crocodile's tongue is wide; it can be lifted from the floor of the mouth and its motion can be delicately controlled, but it cannot be protruded. The crocodile's mouth can be completely isolated from its pharynx by the gular fold, a tissue that can rise from the floor of the mouth to overlap a bony fold on the roof of the mouth. A crocodile is therefore able to keep its mouth open in the water without flooding its lungs and its alimentary canal. It can also breathe when its mouth is filled with water because its valved nostrils, which are on a raised area at the extreme tip of its snout, open into long tubes that lead not to its oral cavity but beyond the gular fold to a cup-shaped cavity in the roof of the pharynx [*see bottom illustration on page 212*]. The valves of the nostrils prevent flooding of the nasal tubes when the animal is submerged.

The crocodile is unique among reptiles in having a series of sensitive pressure receptors between its teeth and its jaws. Pressure receptors of this kind are found in mammals, including man; they gauge the intensity of our bite. If they provide crocodiles with the same kind of information, it could help to explain a number of the animals' remarkable patterns of behavior.

The eyes of a crocodile have bony eyelids; after the lids are closed they can be retracted, thereby streamlining the animal's head. When the eyes are open, they provide excellent fields of vision both to the sides and to the front, where an overlap of the fields may result in binocular depth perception. Immediately behind the eyes flaps of skin cover the oblong ear openings. Both anatomical and physiological studies suggest that crocodilians in general have superior sound perception in the range from about 100 to 4,000 hertz (cycles per second). The crocodile's sense of smell is also advanced. Apparently it can detect odors by flushing air through its nasal tubes; each sniff carries a fresh sample of air into the olfactory chamber adjacent to the olfactory lobe of the brain. The crocodilian brain is more complex than that of any other reptile.

The crocodile's limbs, trunk and viscera also show evidence of adaptation to life in a water's-edge environment. The forelimbs are somewhat longer than the hind ones, but both are stubby. The feet are long and bent, and the toes are tipped with sturdy claws; the toes of the hind feet are webbed. The thick tail, which accounts for some 40 percent of the adult crocodile's length, is almost rectangular in cross section. Its angularity is emphasized by two rows of scales that run along its top edges.

The skeleton of the trunk is powerfully constructed; the partly cartilaginous ribs are long, each consisting of a back, side and belly segment. The vertebrae are solid, and an extra set of cartilaginous abdominal ribs provides reinforcement for the belly. The shoulder girdle retains some flexibility with respect to the trunk; the limbs are attached low along the sides of the trunk and have a marked freedom of forward and backward movement, although vertical movement is restricted.

Most of the space within the bony rib cage is filled by a pair of complexly chambered lungs. The liver is to the rear of the lungs, where it is held in place by a radial array of connective tissue. The stomach and intestines occupy the visceral cavity to the rear of the liver. These organs can stretch well beyond their normal dimensions to accommodate large meals. The skin of the crocodile incorporates a geometrical arrangement of horny plates. Many of the plates have a bony core, and the plates that cover the crocodile's head are actually fused to the skull.

In the past decade a number of investigations of crocodilian physiology have been conducted, for example studies of how the animals digest their food, how they breathe and how their circulatory system is adapted to their amphibious way of life. All such research is useful, but it should be noted that the experimental animals have generally been immature crocodilians, ranging in age from hatchlings to specimens little more than three years old. The life span of any crocodilian is at least 25 years and probably exceeds 50 years, so that making projections on the basis of such studies is the equivalent of basing a physiology of man on tests conducted exclusively with infants.

The studies of crocodilian digestion suggest that the passage of food through the alimentary system is rapid. The food is crushed and torn into large chunks that the animal gulps down. The chunks are further reduced by a kind of milling or churning: stones contained in the animal's stomach are set in motion by regular contractions of the strong stomach muscles. Secretions of hydrochloric acid maintain the digestive fluid at a pH of about 1.5, which is more acid than the optimum for digestive activity. The tissue fluids present in the crocodile's food, however, dilute the digestive fluid in contact with the food to the optimum pH (from 2 to 2.2). The estimated elapsed time from ingestion to elimination is somewhat more than 72 hours.

Even when food is constantly available, the animals' intake is variable. Temperature is one factor that affects feeding; adult crocodilians will not eat when their body temperature falls below a fixed threshold. Juvenile crocodilians are not as sensitive to lowered body temperature and continue to eat when adults will not. Adults will also refuse food when their temperature is at or above the threshold value but the barometric pressure is dropping. This behavior is apparently related to the fact that a falling barometer usually signals a one- or two-day period of lowered air temperature. Cooler weather retards the crocodile's rate of digestion, which is a matter of some significance in the northern and southern extremes of the crocodilian range.

Crocodilian breathing is a complex phenomenon, as might be expected in an animal that moves frequently from land to water and back again. To the rear of the gular fold the floor of the pharynx is normally raised so that the larynx neatly meets the cup-shaped cavity where the long nasal tubes terminate. The cycle of breathing is intermittent: an exhalation, an inhalation and then a protracted pause. The valved nostrils at the end of the snout dilate rhythmically in time with the cycle. In exhalation the space available to the lungs is decreased as the animal's ribs fold slightly and contractions of the abdominal and intercostal muscles shift the liver forward. In inhalation the contraction of the longitudinal diaphragmatic muscles, which connect the liver to the pelvis, pulls the liver backward. The lung volume is increased, and the air is aspirated much as it is when the piston of a syringe is pulled.

The differences between breathing on land and breathing in the water are marked. On land the crocodilian's rib cage is distended as the weight of the animal displaces the flexible ribs outward. In the water hydrostatic pressure tends to fold the ribs inward. The pressure of the surrounding water increases the muscular effort required for inhalation; electromyograms of the inhalatory muscles record a rise in effort as the depth of immersion increases. The crocodilian's inhalations therefore become shorter with depth. At the same time the activity of the exhalatory muscles decreases

MALE CROCODILE holds in its mouth a crocodile egg that is ready to hatch. The male, one of a parental pair, was being tested to see whether its reaction at hatching time was as positive as the reaction of the female parent. The male gently rolled the egg back and forth between its tongue and its palate until the hatchling broke out of the shell. Once free, the hatchling swam ashore to join its crèche-mates.

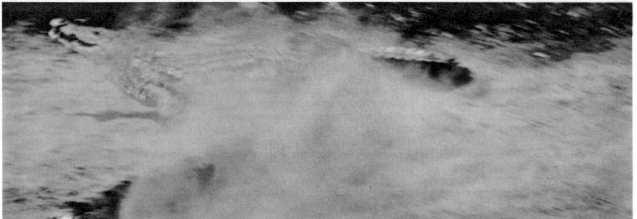

in proportion to the depth. It is as if the animal were recovering some of the energy it has to expend by inhaling against pressure.

The crocodilian's sniff is entirely unrelated to its breathing. The air is flushed through the nasal tubes by a lifting and dropping of the floor of the pharynx as the gular fold and the glottis remain closed. Electroencephalograms show that changes of air in the olfactory chamber stimulate the olfactory lobe of the animal's brain.

The crocodile's heart is analogous to a mammal's. Both direct measurements and cinefluoroscopy confirm that its four chambers can pump blood from the body to the lungs and from the lungs to the body without mixing. A bypass equivalent to one in the heart of the human fetus, however, connects the two ventricles, and it confers a unique advantage. When the crocodile is submerged and its lungs have given up their oxygen, most of the blood bypasses the lungs, so that the heart works primarily to circulate blood through the body.

A behavioral adaptation that contributes to the success of the Nile crocodile is basking. Like other reptiles, crocodilians conserve energy because of their low metabolic rate. The Nile crocodile conserves additional energy when its body temperature is raised by basking. This direct use of solar energy enables large crocodiles to survive even when food is scarce. (At any one time more than 30 percent of the crocodiles in many populations have an empty stomach.) When food is seasonally abundant, the animal can quickly recoup and go on to store the energy needed for growth and reproduction.

These specializations are probably shared by all crocodilians. They provide the basis for the animals' success as predators, a success so outstanding that to judge by the fossil record the order of crocodilians has undergone no significant structural alteration since the end of the Mesozoic era, some 60 million years ago. In considering the animals' predatory adaptations let us return to the Nile crocodile.

The most striking characteristic of crocodilians is their large size. Even the dwarf crocodiles of Africa are more than 1.6 meters long, and the average length for adults of all crocodilian species must be more than three meters. The newly hatched Nile crocodile weighs no more than 125 grams; the female parent weighs 300 to 500

kilograms. Thus the weight gain from immaturity to adulthood is 2,400- to 4,000-fold. The food supply required to achieve this gain might present problems if it were not for the fact that the ecological niche occupied by the Nile crocodile is unusually broad. For example, during its life span it takes prey that is progressively larger; the size of the food objects increases roughly in proportion to the predator's increase in size. The crocodile's predatory behavior undergoes a similar development.

Numerous analyses of the contents of the Nile crocodile's stomach and observations of the capture of prey indicate that the animals are not sit-and-wait hunters but highly active and versatile ones. Juvenile crocodiles necessarily subsist mainly on small prey: insects, snails, frogs and fish fry. Adults can successfully subdue large animals; they routinely capture drinking antelopes, and they can seize and drown Cape buffaloes as heavy as themselves. At the same time the adults will feed opportunistically on lesser prey: frogs, crabs and small fish. Subadults have a similar feeding behavior. As they become larger they seek out the larger prey, but they do not lose the ability to deal with the smaller.

The Nile crocodile probably got its false sit-and-wait reputation from its hiding behavior. Lying in the water, it can breathe, smell, see and hear while only its nostrils and the top of its head are visible above the surface. It regularly enhances this natural concealment by lying next to a stand of reeds or drifting alongside a floating object. Moving in this manner, or swimming underwater, the crocodile can catch fish with a sudden sideways snap of its jaws. All crocodilians have a flattened snout that offers little resistance to lateral movement, and a sideways sweep is mechanically efficient in that most of the crocodile's body remains motionless while most of the mass of its head, the only moving part, is located near the center of rotation. Even so, the slenderer the snout of a species is, the more effective its sideways snap is and the more likely it is that there is a high proportion of fish in its diet. For example, whereas the Nile crocodile eats fish in addition to other prey, its slender-snouted relative *C. cataphractus* feeds mainly on fish and crabs.

The crocodile can also attack by lunging entirely out of the water and onto the land. Swimming crocodilians generally propel themselves by sculling with their tail, keeping their limbs folded against their sides to maximize streamlining. When the crocodile lunges to meet its prey head on, it has usually got close by swimming underwater, perhaps bringing its head to the surface once or twice to check the prey's location. The final lunge may carry the attacker several times its own length. The acceleration imparted by the powerful tail is combined with a simultaneous forward swing of the hind legs as the crocodile touches bottom. The toes and the feet dig into the bank and the powerful legs lever the body upward; if the bank is steep, the crocodile appears to vault

straight out of the water. If the prey is still out of reach, the hind-leg stride may be repeated, and the crocodile's head may lower to hook over the top of the bank and support its body while its legs swing forward for another stride. Juvenile crocodiles have been seen to lunge when they were catching insects. An adult Nile crocodile can also dash across dry land for several yards, moving swiftly in a running stance with its trunk held above the ground.

There are numerous variations on both the land the and water patterns of hunting behavior. For example, a Nile crocodile can stun a small antelope at the water's edge with the same sideways head motion it uses to catch fish. It can also employ its tail for hunting purposes. If there is a nest of weaverbirds in the reeds along the riverbank, the crocodile can bend the reeds down with its tail and flip the nestlings into the water, where they are easily snapped up. In another tail-hunting tactic the crocodile swims slowly parallel to a riverbank with its tail bent toward the bank. The scales at the top of the tail riffle the water slightly, and small fish in the shallows move along ahead of the disturbance. When the crocodile turns its head around to the bank, the fish are trapped and are seized by a sideways sweep of its open jaws. The Nile crocodile is so agile that a fast reverse sweep will even intercept fish that are trying to escape by jumping over its back.

Larger prey animals are crushed by a series of bites along their length. If the prey is too large to be swallowed whole, the crocodile reduces it primarily by jerking and twisting motions. Such manipulations are inertial, with each flip of the crocodile's head moving the food object closer to the gullet; small prey is oriented headfirst to make it easier to swallow. In an alternative tactic the crocodile leaves some unwanted part of the prey outside its mouth and jerks the rest of the tissue away from it; the Nile crocodile uses this maneuver to break off the bony head of an armored catfish. In its hunting behavior the crocodile also makes good use of its adaptation to both land and water. Prey that is caught on land is often killed by drowning and then dismembered in the water; prey that is caught in the water is often flipped into the air, a medium that offers less resistance to jerking than the water does.

The hypothesis that crocodilians have a socially advanced feeding system is supported by observations of Nile crocodiles apparently cooperating in predation. The most commonly observed form of cooperative behavior is one where certain kinds of prey are reduced to pieces of a size convenient for swallowing. The twisting maneuver, in which the crocodile seizes some part of the prey in its jaws and then rolls over repeatedly until the seized part is torn away, obviously does not work with animals that are not large enough: when the crocodile rotates, so does the prey. In such circumstances crocodiles have been observed to

RITUAL CHASE establishes the dominance of a senior male crocodile over a potential challenger. In the top photograph on the opposite page a challenged senior male leaves the water in pursuit of the challenger. The junior male takes flight (*second from top*); the pursuit drives several basking crocodiles off the bank into the water (*third from top*). The junior male now signals its submission by raising its head and exposing its throat (*bottom*). The senior may take one of the junior male's legs in its mouth but will not bite it.

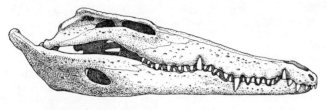

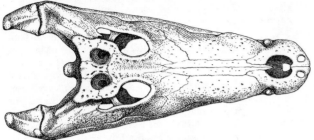

CROCODYLUS NILOTICUS

CROCODYLUS

OSTEOLAEMUS

TOMISTOMA

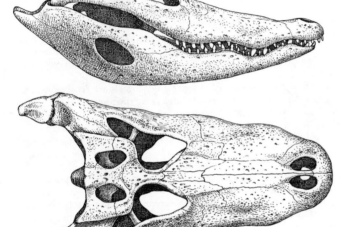

ALLIGATOR MISSISSIPIENSIS

ALLIGATOR

CAIMAN

MELANOSUCHUS

PALEOSUCHUS

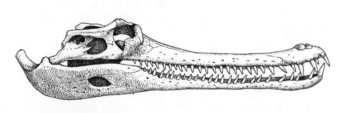

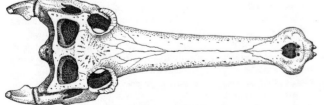

GAVIALIS GANGETICUS

GAVIALIS

move the carcass toward another crocodile. The second crocodile bites the carcass and holds it while the first crocodile rotates, or perhaps both rotate in opposite directions. Each crocodile eats what it tears off without any hostility toward the other.

One of us (Pooley) has witnessed an even more important example of cooperation: two crocodiles walking overland side by side, carrying the carcass of a nyala antelope well off the ground between them. It was not clear whether one or both of the crocodiles had killed the antelope or whether it was carrion. (Crocodiles have been seen to sniff out carrion on land and drag it into the water.) Observations of this kind suggest a high level of nervous integration in the Nile crocodile.

Another example of cooperation may be seen in the early spring, when rivers rise and the water flows into channels leading to pans, or natural depressions, along the river. Subadult crocodiles often form a semicircle where a channel enters a pan, facing the inrushing water and snapping up the fish that emerge from the river. Each crocodile stays in place and there is no fighting over the prey. Any shift in position, of course, would leave a gap in the crocodiles' ranks through which the fish could escape, so that what might be a momentary advantage for one crocodile would be a net loss for the group.

Recent observations have helped to clarify our understanding of the Nile crocodile's pattern of reproduction. In southern Africa the animals evidently reach sexual maturity when they are 12 to 15 years old; they are then between two and three meters long and weigh between 70 and 100 kilograms. It appears that a dominance hierarchy is established among the mature males of a population early in the mating season. The crocodile's aggressive display is a complex sequence that includes its blowing bubbles out of its open mouth (the exhalation bypasses the nostrils and emerges through the gular fold), arching its neck, raising its tail and lashing the water. The bubbling display is accompanied by a grunt or growl. The animal may also partially submerge its head and blow water through its nostrils.

Dominance is established when two mature males display to each other. Whichever animal proves to be the subdominant one

EIGHT GENERA of living crocodilians fall into three broad divisions, illustrated by three skulls shown from the top and from the side: *Crocodylus niloticus* (*top*), *Alligator mississipiensis* (*middle*) and *Gavialis gangeticus* (*bottom*). **The three genera of caimans, here grouped (***color***) with the genus** *Alligator*, **may actually belong in a separate category. Crocodiles have an oversized fourth lower tooth that remains in view even when the animal's mouth is closed because the tooth fits into a prominent gap in the upper row of teeth. This distinguishes crocodiles from alligators, whose oversized lower tooth is concealed behind an overlapping row of the upper teeth.**

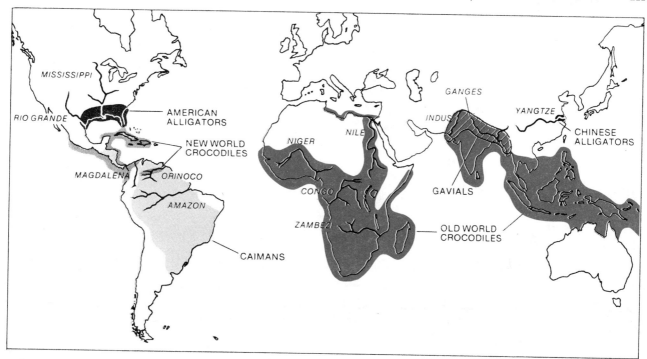

RANGE OF NILE CROCODILE (*center*) is compared with that of other crocodilian species. New World species of crocodiles are found from Cuba and southern Florida, where they overlap the range of the American alligator (*dark gray*), to northern South America, where they overlap the range of the caimans. Old World species of croco- diles are found in Africa and from the Indus River southeastward to New Guinea and northern Australia; their range in Asia overlaps the range of the gavial. A small population of alligators is also found in China. The Nile crocodile no longer ranges as far along the Mediter- ranean coast as is shown here and has become rare in the lower Nile.

eventually turns and swims off at high speed; the dominant male, usually the larg- er of the two, pursues, threatening to bite [*see illustration on page 208*]. If the sub- dominant animal slows in its flight, it raises its head almost vertically to expose its throat. The pursuing male may then seize one of the subdominant animal's limbs in its jaws, but it will not bite. In- teractions of this kind result in the dis- placement of subdominant males from the territory occupied by mated pairs.

We are beginning to obtain evidence that paired Nile crocodiles stay together for some time. The species appears to be mo- nogamous at least during the sexually active part of its annual reproductive cycle. The male and the female join in an elaborate courtship ritual that includes a mutual lift- ing of heads and rubbing of jaws. The ritual is also marked by a wide gaping of the jaws, but there is no biting. Within two or three days following the courtship ritual the ani- mals copulate at least twice while they are in the water.

Crocodilians, like most reptiles, lay eggs. Nile crocodiles lay theirs some five months after fertilization. The clutch varies from 16 eggs to more than 80. The larger the female is, the larger are both the number of eggs and the weight of the individual egg (from 85 to 125 grams). When egg-laying time approaches, the female selects a nesting site. After the site is first used it becomes a per- manent one; the female returns to it year after year, resting at it after mating and

defending it against the approach of other females until egg-laying time.

When the nesting site is being used for the first time, the female occupies it for a few days. Then one night she digs a hole with her hind legs, deposits her eggs and buries them under 30 to 45 centimeters of soil. She now remains on the nest or in its immediate vicinity. She defends a fairly large territory around the nest by charging at intruders.

The incubation period lasts 84 to 90 days. During this period the female evidently does not feed at all, and she has become quite inactive by the time the young begin to hatch. The male parent also remains in the general vicinity of the nest, although he does not approach the nest itself and goes off at intervals to feed. One student of the Nile crocodile, M. L. Modha of the Ken- ya Game Department, observed an entire breeding season on the shores of Lake Ru- dolf; his observation blind overlooked an area that included several nesting sites. He reported that the largest males in the local population patrolled the beach in the vicini- ty of the nesting sites and that no subdomi- nant crocodiles were to be seen there.

At hatching time the young crocodiles begin to call from within the egg. The sound they make is loud enough to penetrate the overlying soil and to be heard as much as 20 meters from the nest. On hearing the sound the female moves to the nest and starts to excavate it, working with her forelimbs and scraping and biting with her jaws. When the

nest is opened, a remarkable event in the crocodile's life cycle takes place: the female picks up the hatchlings one by one until they are all in her mouth and then carries them down to the water.

To do this the female rotates her head 45 to 90 degrees, gently picks up one of her tiny offspring with her teeth and flips it into the back of her mouth. To make room for all the hatchlings she depresses her tongue and the entire floor of her mouth, forming a pouch. As the young crocodiles enter the female's mouth their call changes to a softer chirping. Those hatchlings that wander away from the nest while the female is busy collecting the others soon begin to emit dis- tress calls. Some turn to approach the fe- male; others move around at random until she locates them by their calls and puts them in her mouth.

When the entire brood has been collect- ed, the female enters the water and releases the young by opening her mouth and swing- ing her head from side to side in the shal- lows. Washed clean of sand from the nest, the young swim ashore and mill about at the water's edge, emitting pulsating chirps. This chorus elicits a vocal response from adult and subadult crocodiles in the vicin- ity. The parental male now approaches the female and is greeted by a low warble.

The hatchlings do not disperse for some six to eight weeks. Both parents remain near the crèche during that time, defending their young against other crocodiles and against predators in general.

Experiments conducted by one of us

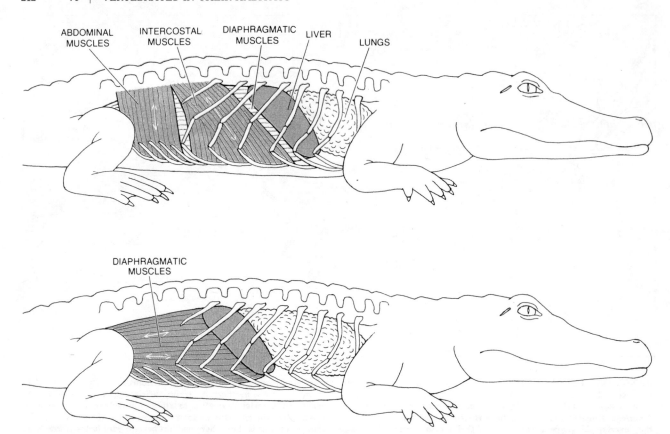

CROCODILE'S BREATHING is assisted by the movement of its liver, which lies just behind its lungs. In exhaling (*top*), space for the lungs is decreased as the ribs fold slightly and contraction of certain muscles (*color*) shifts liver forward. The inhalation that follows (*bottom*) begins when contraction of other muscles (*color*) shifts the liver backward; the air is drawn in as if it were pulled by a syringe piston.

(Pooley) show that if the parental male has the opportunity, he too will collect the hatchlings in his mouth and deposit them in the water. The male will even break eggs to free the hatchlings. This he does by taking the egg in his mouth and rolling it back and forth between his tongue and palate. If one remembers that the weight ratio of the adult to the hatchling can be as much as 4,000 to one, both the male's grasping action and his palpation of the egg are demonstrations of a spectacular oral sensitivity and muscular control. The same jaws that can crush the femur of a Cape buffalo can pick up an egg without harming the little crocodile inside.

If one of the juveniles in an aggregation is disturbed or encounters any other kind of trouble while it is still a member of the crèche, it emits a long, loud, high-pitched distress call; the other hatchlings may join in, chorusing for as long as 30 seconds.

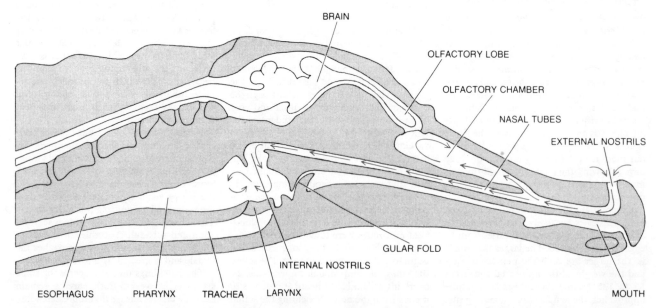

SNIFFING, an action that is independent of respiration, is made possible by the crocodile's ability to shut both its glottis and its gular fold, thereby making a closed system out of the nostrils and pharynx. By dropping the floor of its pharynx the crocodile is able to draw a sample of fresh air (*color*) into the olfactory chamber. Encephalograms show that the change of air that is produced by a sniff stimulates the olfactory lobe of the crocodile's brain; the mechanism may account for the crocodile's ready location of carrion in its vicinity.

Simultaneously the hatchlings take shelter, fleeing to the water or hiding under vegetation. All the adult crocodiles in the vicinity respond to the distress call by moving immediately toward the sound, even leaving the water to do so. The subadult crocodiles, however, do not react. The adults respond to distress calls even after the crèche disperses, although their reaction is less lively.

At the end of the crèche phase the young crocodiles gradually disperse, seeking out pools and streams that are not inhabited by subadults or adults. Their earlier positive response to the presence of an adult crocodile is reversed; when they encounter a crocodile larger than themselves, they turn and move away. Young crocodiles soon use their claws and teeth to dig a tunnel in the riverbank; the burrowing is sometimes a communal enterprise, with several young sharing the work. The tunnels are up to three meters long, and the animals use them as a shelter until they reach the age of five years or so. The tunnels afford the juveniles

protection against various predators, including adult crocodiles, but it may be that the warmth they provide is equally important. During the winter months in southern Africa the air temperature in the tunnels can be eight to 10 degrees Celsius above the temperature of the river water.

There is almost no information on the life span of the Nile crocodile in its natural environment. This is hardly surprising; it is only in the past few years that systematic observations of crocodilians in the wild have been undertaken. We can confirm the impression that the young and the subadults have the highest mortality rate. Relatively few live to reach sexual maturity, let alone maximum growth. Predators that take hatchling and juvenile crocodiles include crabs and large fishes, other reptiles, herons and storks and such mammals as the mongoose and the hyena. A crocodile that survives to sexual maturity may play host to a few parasites but rarely encounters predators.

It seems clear that the 12-week crèche period, when not only the parental pair but also other adults behave protectively toward the hatchling crocodiles, must greatly reduce predation at a time when the young animals are at their most vulnerable. At the same time simple arithmetic—the 20- to 40-year reproductive span of the adult female crocodile multiplied by the annual production of 16 to 80 eggs—provides persuasive evidence that the annual loss of immature crocodiles to predators is a substantial one. The same arithmetic also suggests the crocodile's potential for the invasion of new habitats and for the renewal of populations after catastrophes.

Bearing all this in mind with respect both to the Nile crocodile and to crocodilians in general, one can briefly summarize the nature of the Nile crocodile's successful adaptation. Any food chain has a bottom level occupied by animals that feed on plants, the primary producers. Above the

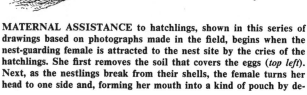

MATERNAL ASSISTANCE to hatchlings, shown in this series of drawings based on photographs made in the field, begins when the nest-guarding female is attracted to the nest site by the cries of the hatchlings. She first removes the soil that covers the eggs (*top left*). Next, as the nestlings break from their shells, the female turns her head to one side and, forming her mouth into a kind of pouch by de- **pressing her tongue, gently gathers them in (*top right*). When all the hatchlings are in her mouth (*bottom left*), she moves from the nest site to shallow water. There she opens her mouth (*bottom right*) and deposits the hatchlings in the water by swinging her head from side to side. After the little crocodiles swim ashore they congregate; both parents guard the crèche during the weeks before the young disperse.**

herbivores are primary predators and above them are secondary and tertiary predators. The crocodile, as a tertiary predator, can feed on any of the animals below it in the chain, and in addition it can play the role of a scavenger. Moreover, as it grows larger it can add larger prey to its diet without losing the ability to secure small prey.

The crocodile's keen senses and advanced nervous system, which enable it to track live prey and to locate carrion, also facilitate cooperation with other crocodiles in the capture and dismemberment of prey. Its low rate of metabolism makes it possible for the female to endure the long fast while her eggs are incubating and also enables an entire population to survive when energy sources are limited. The crocodile's ability to maintain its optimum body temperature by basking means that a larger share of its total energy input can be allocated to growth and reproduction. The net effect is that crocodiles extract enough energy from a variety of waterside environments to maintain large populations of large animals.

The crocodile's large size confers its own selective advantage. The mature crocodile combines formidable power and speed with remarkable agility; the larger the crocodile is, the less likely it is to be the victim of predation. At the same time the larger males have the advantage in dominance confrontations, and the larger females produce more and larger eggs and are probably able to defend their offspring more effectively. That ability in turn increases the hatchlings' chances for survival. In this connection, it was once the custom to assess an animal's "reproductive effort" on the basis of the gross weight of each litter of its offspring. The application of this standard to the Nile crocodile would neglect its behavioral investment in the next generation: the female's three-month fast while she is guarding the nest. The weight of prey the female forgoes during her abstinence far exceeds the 10-kilogram mass of the largest clutch of eggs.

The burrowing habit of juvenile crocodiles, together with their ability to feed at body temperatures below the normal adult threshold for ingestion, are mechanisms that protect the young animals and accelerate their passage through the hazardous small-size stage. The same is true of the postcrèche dispersal patterns of the young crocodiles. The dominance hierarchies among the adults also do much to explain how crocodile populations manage to live with little overt conflict in locales where the area available for basking sites may be limited.

The Nile crocodile has adjusted successfully to a wide variety of African water's-edge habitats. It is nonetheless in danger of extinction, not because it has failed to find its place in nature but because it is the prey of human hide-hunters. One might not want to have a crocodile in every water hole, but it will be a sad day when a few stuffed or pickled specimens are all that remain of these splendid animals.

Kangaroos

by T. J. Dawson
August 1977

*These hopping marsupial mammals have evolved in
relative isolation for some 25 million years. Their
adaptive strategies closely parallel those of the hoofed
mammals of the semiarid Old World grasslands*

Kangaroos have a certain fascination because they differ so much from the usual notion of what constitutes a mammal. They stand apart from other large mammals because they rear their young in a pouch and because they hop. The explanation often given for these odd characteristics is simply that kangaroos are marsupials and marsupials are primitive mammals. This is no answer but reflects an impression that seems to have arisen, particularly among people who live in the Northern Hemisphere, from statements made about the Virginia opossum, the only marsupial in North America. The opossum is frequently described as an archaic primitive mammal, little changed since the time of the dinosaurs. There is perhaps some basis for such statements: fossils of marsupials that coexisted with dinosaurs toward the end of the Mesozoic era, some 100 million years ago, have been classified as belonging to the same genus as that of the Virginia opossum (*Didelphis*). But even if the opossum can in some ways be considered a "living fossil," the same cannot be inferred for the kangaroos. These Australian marsupials represent a recent and successful specialized adaptation to a changing world environment.

The family Macropodidae (literally "big feet") consists of some 45 species, including not only the five or six species of large kangaroos but also smaller hopping relatives ranging from the wallabies, the next largest, down to the rabbit-sized rat kangaroos. The group has evolved from small forest-dwelling browsing animals into predominantly open-range grazing ones. This development represents a recent adaptive radiation in Australia comparable to the radiations of nonmarsupial mammals elsewhere in the world. The rise of the macropodids, possibly the most successful marsupial group among the animals of Australia, was probably related to the spread of grassland in the interior of Australia in the Miocene epoch, between 10 and 25 million years ago.

Similar events were taking place in other parts of the world in the Miocene.

It was a time of expansion for the even-toed ungulates: the mammalian order Artiodactyla. We know a good deal about the biology of the artiodactyls, particularly those in the suborder of ruminants, because they include such domestic animals as sheep, goats and cattle. The adaptations found in kangaroos represent solutions to the problems of life in a grassland environment that differ from the solutions arrived at by the artiodactyls. Moreover, even when the solutions arrived at by the marsupials and the nonmarsupials are similar, they embody many differences in detail. A major theme of this story is how these differences in evolution and adaptation to life on the open grasslands bear on the supposed primitiveness of the marsupials.

A brief review of the evolution of mammals will help to establish the relations between the marsupials and other mammals. It is believed all mammals are descended from a mammal-like reptile that lived late in the Triassic period, more than 200 million years ago. Soon thereafter, about 180 million years ago, the stocks leading to the two principal groups of mammals separated. One group was the prototherians, which consist of the egg-laying mammals of the order Monotremata; among the modern representatives of the order are the platypus and echidna of Australia. The other group was the therians, which consist of the marsupials and the placental mammals, that is, mammals characterized by a "true," or chorioallantoic, placenta. Thus the marsupials are more closely related to the placentals than to the monotremes, which retain many reptilian features in addition to egg-laying.

The initial radiation of the therian mammals early in the Cretaceous period stemmed from tiny insect-eating animals and was undoubtedly based on their adaptations to the newly developing flowering plants of the period and their pollinating insects. The separation that gave rise to the marsupials and the placentals took place at this time, some

130 million years ago. Information from a variety of sources suggests that the marsupials developed in North America and were the dominant therians in that region for most of the Cretaceous period, or until about 70 million years ago.

The placentals apparently developed initially in Asia and only reached North America late in the Cretaceous, when there was no oceanic barrier between the two land masses. This invasion of North America was followed by a major evolutionary radiation of placentals that coincided with the extermination of all but one of the previously abundant marsupial lines. Competition with the invading placentals has been suggested as the reason for the great marsupial extinction, but other causes may have been involved, since this was also the time when the dinosaurs finally died out.

Tracing the evolutionary pathways from the tiny insectivorous marsupials of the Cretaceous in North America to the kangaroos of the grasslands and open forests of Australia has long presented a problem. The emergence of evidence supporting the theories of continental drift and plate tectonics, however, has now given many clues to how the marsupials reached Australia. It appears that even though marsupials were disappearing from North America at the end of the Cretaceous, they had by that time spread widely through South America. There the marsupials radiated further. It is from this fauna that the Australian marsupials were apparently derived, at a time when South America was linked to Australia by Antarctica. The latest possible date for this further spread is about 50 million years ago, when Australia and Antarctica became separated.

The fossil record in Australia in this period is impoverished, so that we have no direct information about the original marsupial immigrants. Michael Archer of the Queensland Museum has suggested, however, that they came in several waves. His proposition is based on similarities between the teeth of Australian marsupials and those of South American marsupials, both living and extinct. Whatever the form of the earliest Australian marsupials, by the start of the Miocene, 25 million years ago, the process of differentiation had established their major groups. One of these groups was the macropodids. Their earliest-known fossil representative is a small rat kangaroo from a Miocene formation in the region of Lake Eyre.

One of the most significant aspects of faunal change in Australia over the past 10 to 15 million years has been the increase in abundance and diversity of macropodids. Several lines of kangaroos resulted from this relatively recent radiation. For example, one fossil line consists of giant, short-faced kangaroos with a skull and teeth specialized for heavy browsing. Another line had comparatively high-crowned cheek teeth, adapted for grazing. It was this line that gave rise to all the living groups of large kangaroos.

In short, the kangaroos are far from being primitive and unchanging mammals. In terms of the geologic time scale they have only recently arrived on the scene, evolving in response to new and changing environments, notably the spread of semiarid grasslands. Given the genetic heritage of their insectivorous ancestors of 130 million years ago, how did the two groups of grazing therian mammals, the kangaroos and the ruminants, differ in their adaptation to the same kind of specialized environment?

The rearing of young in a pouch is one obvious difference between the kangaroos and their placental counterparts. To understand why the kangaroos retain this basic marsupial feature it is necessary to consider marsupial reproduc-tion. The marsupial female rears her young after a pregnancy that is shorter than that of placental females of comparable size. As a result her young are born at a virtually embryonic stage. Not all marsupials have a marsupium, or pouch, but all of them have young that are firmly attached to a teat for a period after birth that corresponds roughly to the latter period of intrauterine development among placentals.

There is some controversy over the evolution of the bearing of live young among both the marsupials and the placentals. Since the monotremes still lay eggs, however, it is assumed that the bearing of live young evolved only after the early mammals had become well established. The discussion therefore revolves around whether the bearing of live young evolved once, before the separation of the marsupials and the placentals, or evolved independently in the marsupials and the placentals. Actually in the monotreme anteater *Tachyglossus aculeatus* the leathery egg remains in the uterus (and grows markedly) for 27 days and is then laid into a pouch, where it hatches after 11 days. The intrauterine component of this form of reproduction therefore bulks quite large. It is not hard to visualize the next evolutionary step to the deposition in the pouch not of eggs but of small young. This speculation is supported by Jason A. Lillegraven of the University of Wyoming, who suggests that all primitive therians were born helpless and needed intensive maternal care. Marsupials have retained this pattern, but most placentals have extended the period of internal gestation and give birth to more developed young. The ungulates have fully developed the option: their young are born active.

The reason for the retention of this ancestral birth strategy in marsupials and for its high development in kangaroos is something of a puzzle. Among the proposed explanations are the lack of a true placenta and the lack of a mechanism to extend the active life of the corpora lutea of the ovary: the bodies that secrete progesterone, the hormone that is largely responsible for the preparation and maintenance of the uterus in pregnancy. Neither explanation is acceptable. In some marsupials, notably the rabbit-sized bandicoot, there is both a true placenta and a prolonged active life of the corpora lutea.

Another proposal is that the crucial adaptation leading to prolonged gestation in placental mammals is the mechanism that keeps the pregnant animal from mounting an immune reaction that would reject the "foreign" tissue of the embryo. (The developing embryo is foreign in the sense that it consists of tissues derived not only from the mother but also from the father.) Lillegraven has suggested that in placentals the major barrier between the fetal antigens and the maternal antibodies is an anatomical one, consisting of a special component of the placenta, the trophoblast. The marsupials have not evolved this special barrier and retain the ancestral condition: an inert eggshell (which is, of course, of maternal origin) is maintained until late in pregnancy and apparently blocks the maternal immune response. The retention of the eggshell limits the size of the young because once it breaks down, the embryo has to be born before a full-fledged immunological attack can be launched on it.

The marsupials have retained the basic ancestral therian reproductive pat-

CHANGING GAITS of kangaroos range from pentapedal locomotion (*top*) typical of slow movement to high-speed hopping in excess of 40 kilometers per hour (*bottom*). In pentapedal locomotion the tail of the kangaroo serves as the third leg of a tripod, helping the animal to support itself on its forelimbs while the large hind limbs move forward. In low-speed hopping (*middle*) the tail assists balance but no longer acts as a support. Number of hops per minute is almost constant at speeds between 15 and 35 k.p.h. (about nine to 22 miles per hour); more speed comes from longer strides. High-speed hopping, from 40 to 50 or more k.p.h. (25 to 30 or more m.p.h.), requires still longer strides and faster hopping. These schematic representations are based on silhouettes of red kangaroos prepared by Frank Knight.

tern, but the kangaroos have modified it so that in some respects it even parallels the ungulates' advanced pattern of bearing fully active young. For example, the kangaroo pouch is much larger and more baglike than the average marsupial pouch, and the time the young kangaroo spends in it is markedly extended. In the red kangaroo (*Megaleia rufa*) pregnancy lasts for only 33 days, two days less than the estrous cycle. The offspring weighs only about .8 gram at birth, but it is larger, both relatively and absolutely, than the offspring of any other marsupial. The newborn red kangaroo still has a lot of growing to do after it makes its unaided climb from the birth canal to its mother's pouch. It does not leave the pouch permanently until 235 more days have passed.

When the young kangaroo, or joey, finally does leave the pouch, it can be regarded as having been born again, and it is quite comparable in its development to the newborn young of such placentals as deer and sheep. It continues suckling for a period comparable to the suckling time of ungulate young. During this additional suckling period, however, the nature of the mother's milk changes. The ability of the kangaroo mammary gland to secrete two different types of milk is an adaptation that allows the female to suckle a large "young-at-foot" offspring while simultaneously nursing another in her pouch. The young-at-foot puts its head back into the pouch to suckle from its original teat; the joey in the pouch is attached to another one. The two milks secreted by the separate mammary glands are quite different in volume and composition. How this can be achieved under the same hormonal conditions is an intriguing question.

One peculiarity of the reproduction of most species of kangaroos is the phenomenon known as embryonic diapause. During this period of reproductive quiescence a viable embryo is carried in the uterus with its development entirely arrested at the blastocyst stage, when it consists of from 70 to 100 cells and is about .25 millimeter in diameter. What is the function of this adaptation? To understand it one must consider the kangaroo's reproductive cycle.

Fertilization does not generally alter the course of the kangaroo's estrous cycle, and the cycle is not suppressed during pregnancy. It is halted only by the stimulus of lactation after birth. In most species estrus and mating occur again very soon after birth, that is, at the same time they would have occurred if the female had not become pregnant. The fate of the ovum fertilized at this immediately postbirth mating was first described in 1954 by G. B. Sharman of Macquarie University in Australia.

The ovum develops to the blastocyst stage but then becomes dormant. In the red kangaroo if the newborn young in the pouch survives normally, the blasto-cyst will remain dormant for about 200 days. Thereafter development resumes, and within 30 days birth takes place; the previous inhabitant of the pouch is evicted shortly before the new offspring enters its shelter. Birth is again followed by estrus and mating, and another blastocyst may result. Since the evicted joey may continue to suckle for another four months, the female red kangaroo may have three offspring in the "pipeline" at any one time: a dormant blastocyst, a small joey nursing and developing in the pouch and a larger young-at-foot still suckling.

The adaptive significance of embryonic diapause has been debated at length because it takes different forms in the macropodid family. For example, among red kangaroos its significance would appear to be associated with a rapid rebuilding of the kangaroo population after a reduction in numbers due to a drought in the arid Australian interior. Unlike the placental embryo, which is plugged into the maternal blood supply for nourishment, the joey is fueled only by milk. During a drought the young in the pouch and even the young-at-foot may perish if the milk supply decreases. When this happens, the blastocyst resumes development, and after the embryo is born the mother mates and produces another blastocyst.

During a drought the stress would be greater on the larger young, and in a prolonged drought the young in the

FEMALE RED KANGAROO rests in the shade of an acacia bush on a hot day with the air temperature 46 degrees Celsius (115 degrees Fahrenheit). Offspring at her side is still nursing but has long since left the pouch. In the female's pouch is another maturing offspring that will soon join the older "young-at-foot." The female's posture, with tail tucked under, helps to reduce radiant-energy input.

pouch die at progressively earlier ages until the female stops breeding altogether. One can thus see the adaptive advantage in an arrangement whereby the female will, except under the severest conditions, harbor a developing offspring. The significance of this adaptation to a widely dispersed population can be further appreciated in view of an additional fact. As A. E. Newsome of the Wildlife Research Division of the Commonwealth Scientific and Industrial Research Organization has shown, male red kangaroos may become temporarily infertile under hot drought conditions.

Knowledge of the reproductive physiology of kangaroos has been slowly accumulating for many years. It is only quite recently, however, that the characteristics of kangaroo locomotion have become appreciated. The hopping of kangaroos intrigued the earliest European visitors to Australia. One of the first comments was made by Sir Joseph Banks in the journal of his voyage with Captain James Cook on the *Endeavour* from 1768 to 1771:

"Quadripeds we saw but few, and were able to catch few of them that we did see. The largest was called by the native *Kangaroo*. It is different from any European and indeed any animal I have heard or read of except the Gerbua of Egypt, which is not larger than a rat when this is as large as a middling Lamb; the largest we shot weighed 84 lb. It may however be easily known from all other mammals by the singular property of running or rather hopping upon only its hinder legs carrying its fore bent close to its breast; in this manner however er it hops so fast that in the rocky bad ground where it is commonly found it easily beat my grey hound, who tho he was fairly started at several killed only one, and that quite a young one."

Banks had every reason to be surprised by the sight of a large hopping animal; small hoppers such as his Egyptian "Gerbua" (the jerboa, a hopping rodent) abound in the deserts of the world but large ones were unknown. Indeed, if he had known what we know today he would have been even more surprised: the kangaroo seems to represent the only instance in evolutionary history of any animal weighing more than three to five kilograms having adopted a hopping gait.

It was once believed bipedal dinosaurs hopped, but there is no evidence to support this view. Why, we may ask, is hopping such an unusual method of locomotion in larger animals and why did the kangaroos develop it? As we now begin to understand such matters as the relative energetic cost of various forms of locomotion and the relative agility associated with different gaits, the answers to these questions are becoming more apparent.

An understanding of the characteris-

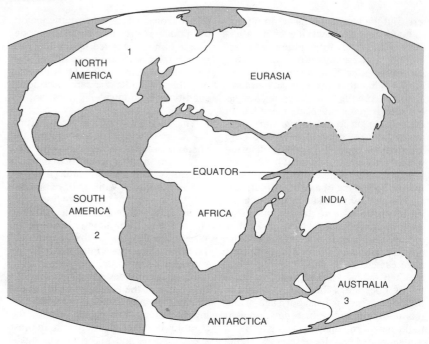

MARSUPIAL EXODUS to Australia began at the end of the Mesozoic era when some of these successful early mammals left their North American homeland (*1*) for South America (*2*), which was still joined to Antarctica and Australia (*3*) in the early Cenozoic. North and South America were separated for much of the Cenozoic. When they were rejoined, advanced placental mammals from the north competed with the marsupials of South America, leaving Australia the major marsupial reservoir. Marsupials died out in North America by the middle Cenozoic. The opossum is not a Mesozoic leftover but a migrant from South America.

EPOCH	AUSTRALIA	SOUTH AMERICA	NORTH AMERICA	EUROPE	ASIA
RECENT					
PLEISTOCENE					
PLIOCENE					
MIOCENE					
OLIGOCENE					
EOCENE					
PALEOCENE					
LATE CRETACEOUS					
EARLY CRETACEOUS					
JURASSIC					

EVOLUTIONARY RECORD of the marsupials is outlined in this diagram, beginning early in the Cretaceous with the divergence between placental and marsupial mammals. The first marsupials flourished in North America and also entered South America at this time. During the Paleocene some marsupials from North America reached Europe; in the Miocene both groups died out. Now the majority of marsupial species are found in Australia and its vicinity.

tics of hopping has been difficult until recently because there was no general framework of information about the energetic cost of locomotion with which the cost of hopping could be compared. C. Richard Taylor of Harvard University, Knut Schmidt-Nielsen of Duke University and their collaborators have now begun to erect such a framework. It is apparent from their studies that for the different types of running there is a regular and predictable energetic cost that depends primarily on the size of the animal and the running speed. The cost of running differs between quadrupeds and bipeds, but it is largely consistent and predictable, at least up to moderate speeds.

A joint interest in several aspects of kangaroo energetics and temperature regulation led to a collaboration between Taylor and me. Four kangaroos were flown from Sydney to Boston. We were able to train two of them to hop on a large treadmill as their oxygen consumption (and consequently the energetic cost of hopping) was measured. These studies showed that the energetic patterns of hopping diverged markedly from those of both bipedal and quadrupedal running. The most significant difference is that a kangaroo traveling at moderate speeds does so more economically than running bipeds or quadrupeds of similar size do. This is obviously advantageous to the kangaroo.

Why, then, is hopping so rare in large animals? The answer probably lies in the locomotor pattern of kangaroos at low speeds. At speeds below six kilometers per hour (about 3.7 miles per hour) kangaroos do not hop. They move in a rather odd way, using their heavy tail as an additional support. This gait has been called pentapedal because the tail acts as a fifth leg that helps the forelimbs to support the animal as the large hind limbs are moved forward together. (Contrary to a widespread belief, kangaroos can move their hind legs independently. They are good swimmers, and as they swim their hind legs alternate.)

The pentapedal gait is clumsy and energetically costly. The reason kangaroos resort to this odd mode of locomotion at low speeds seems to be that it would be even more costly to hop. A simple physical analysis of the various energetic components of hopping supports this view. The same kinds of factors explain why human beings shift from walking to jogging when they want to move faster, and why it is costlier for large animals to climb trees or run up slopes than it is for small animals. Moreover, shifting to quadrupedal locomotion at low speeds, as small hopping animals do, seems to be precluded in kangaroos by the gross anatomical specialization of their hind limbs and the reduction in size of their forelimbs. That these specializations are related to increased size is also indicated by the fact that some small members of

the macropodid family do employ quadrupedal locomotion at low speeds.

In short, at low speeds large kangaroos have major locomotion problems but at high speeds their hopping gait has marked advantages over running. Once they start hopping the energetic costs do not change over a wide range of speeds. In the treadmill experiments when the kangaroos hopped at speeds above 15 k.p.h. (about 9 m.p.h.), they traveled more economically than running quadrupeds or bipeds of similar size. The pattern of hopping energy expenditure appears to be explained by the combination of a fixed hopping rate and the elastic storage of energy in tendons. Energy can be stored in such elastic fibrous tissues much as it is stored in the spring of a pogo stick. The treadmill measurements of the energetic costs of hopping were not practical at speeds above 22 to 23 k.p.h. (about 14 m.p.h.), but some overall comments can be made on the basis of knowledge gained from observations of small hopping animals and from field observations of kangaroos.

The proposition can be advanced that unhindered animals normally travel at their most comfortable speed, that is, their economical speed. For kangaroos this appears to be between 20 and 25 k.p.h. Energy costs probably increase above that speed because the elastic storage of energy cannot increase indefinitely. Hopping frequency, however, remains constant up to about 40 k.p.h. (25 m.p.h.); the increase in speed is achieved mainly by increasing the length of the stride. If a kangaroo is pressed, it can maintain this speed for a couple of kilometers. Kangaroos can even increase their speed above 40 k.p.h., but they rarely do so. They hop in such high-speed bursts only in emergencies and then only for a few hundred meters. I have monitored kangaroos traveling at speeds of up to approximately 50 k.p.h., and a speed of 65 k.p.h. (about 40 m.p.h.) has been reported. At these maximum speeds there is a noticeable increase not only in stride length but also in stride frequency. It is certain that an increase in stride frequency is accompanied by a marked increase in energy consumption, but how the energetic cost of hopping at maximum speeds is related to the cost of high-speed quadrupedal locomotion is not yet known.

It is clear that the kangaroo mode of locomotion has advantages in spite of its shortcomings at low speed. Kangaroos have radiated widely throughout Australia and have survived the coming not only of aboriginal man and his semidomesticated dog, the dingo, but also of European man and his dogs. A biomechanical analysis of hopping indicates, however, that its advantages diminish significantly with increasing size. It is perhaps pertinent that the largest kangaroos known to have evolved, some of

them more than three meters tall, vanished along with many other large marsupials in the Pleistocene epoch.

Why did hopping come into existence at all? The answer may lie in a conservative or perhaps primitive marsupial characteristic. For many years one of the factors that was thought to support the notion that kangaroos are primitive was the reported low metabolic rate and the low and fluctuating body temperature of marsupials in general. The basal metabolism of kangaroos and other marsupials is only about 70 percent of that of advanced placentals. Their body temperature too is slightly lower, 35.5 to 36.5 degrees Celsius (96 to 98 degrees Fahrenheit). They do, however, maintain a constant temperature level.

How can a low metabolic rate be reconciled with energetic hopping? Here two aspects of locomotion must be considered. One is maximum sustained speed for escaping danger. The other is economy in long-distance travel. The investigations by Taylor and his co-workers suggest that the costs of quadrupedal locomotion are more or less fixed. Maximum sustained energy consumption may be estimated from maximum oxygen uptake, which in turn appears to be related in some way to basal metabolism. Since marsupials have a low metabolic rate, it is possible that they have a low maximum energy output and consequently a reduced potential for speed of locomotion. The bipedal hopping of kangaroos and other macropodids could be a mechanism that helps to overcome this limitation. Also important for a metabolically limited animal, particularly for a kangaroo ranging over the semiarid interior of Australia, would be economy in long-distance travel.

Past assumptions about the kangaroo's limited ability to regulate its body temperature were based not only on the slightly lower body temperatures of marsupials but also on observations of a kangaroo behavioral trait: thermoregulatory licking. When kangaroos are chased on hot days, they occasionally stop hopping and lick themselves, particularly on the forelegs. Because licking has been generally accepted as a primitive mode of cooling, this behavior helped to sustain the notion that kangaroos cannot maintain a constant body temperature. Such a view would have been difficult to hold if those who held it had ever observed desert kangaroos such as the red kangaroo in their natural habitat.

It is hard to conceive of an inefficient temperature regulator spending long summer days in the sparse shade of a small desert tree when the air temperature can be in excess of 45 degrees C. (113 degrees F.) and the radiation heat load makes the effective environmental

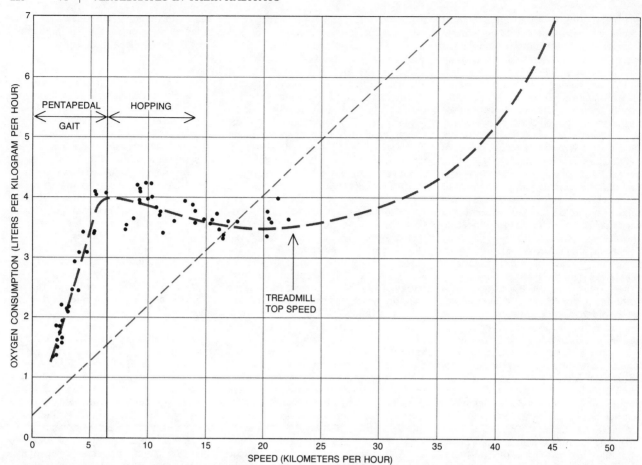

ENERGETIC COST of locomotion, measured in terms of oxygen consumption at various speeds, was determined by placing kangaroos on a treadmill that made it possible to simulate natural gaits at velocities up to 23 k.p.h. (14 m.p.h.). The energetic cost of pentapedal locomotion increased rapidly, but once the animals began to hop the cost leveled off and even decreased. The straight line (*color*) plots the pre-dicted cost increase with speed for a quadruped of the same body weight. Beginning at a speed of about 17 k.p.h. (11 m.p.h.), hopping appears to be more economical of energy than quadrupedal locomotion is. Projecting the curve for the kangaroo (*broken line*) beyond the maximum treadmill speed suggests that this may be true up to the top limit for sustained hopping: approximately 45 k.p.h. (28 m.p.h.).

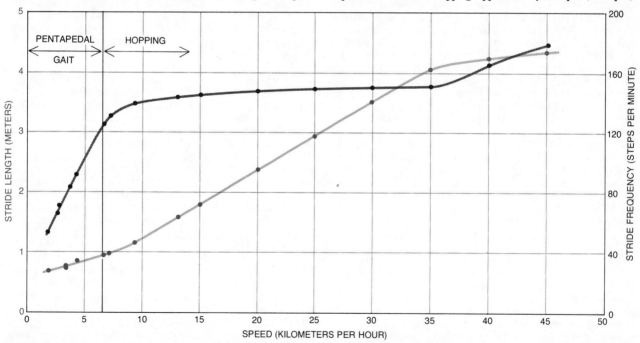

HOPPING SPEED is a function of frequency and length of stride. The frequency increases rapidly as the speed of pentapedal locomotion rises, reaching a rate of two steps per second before the kangaroo begins to hop. From a hopping speed of 15 k.p.h. (10 m.p.h.) up to a speed of 35 k.p.h. (22 m.p.h.), however, the increase in stride frequency is trivial. In this range the main factor in increasing speed is an increase in length of stride (*color*). At speeds above 35 k.p.h. the kangaroo increases both the length and the frequency of its stride.

temperature even higher. It has been found that far from being inefficient these kangaroos have possibly the most efficient and best organized thermoregulatory defense against overheating to be found in any mammal. Kangaroos not only lick their forelimbs during heat stress but also pant and sweat. The interrelation of these different methods of evaporative heat dissipation is seen only during exercise. When the kangaroo is resting, it dissipates heat primarily by panting; it does not sweat while resting, but under severe heat load it may lick its forelimbs.

The kangaroo's licking behavior was long a puzzle to my colleagues and me because the forelimb area usually licked is small. It seemed that the overall heat-dissipation benefit would be doubtful in view of certain possible disadvantages. Probably more in frustration than for any other reason we undertook an examination of the blood supply to the forelimbs. Injecting the blood vessels in the forelimbs of a dead kangaroo with liquid latex, we found that there was a dense and intricate network of superficial blood vessels in the region the animals usually lick. Further study revealed that during heat stress the blood flow to the region is greatly increased. The forelimb region is thus a site of significant heat transfer. Indeed, by spreading saliva on their forelimbs kangaroos may well be making the most efficient use possible of an overflow of fluid from their respiratory system, the principal site of evaporative heat dissipation in resting animals.

The pattern of heat loss in exercising kangaroos is very different from that in animals at rest. The reason is that exercising kangaroos sweat. Early in the course of our investigations it was suggested that sweating replaced panting during exercise because panting is not possible when the animal is exercising. We soon realized, however, that evaporation from the respiratory system is still highly significant during exercise because when the animal is exercising, water loss from these surfaces is greatly increased. In fact, the increase raises respiratory evaporation well above the levels the animal achieves by panting when it is at rest. This increased respiratory ventilation is probably primarily in response to the increased demand for oxygen during exercise and is therefore probably limited by the level of oxygen intake. Body sweating, then, appears to bring into play an additional surface area for further evaporative heat loss. Licking is not possible when the animal is hopping, although hopping kangaroos do stop occasionally and spread saliva on their forelimbs. Licking after exercise seems to assist the rapid return to normal body temperature.

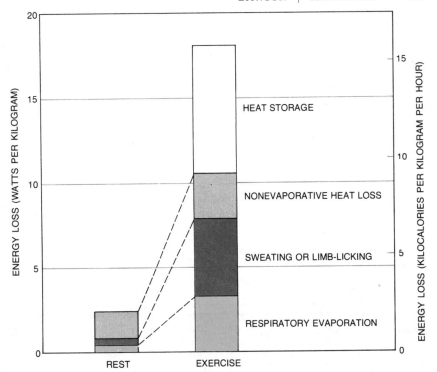

HEAT BALANCES of kangaroos at rest and exercising strenuously at the moderate air temperature of 24 degrees C. (75 degrees F.) are compared. Nonevaporative channels of heat loss suffice to maintain a stable balance in the resting kangaroo; skin cooling and respiratory-tract water loss combined are only half as important. The exercising kangaroo cannot significantly increase nonevaporative heat loss, and even sharply increased evaporative heat loss is not sufficient to dissipate the load. The "stored" excess is dissipated when the kangaroo stops exercising.

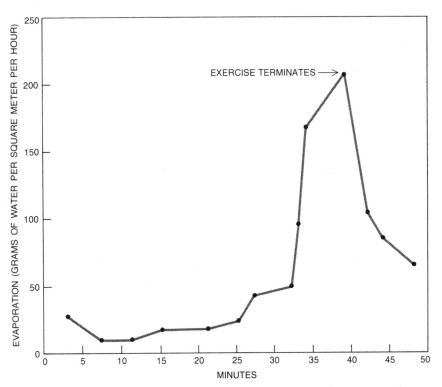

SWEATING not only is directly correlated with exercise in kangaroos but also requires a warm-up period. The graph records the outbreak of sweat on the face of a kangaroo forced to engage in pentapedal locomotion on a treadmill moving at a rate of 4 k.p.h. (2.4 m.p.h.). Some 30 minutes after starting to exercise the kangaroo began to sweat copiously. When exercise ended some 10 minutes later, sweating also diminished, virtually ceasing in 10 more minutes.

An aspect of this picture that sets kangaroos apart from other mammals is that sweating stops as soon as a kangaroo stops exercising, even when its body temperature is elevated and it is panting and licking its forelimbs. Other mammals capable of sustained exercise, such as horses and men, not only sweat to dissipate heat when they are exercising but also sweat to dissipate it when they are at rest. A few mammals, such as cattle, sheep and other bovids, sweat and pant simultaneously. What does the kangaroo achieve by ceasing to sweat when it is at rest and starting to pant? The answer lies in water economy. An animal that lives in an arid environment must husband its body water, and water is better conserved when it is evaporated from the respiratory tract than when it is evaporated from the skin. Evaporation from the skin results in a lowered surface temperature, which leads in turn to a greater inflow of heat from the environment and the utilization of additional water to eliminate the additional heat. Panting, of course, does not lower the body surface temperature.

On a very hot summer day the red kangaroo is an example of an animal doing the most it can to both regulate its body temperature and conserve its water. When the temperature is higher than 45 degrees C., the red kangaroo does not lie down in the shade but stands hunched, thus presenting the smallest amount of surface area for the uptake of heat from the environment. The animal's dense fur provides an almost ideal insulation against the ambient heat. To further minimize the exposure of surface area to heat radiation the kangaroo's long, thick tail is pulled between its legs. The tail too has a complex network of superficial blood vessels that are involved in heat dissipation, but when conditions favor heat flow into the body, this system is apparently nonfunctional.

One stratagem for conserving water under arid conditions that has not evolved among kangaroos is the ability to let the body temperature rise during the day. The oryx, a ruminant of Africa and the Middle East, has a special arrangement of blood vessels in the base of the brain that allows the brain to be maintained at a tolerable temperature even when the temperature of the rest of the body rises as high as 46 degrees C. (115 degrees F.). Camels presumably have a similar arrangement. Hence both of these desert animals are able to store heat, which they can then dissipate in the cool of the evening without expending water in evaporative cooling. Kangaroos do not have this vascular arrangement. Under conditions of heat stress and dehydration, however, the kangaroo lowers its temperature overnight, so that it can start the next day with its body temperature from two to four degrees below the normal level.

As a result of these combined adaptations for water economy the minimum water requirements of kangaroos are similar to those of the desert ungulates of Africa. Field studies indicate that arid-zone kangaroos turn over only a fourth of the water needed by sheep and wild goats. The oryx is completely independent of drinking water; it can get the water it needs from the plants it eats and from the oxidation of foodstuff. The kangaroo, like the camel, does need to drink occasionally under severe heat conditions.

The persistent notion of primitiveness colors many perceptions of the physiology of marsupials. Consider the fact that the kidney of the arid-zone kangaroo has concentrating abilities superior to those of the kidney in comparable placental mammals. This should not really surprise anyone; the fundamental characteristics of the mammalian kidney appear to be very old. Even monotremes have a good kidney of the mammalian type, and it may well be that the reptilian ancestors of the mammals, the therapsids, also had such a kidney. Kangaroos too compare well with placental mammals in other areas of water economy. For example, kangaroos and camels resemble each other in their cardiovascular response to dehydration: when they lose water, they adjust the volume of the vascular system so that the functioning of the heart is not jeopardized.

Adaptations involved in feeding and digestion provide interesting evolutionary comparisons between kangaroos and ruminants. Both lineages have arrived at a similar solution to the problem of living off plant tissues. It has often been suggested that the evolution of the mammals was closely associated with the simultaneous evolution of flowering plants. Early in the Cenozoic era these plants were largely woody types. By the Eocene epoch, however, herbs and grasses had begun to spread, and in the Miocene, beginning some 25 million years ago, large areas of grassland covered the drier interiors of most continents. The evolution of grazing kangaroos appears to be associated with this spread of grassland in Australia, as does the expansion of the artiodactyl group of placental mammals elsewhere in the world.

No mammal can synthesize the enzymes that break down cellulose, the principal carbohydrate of grasses. The utilization of this food source calls for the evolution of a symbiotic association between the mammal and species of bacteria and protozoa that have such enzymes. The microorganisms take up residence in various expanded sections of the mammalian gut. In horses and rabbits they live in parts of the hindgut, such as the caecum. In ruminants they live in sections of the foregut and specialized enlargements of the esophagus and the stomach. There are marked advantages in allowing the plant material to ferment in such chambers before it reaches the stomach: both the products of fermentation and the microorga-

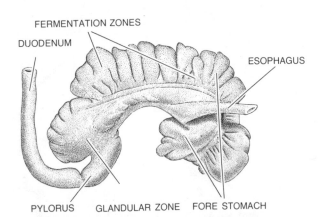
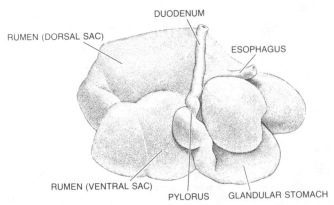

DIGESTION OF CELLULOSE, made possible by the action of enzymes secreted by stomach bacteria, is a capability independently evolved both by the marsupial kangaroos and by the placental ruminants. At left is the stomach of a red kangaroo, at right the stomach of an ox. The cellulose present in the diet of both grass eaters is largely broken down by fermentation before it passes into the intestine.

nisms can be digested and absorbed along the entire length of the small intestine. The greater efficiency of this system may account at least in part for the fact that whereas the dominant herbivores of the Eocene were hindgut fermenters such as horses, these animals were displaced in the Miocene by a massive radiation of fore-stomach fermenters such as ungulates.

Kangaroos are fore-stomach fermenters, so that they share with such modern ruminants as cattle, sheep, goats, deer and camels the benefits of this type of digestion. Those benefits include a lower requirement for protein. Urea, formed by the breakdown of protein in the body, is returned in the saliva to the fore-stomach, where it is resynthesized into nutritive protein by the resident microorganisms instead of being entirely excreted in the urine and lost. This recycling enables kangaroos and ruminants not only to gain nourishment from the dry, low-protein grasses of the arid zone but also to conserve the body water that would otherwise be needed for the excretion of urea. The combination of this efficient mode of digestion and the low rate of marsupial metabolism gives the kangaroo a formidable array of characteristics for the successful exploitation of a dry grassland environment.

The picture of the kangaroo that has now been developed is one of a group of mammals that are not at all primitive but have adapted and radiated rapidly in recent times in response to a new and changing environment. That kangaroos still possess some characteristics that may be considered primitive is true, but it is also true that they have adopted ways of circumventing the limitations of these ancestral traits. Moreover, kangaroos, like other marsupials, have evolved abilities that surpass those of comparable placental mammals. Marsupials and placentals are close relatives; their separate evolution over 130 million years illuminates how there can be different evolutionary solutions to similar problems. It is a large jump from the tiny shrewlike mammals of the early Cretaceous to the hoofed mammals of the Old World and the Americas on the one hand and the kangaroos of Australia on the other. Historical records suggest that there are more kangaroos in Australia now than there were before the Europeans arrived. Thus it may be that, far from traveling a road to extinction, at least the large kangaroos of the arid zone are thriving in spite of competition with man and the ruminant stock he has introduced. This is probably attributable to an extension of a suitable arid habitat. That extension, however, has simultaneously reduced the habitat suited to the kangaroos' smaller relatives, the wallabies and the rat kangaroos. It is these marsupials that have largely disappeared from the interior of Australia.

21 The Archer Fish

by K. H. Lüling
July 1963

This remarkable animal of southeast Asia downs its insect prey with a powerful spout of water. A discussion of the anatomical and behavioral adaptations that account for its unique ability

In 1764 John Albert Schlosser, a Fellow of the Royal Society of London, described to his colleagues an extraordinary animal: the East Indian "jaculator, or shooting fish." Schlosser reported that "this cunning fish," sighting a fly on a plant above the water, "ejects out of its tubular mouth a single drop of water, which never fails striking the fly into the sea, where it soon becomes its prey." Although Schlosser's story was inaccurate in some of its details, it was essentially correct. Scholars nonetheless discounted it and many similar reports. Only at the beginning of this century did the Russian ichthyologist N. Zolotnisky take the trouble to observe some archer fish that had been imported and placed in an aquarium. He confirmed Schlosser's account of their shooting prowess and noted several other aspects of their behavior. The archer fish became a popular aquarium specimen, but it was not until the 1930's that a U.S. ichthyologist could describe the anatomical structures that enable it to function as a living water pistol. Ten years ago I undertook to investigate the remarkably specialized behavior of the archer fish—behavior that so far as I know has no parallel in the animal kingdom.

The habitat of the genus *Toxotes* (from the Greek word for "archer") ranges from India to the northeastern tip of Australia. *Toxotes* is adapted to various environments at the edge of the sea: coastal salt water, the brackish waters of mangrove swamps, bays and estuaries; it even moves some distance upstream in rivers. Of the five species that have been distinguished, *T. jaculatrix* (Schlosser's "jaculator") is the best known and the one with which I have usually worked. It is a small fish, attaining a maximum length of about seven inches, and is silvery in color with dark patches and bars on its flanks. As one might expect, the eyes of this sharpshooter are quite distinctive: large, rather protruding and mobile. Most observers have been struck by their brightness and even an appearance of keenness not usually associated with the eyes of fishes.

One other anatomical detail should be mentioned because of its apparent adaptive significance. The bodies of juvenile archer fish shine with luminous spots that at times are so bright they resemble tiny greenish fluorescent lamps. The light, however, is not produced by the fish itself; it apparently results from the enhanced reflectivity of the tissue between the dark dorsal patches, which is particularly soft in the young fish. Juvenile archer fish are gregarious, and it seems likely that their luminosity helps them to keep contact with one another in the dark and muddy waters they so often frequent.

In its native habitat the archer fish swims at the surface or just beneath it, the great eyes peering upward in what appears to be a purposeful search for prey. When it spots a likely insect on a water plant or a mangrove root, it takes up a characteristic position with its snout just breaking the surface and lets fly a jet of water. Its aim is remarkably good: the first shot is usually successful at distances up to two or three feet. The insect, its wings doused with water, falls to the surface, whereupon the fish quickly eats it. In captivity *T. jaculatrix* retains its spouting behavior. Once it has become acclimated to the aquarium it will readily shoot at a meal worm held above the surface with a pair of tweezers or at a cockroach running about on the glass cover of the tank. It may even shoot the aquariist in the eye as he peers into the tank, the movement of his eyelids having stimulated the act of spouting.

One of the most interesting things about *T. jaculatrix* is that its highly specialized water-shooting is not its primary means of food-gathering; a number of restrictions are placed on this behavior by external factors and by the nature of the fish itself. First of all, the spouting is only possible where the surface is calm and there is vegetation; it is obviously impossible in open coastal waters where the fish spend some of their time. Moreover, the aerial targets must be within a reasonable distance of the surface. Very young fish can manage a jet only two or three inches long; in fact, they shoot down hardly any of their food. Adults are rarely successful beyond three or four feet. Another restriction is that the fish tend to fire with more vigor than is necessary; a powerful squirt may drive a nearby insect out of reach. Perhaps most important, the fish are not squirting machines but living things; they get tired. Even if insects are still available, a fish may weary of squirting and shift to another form of food-gathering if it is still hungry.

For all these reasons *Toxotes* depends largely on food it finds on or below the surface. It prefers insects that have fallen to the surface, but it will also take food that has sunk a few inches into the water. This raises an interesting question for evolutionary theory: Spouting, if it is so unimportant, can hardly have been a significant factor in the survival of the species or in selection and differentiation within the species.

Leaving this question aside, it is true that the archer fish does spout and knock down insects. The main reason for the long refusal of scholars to accept this fact was the lack of any obvious water-propelling mechanism in

ARCHER FISH is a native of coastal waters and rivers from India to northeastern Australia. This photograph shows *Toxotes jaculatrix*, the most common species, scoring a direct hit on an insect on a twig above the water. The fish spouts a stream of water, which breaks up into drops that strike the prey. The photograph was made by Heine Hediger and H. Heusser of the University of Zurich.

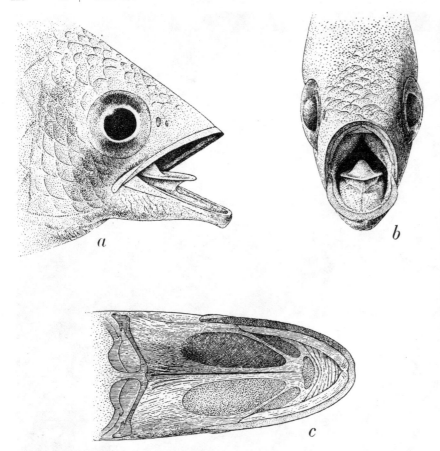

ARCHER FISH'S MOUTH, while superficially similar to that of many other fishes, has structures that explain its ability to spout. The bony tongue is thin in front but swells into a ridge toward the rear (*a*). When the mouth closes, the tongue almost fits into a V-shaped depression in the roof of the mouth (*b*) but leaves a small channel through which water is forced by compression of the gill covers. The roof of the mouth is seen from below at *c*.

the archer fish. As the British ichthyologist Francis Day wrote in 1889, "the action is one which the mouths of these fishes appear incapable of effecting." This objection was finally disposed of by the U.S. ichthyologists Hugh M. Smith and George S. Myers. Smith found that the propulsive force came from a sudden compression of the fish's gill covers; in fact, he was able to simulate the natural shooting act by holding a fish in his hand at the water surface and forcefully squeezing the gill covers. Myers then searched for some structural peculiarity of the mouth parts that would explain the formation of a jet of water. He discovered a narrow groove running along the roof of the mouth. When the fish closes its mouth, its tough tongue is pressed up against this groove, creating a tube like the bore of a gun. Pressure from the gill covers forces water through the tube and out of the slightly open mouth; the stream may be regulated in part by the thin tip of the tongue [see *illustration above*].

The work of Smith and Myers left un-explained one important question: How does a fish, with its eyes below the surface, discern small targets above the water and sight them accurately enough to score hits? A fish's-eye view of objects above the surface is notoriously poor. Moreover, the refraction of light rays as they enter the water makes an object above the surface appear to be where it is not. These difficulties prompted some people to insist that the archer fish must raise its eyes above the water surface at the moment of spouting. When this turned out not to be true, one ichthyologist suggested that the fish must have "a truly remarkable trigonometric range finder in its brain."

Neither this visual problem nor other details of the unique behavior of the fish could be satisfactorily investigated by casual observation in aquariums or in the field. I decided in 1952 to make a careful study of the spouting act with motion picture photography and to supplement this analysis with an anatomical study of the archer fish's eye structure.

I obtained seven specimens of *T. jaculatrix* ranging from two to four inches in length. Four of these I dissected for the anatomical investigation. I put the others in a tank, in which I stood a pane of frosted glass that projected 20 inches above the surface. When my fish squirted at cockroaches or ants placed on the glass, the location and shape of each broadside was recorded by a drenched area on the glass. With a soft pencil I traced these areas for later analysis. I used a similar setup for the filming operation, placing the lens of my camera level with the water surface. In this case I offered live targets on a pane of glass and also various dummies either fastened to wires or placed in flat-bottomed glass tubes fixed to the pane. In order to measure the angles at which the fish spouted I wanted to catch a fairly long stream of water on each frame, so I shot the film at 24 frames per second rather than at a very high speed.

My observations confirmed reports from the field that the archer generally keeps to the upper few inches of the water, often swimming with its head in contact with the surface. In the turbid water it often frequents this is a necessity if it is to catch sight of its prey. A fish that is in a squirting mood (which is to say a fish that is hungry and has not spouted for a while) keeps looking upward and, when it sees a target, heads for it in a manner so characteristic that I could take it as a signal to start my camera. The fish gives the impression of stealth and determination as it closes in on the prey slowly and deliberately. The archer has binocular vision and its great eyes can be seen to range horizontally as they focus on the target. If the fish is high in the water, it moves toward its prey directly beneath the surface; if the fish is deeper, the approach path is diagonally upward [see *illustration on opposite page*].

In either case the fish swims until it is almost directly below its prey. The reason is immediately apparent, and it is important: The refraction of a ray of light decreases as the angle of incidence increases; light striking the water surface at right angles is not bent at all. When the archer fish is directly below its prey or nearly so, there is no refraction, or extremely little, of the light rays reflected from the quarry and reaching the archer's eyes. The eyes, in any case, are so close to the surface of the water that refraction can hardly be much of a problem.

Once it has arrived below the target, the archer fish makes a peculiar movement that occurs so quickly it can hardly be perceived except in a series of photographs. The fish shifts from a diagonal position, with the top of its head touching the surface, to a more vertical position. This characteristic movement was repeated in every spouting performance I recorded. Once it has reached its near-vertical firing position, the fish may move slightly forward or even backward, apparently adjusting its aim and making its final distance calculations. There is a noticeable increase in tension, visible as an extension of the pelvic and dorsal fins, and then the fish lets go its jet of water.

There would seem to be two reasons for the pivoting movement of a fish that is about to fire. One is that the shape of the archer fish's snout sets limits to the angle between the axis of its body and the line of fire. As will be seen, however, this angle is subject to quite some variation. A more important reason, I think, is that the near-vertical position minimizes parallax—the angle between the eyes-to-target line of sight and the mouth-to-target line of fire—and thereby eliminates what could be a serious obstacle to accurate aim [see illustration on page 229]. Parallax would still exist for short-range shooting, but as a matter of fact prey very near the surface are almost always seized by jumping rather than shooting. Accurate jumping is

made possible by the archer's binocular vision, and particularly by its ability to direct its eyes forward to see prey close to its snout.

One of the details I sought to pin down was the precise angle between the longitudinal axis of the fish's body and the line of fire. I found that the archer fish is able to score hits when this angle is anywhere from 140 to 170 degrees [see illustration on page 229]. In other words, the fish can adjust its direction of spouting within these limits. Actually this is rather surprising. One would have thought that the anatomy of the animal would prescribe a constant angle, in which case the archer fish, like a fighter plane with a fixed machine gun, could aim only by pointing its entire body in the proper direction.

One of the most difficult things to determine was the exact nature of the spouted stream of water. Earlier observers had spoken of single drops of water or a succession of drops. Even my film did not yield clear evidence whether there is in fact a single jet of water or a spray. It often appeared to me that there was a spray made up of a large number of individual drops discharged in rapid succession. It remained for some later investigators to clear up this question. Two years ago Heine Hediger and H. Heusser of the University of Zurich

made a careful study of the archer's spouting behavior with the aid of very high-speed motion picture photography. They found that the fish actually ejects a single jet of water that travels between two and four inches and then breaks up into a fine spray and a few large drops. It is these drops, which quickly pass the spray, that strike the insect prey.

I noticed that the area of frosted glass wetted by the stream of water was substantial and was usually larger in the vertical dimension than the horizontal. This is partly because of the fringe of spray that falls somewhat short of the target and partly, as Hediger and Heusser observed, because the fish is not absolutely still in the water as it shoots. It begins to spout while it is still "rocking" toward a vertical position, so that the beginning of the jet leaves its mouth at a smaller angle to the water than the end does.

A final reason for the large size of the area wetted by a stream is the force with which it is ejected, which is always the same regardless of the target's distance. This constancy in the expenditure of energy is a conspicuous example of rigidity in instinctual behavior. The amount of energy in a jet becomes apparent if the stream misses its target: Myers noted that some drops spouted by a three-inch archer at an insect three feet away missed and traveled more

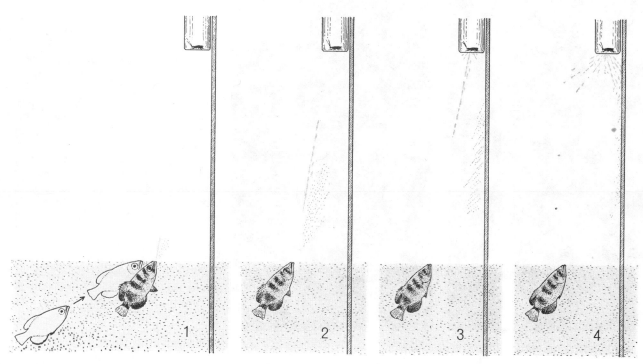

ACT OF SPOUTING is diagramed in these drawings based on a film made by the author. A characteristic approach is shown by the two fish drawn in outline (left). Once in position the fish pivots to a more vertical attitude, tenses its fins and ejects a stream of water (1). The stream breaks up (2) and strikes the target (3) as the fish relaxes its fins (4) and waits for the prey to fall.

than 15 feet! I observed that the jets carried at least 20 or 22 inches on a straight trajectory. An archer fish that fails to hit its prey with its first volley will immediately try again, once or several times, if it is not too tired.

Sheer force, however, is not always enough to knock the archer fish's prey from its perch. If the insect is on a vertical surface (my pane of glass or, in natural conditions, a twig or a branch), a jet of water striking its back would often "fasten" it more firmly instead of dislodging it. In this case, I was fascinated to observe, the fish apparently aims not directly at its prey but at a point just below it. The water ricochets from the vertical surface and catches the insect in the abdomen, lifting it neatly from its perch.

The archer fish's aim is not matched by selectivity in choice of target. When it is ready to spout, the fish will let fly at almost anything; in my experiments the fish spouted eagerly at drops of water on the frosted glass, paper cutouts in the shape of insects, a large paper rectangle and a real wasp that was much too big for the fish to swallow. When the target—any target—falls to the water, the fish seizes it. Only then does the fish decide whether or not the fallen object is edible. When I dropped a pin on the surface immediately after a fish had spouted, the fish rushed to seize it and then, of course, released it at once. This impetuous retrieving behavior immediately after shooting is probably dictated by the fact that other archer fish nearby will snatch the marksman's prey if they can.

A fish that has a lot of what I call "reaction-specific energy" stored up will spout even when it sees food that does not call for spouting at all. For example, I have placed a mealworm on the surface of the water near a hungry fish; the fish darts forward, stops short a fraction of an inch from the worm and squirts water into the air. A droller version of this behavior may rarely take place at the bottom of a tank. A fish rushes toward some food lying on the sandy bottom and then suddenly stops. As it does so the food zooms' off a few inches along the bottom, struck by an underwater jet from the fish. One is reminded of the circus clown who keeps stooping to pick up a ball but can never quite get it because he kicks it along the ground just as he reaches down.

Successful spouting seems to occur only when a fish is hungry and has enough energy available. The drive to spout is strongest in fish that are hungry and have not spouted for some time. If an archer fish is hungry but in a state of reduced energy because it has been spouting a lot, its efforts to hit a new target are unsuccessful. The stream of water simply falls short [see bottom illustration on page 229]. The fish may continue to squirt desultorily for a while; eventually it ceases even to try.

 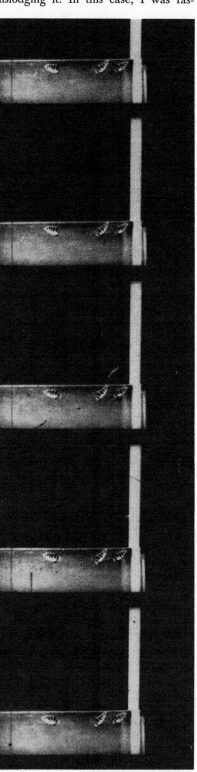

MOTION PICTURE SEQUENCES show two spouting acts. In each case an archer fish is seen (*top to bottom*) shifting to the near-vertical shooting position and then squirting at a target. The strip at right shows a fish spouting weakly because it has insufficient energy. These are frames from two segments of the author's film, shot at 24 frames per second.

A little should be said about the relation between the archer fish's spouting behavior and its eyes and eyesight.

The importance of binocular vision is indicated by the ineptitude of an archer fish that has only one good eye as a result of an injury or a parasitic infestation. It is unable to judge either distance or direction correctly. One of my specimens, which was blind in the left eye, regularly aimed its jet too far to the right and was also unsuccessful in jumping at a target.

The archer fish's eyes are not only large and capable of binocular vision but also structurally more highly developed than the eyes of most other fishes. The retina is particularly advanced. It has a very large number of cones and rods: I have counted eight or nine cones (for daylight vision) and 217 rods (for vision in semidarkness or in turbid water) ranged along a line .08 millimeter long. (The eel, by comparison, has in the same distance seven much smaller cones and 143 rods.) This retinal structure gives the archer fish excellent vision in dim light as well as bright. Hugh Smith, who squeezed the gill covers of archer fish to demonstrate how they could fire a jet of water, asserted that they could extinguish cigarettes in total darkness. There is good reason, in fact, to believe that an archer fish can shoot down a luminescent insect at night.

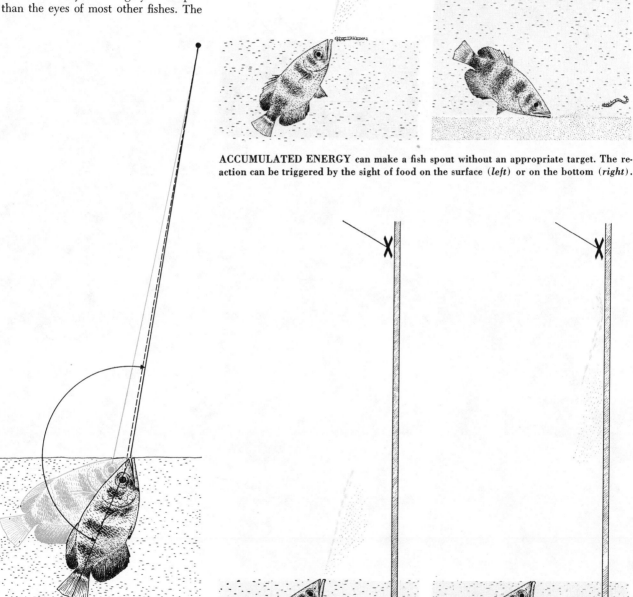

ACCUMULATED ENERGY can make a fish spout without an appropriate target. The reaction can be triggered by the sight of food on the surface (*left*) or on the bottom (*right*).

PIVOTING MOVEMENT before spouting begins brings the fish from a nearly horizontal (*color*) to a nearly vertical (*black*) position. This has the effect of reducing the parallax between the direction of sighting (*colored and broken black lines*) and the mouth-to-target direction of spouting (*solid black line*). The angle (*arrow*) between the spouting direction and the long axis of the body is between 140 and 170 degrees.

ENERGY IS REDUCED after the fish has spouted repeatedly. If it is still hungry and sees a target (in this drawing a paper cross on a wire), it spouts weakly and falls short.

The Weddell Seal

by Gerald L. Kooyman
August 1969

*This Antarctic mammal can swim under the ice for
more than an hour without coming up for air. How is
it able to find its way back to its breathing hole,
particularly during the long Antarctic night?*

A century and a half ago the explorer James Weddell, during a record penetration into the Antarctic, discovered a remarkable mammal that spends its life in rigors of cold few other mammals are known to endure. The Weddell seal, as it has come to be known, lives in the coastal waters of the Antarctic Continent, where the air temperature drops as low as −70 degrees Fahrenheit, the water temperature is generally about 28 degrees and the sea surface is frozen to a depth of several feet for about eight months of the year. The animal has to forage for its food under the ice. Yet the Weddell seal manages to thrive in this icy fastness the year round, and during the long winter night it actually grows fat.

Weddell described the seal in his book *A Voyage Towards the South Pole* in 1825, but little more was learned about the animal until major expeditions began to visit the Antarctic early in this century. In an expedition of 1901 to 1904 the English physician Edward A. Wilson (who later lost his life in the Scott expedition to the South Pole) made a general study of the Weddell seal's habits. The animal also received the attention of investigators in later expeditions, and it has been studied intensively since the establishment of permanent bases in the Antarctic in the 1950's. Probably more is now known about the Weddell seal than is known about any other marine mammal. The reasons for the great

interest in this animal are twofold: it provides a unique opportunity for study of a mammal's adaptation to extreme conditions, and both its habitat and the animal itself furnish an ideal laboratory situation for investigation of the animal's behavior.

By virtue of its isolation in an environment too harsh for human invasion (except by scientific expeditions) the Weddell seal has no fear of man. Whereas seals in other parts of the world have been hunted so aggressively that they can be approached only by stealth, the Weddell seal does not flee from men either afoot or in vehicles. One can drive up to the seal in a truck and capture it easily. The ice platform over the deep sea, particularly at McMurdo Sound, where the ice remains firm into the Antarctic summer, provides a stable base for a marine laboratory (in contrast to a tossing ship). And the water under the ice is the clearest oceanic water in the world, so that the seal's movements in the water can be seen from an observation chamber under the ice. These conditions, together with the presence of large populations of Weddell seals in the sound, have drawn biologists to various specific investigations of the animal. Among others, investigators from the Woods Hole Oceanographic Institution have made a comprehensive analysis of underwater vocalization by the seals, physiologists from the Scripps Institution of Oceanography have studied their defense mechanisms against asphyxiation during dives, and zoologists from the University of Canterbury in New Zealand have analyzed the seals' population structure.

My own investigations have focused on the seals' diving behavior and their capacity for doing without oxygen and withstanding pressure in lengthy or deep dives. Thorough studies of these

matters had never been made on a marine animal in its natural environment, and the Weddell seal provided an outstandingly suitable subject for experimental examination of its underwater behavior.

There are small populations of Weddell seals in various island groups of the South Atlantic and sub-Antarctic regions, but the main habitat of this animal is the icy waters off Antarctica. It is known to breed at least as close as 10 degrees from the South Pole. The Weddell seal lives on fish, and for most of the year it must hunt them under the ice. For access to the water, and to return to the air for breathing, it depends on breaks in the sea ice produced by the wind, tides and currents. The animal has unusual canine and incisor teeth that it uses to ream out breathing holes in the thin ice when the breaks freeze over. One of the principal mysteries about the Weddell seal is how it manages to find the widely dispersed breaks in the ice when it is underwater, particularly in the darkness of winter.

For our studies of the animal's underwater performances we set up a station (a heated hut) on the sea ice in a location where the water was approximately 600 meters deep. We chose an area where the ice was free from cracks and seal holes for some distance around the station. At the station itself we cut a hole in the ice through which we would release our seals into the water. Our procedure was to capture a seal on the ice several miles away, bring it to the station, attach a small pack of instruments to its back and then obtain a record of its movements after it plunged into the water through the hole we had provided. We used a variety of instruments: a manometer for measuring the maximum depth of dive, other devices for recording the depth of the dive as a function of time,

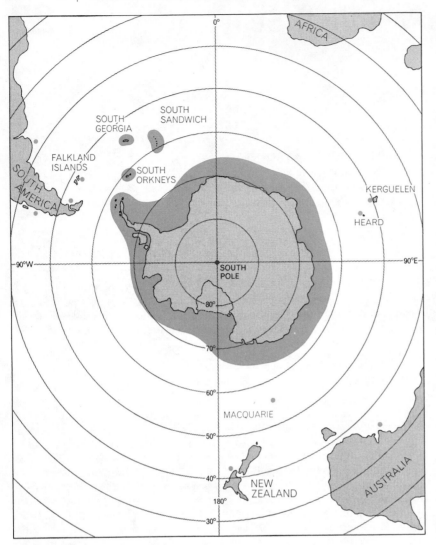

RANGE OF WEDDELL SEAL is shown on this map. The animal is primarily a coastal dweller of the Antarctic Continent; its range is shown in color. Dots indicate areas where the seal is occasionally found. Its most northerly appearances have been in sightings from the Falkland Islands and the coasts of Uruguay, Chile, New Zealand and southern Australia.

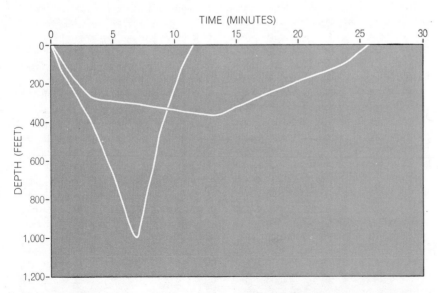

TIME AND DEPTH of exploratory and deep dives of the Weddell seal are portrayed. The patterns were obtained partly from watching seals from an underwater chamber after they were released in a new hole and partly from recording equipment strapped to the seals.

and an underwater telemetry device that supplied us with running records during the dive. With these aids we were able not only to determine the depth and duration of the seals' underwater swims but also to estimate the swimming speed.

We observed more than 1,000 dives, and we found that the Weddell seal typically engages in three types of dive with different purposes. In each kind of dive the animals' performance in the water raised interesting questions for investigation.

The first type of dive is a brief foray of local exploration. When we released a seal into the water at the dive station for the first time, it would swim about at shallow depths, no more than 100 meters, and soon return to the breathing hole, usually within five minutes. In these initial, local dives, which might be repeated several times, the seal evidently occupied itself in testing the immediate surroundings and searching out other occupants of the water. If there were several seals in the water at the same time, they often engaged in competitive fighting, apparently for privileges such as being the first in line at the breathing hole.

After exploring the local scene the seal usually proceeded to a second stage of much more extensive exploration. These dives were somewhat deeper, down to 130 meters or so, and lasted from 20 minutes to an hour or more. The animal would range over considerable distances under the ice, often for miles, looking for other breathing holes. If it failed to find one, it would return to the dive station to rest. It sometimes repeated these long, exhausting swims for several hours with a few short, local dives after each prolonged dive.

When a seal did find another breathing hole, it might or might not come back to the station hole. One seal returned from its expeditions again and again and made its home at our station for more than a month. We identified it easily because its head was hairless, and we developed a fond feeling for it. When a seal did not return to us after finding another hole, we would hunt for it to recover our instruments. With the aid of our rapid means of transportation, including a helicopter, we succeeded in finding nearly all the errant seals. In one case we recaptured the animal on the ice 15 miles from the dive station.

The eminent marine physiologist P. F. Scholander, of the Scripps Institution of Oceanography, has investigated the extraordinary mechanisms that enable diving animals such as seals to spend long periods underwater without as-

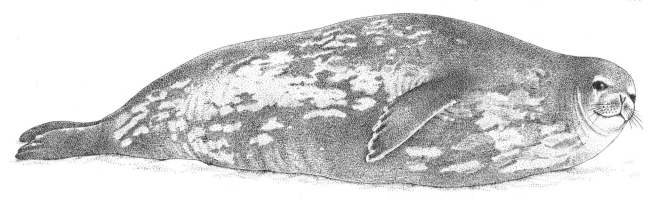

ADULT WEDDELL SEAL can attain a length of about 10 feet and a weight of as much as 1,000 pounds. Most of the animal's fur is black or dark brown; the lighter spots are almost white. Seen from underwater the seal looks black and, like the normal pattern of light near the surface, rather mottled. When the seal is out of the water, the fur is seen to be predominantly brown after it dries.

phyxiating. In laboratory experiments he found that during a dive a seal slowed its heartbeat to a tenth of the normal rate, reduced its rate of metabolism and drastically restricted the circulation of the blood to all tissues except the heart and the brain [see "The Master Switch of Life," by P. F. Scholander; SCIENTIFIC AMERICAN December, 1963]. In this way the animal conserved the supply of oxygen in its blood during its dive. Scholander showed in the laboratory that a seal could stay underwater for 20 minutes or more.

In our observations of unrestrained Weddell seals in their natural environment we found the animals to be capable of performances considerably better than this. They sometimes spent up to an hour in the water without surfacing for air, as our instruments showed. How far did they swim? We can estimate from indirect indications that they frequently covered several miles. We found that a seal normally took about 12 minutes to swim from one hole to another approximately a mile away; if we assume that it swam a straight route (and our observations indicated that the seals rarely deviate from a fairly straight line in such a journey), this meant that the animal's swimming speed was about five miles per hour. It is obvious, therefore, that in an underwater stay of half an hour or more a seal swims several miles.

On occasion a seal would swim away from our station and return after 30 to 60 minutes without having surfaced elsewhere—an indication that it had failed to find another breathing hole. This raises a fascinating question: How does the animal know when it must turn back? What manner of internal signal induces the seal to turn back, and is this control mechanism sensitive to the half-life of the seal's submersion capacity? Research on the seal's ability to survive underwater has not yet supplied an answer to this puzzle.

There is a corollary question: What gift of farsighted perception informs the seal when a breathing hole lies ahead within reach so that it can safely pass the point of no return?

Certain observations of the seal's behavior during long trips offer grounds for speculation. On such a trip the seal seemed to maintain a constant heading: in almost 90 percent of the cases the seal would swim away from the station in a particular direction and return from the same direction. This practice makes it comparatively easy for the seal to find its way back to the breathing hole it has left. Secondly, in their long trips of exploration the animals swim at relatively shallow depths, and in the daylight conditions of the Antarctic summer the frozen sea surface (and any discontinuity in it such as a crack or hole) is readily detectable from the normal depth of their prolonged dives.

How does the seal find its way, however, in the winter darkness? During the long winter night the only light in the Antarctic seas, other than that from the moon and stars, is the bioluminescence of certain living things in the water. Does the Weddell seal use sound (perhaps echo location) or other external cues to navigate in the dark? It would seem that the animal must then depend on a navigational aid of some kind other than vision, not only for its shallow exploratory journeys but also in the third type of dive in its repertory: deep dives in pursuit of food.

In these dives the Weddell seal commonly descends to depths of 300 to 400 meters, and we have found that it can go as deep as 600 meters. The deep dive is relatively brief, seldom lasting more than 15 minutes. With three-minute rests at

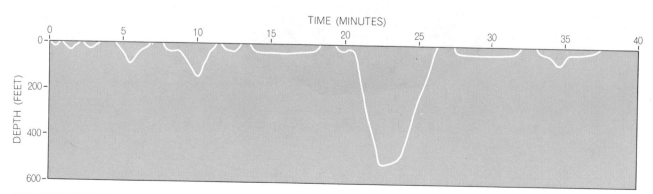

DIVE PATTERNS of the Weddell seal include a series of short, shallow explorations immediately after the seal is released in an unfamiliar hole. Spaces between dive profiles represent times when the animal returned to the breathing hole for air or rest or both.

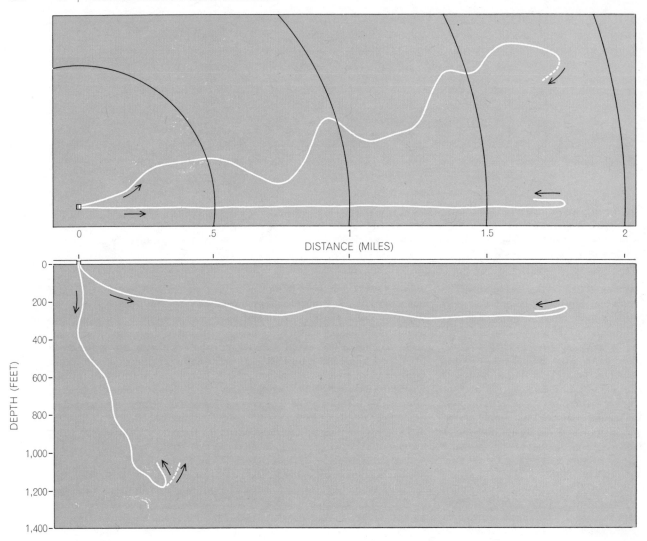

DISTANCE (MILES)

DEPTH (FEET)

EXPLORATORY AND DEEP DIVES are portrayed. At top are plan views of exploratory dives; headings are hypothetical. Random deviations (*upper curve*) would be hazardous because the seal might lose track of its breathing hole; the actual heading appears to be straight out and back (*lower curve*). At bottom are elevation views of two dives. In an exploratory dive the seal retains visual contact with the ice; in a deep dive random deviations are possible because the seal can find the breathing hole merely by going up.

BREATHING HOLE surrounded by a great expanse of ice is characteristic of the Weddell seal's environment. The seals have adapted to the topography by developing navigational skill at returning to familiar holes and skill at reaming new holes in thin or cracked ice.

the surface between the dives, the seal may continue its feeding dives for several hours. The problem of returning to the breathing hole is much simpler than it is in an exploratory dive; in the deep dive the seal remains in the vicinity of the hole, the direction for the return (more or less straight up) is reasonably clear, and the short duration of the dive gives the animal a large margin of safety within its capacity for underwater endurance.

Our studies of the seal were conducted during the season when it has light to assist its navigation. Future investigations in the dark season may well prove highly interesting. Perhaps the animal's diving behavior changes to a different pattern during the winter. At all events, the Weddell seal's method of navigation under the ice, in summer or winter, is no doubt both effective and complex, utilizing an array of external cues and internal receptors. The analysis of its guidance system should be an inviting subject for study.

No less interesting is the seal's ability to dive to great depths without suffering ill effects from the increase in pressure or undergoing the effects of decompression ("the bends") on its return to the surface. The Weddell seal adjusts to rapid changes of pressure: it can descend to several hundred meters at the rate of 120 meters per minute without harm. How does it withstand the high pressure at such depths?

The seal's anatomy indicates that the animal copes with the pressure not by resisting it but by yielding to it. Its main body cavities are so constructed that they readily compress. The rib cage is very flexible, and the diaphragm underlying the lungs is at an oblique angle to the chest cavity. Consequently external pressure can cause complete collapse of the lungs, and the gases are pushed into the bronchi and the trachea. In the Weddell seal and other Antarctic seals the trachea has an unusual shape: the tube is not round but has the form of a flat bow. Hence the trachea in turn is easily squeezed by pressure to complete collapse. Gases may also be compressed in the cavity of the middle ear; it has been observed that in some seals, sea lions and whales this cavity is lined with a network containing expandable veins, which probably swell with blood under pressure and so reduce the volume of the cavity. These anatomical characteristics suggest that during a deep dive the external water pressure may successively drive the gases out of the lungs and bronchioles and into the bronchi and the

trachea. The lungs collapse, and the trachea is compressed; thus the animal can avoid dangerous differences between external and internal pressure. If this hypothesis is correct (it has not yet been verified), the seal's tolerance of pressure in a deep dive depends on the extent to which its volume of internal

gases can be compressed in the lungs, trachea and middle ear.

The avoidance of the bends is a more complicated matter. In mammals high pressure forces nitrogen from the air in the lungs into solution in the blood and tissues, and if the body is rapidly decompressed, the bubbling of the nitrogen

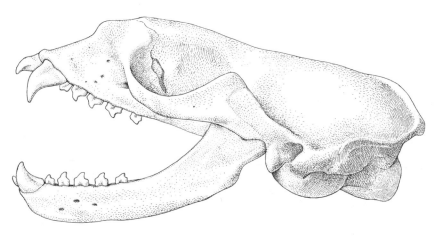

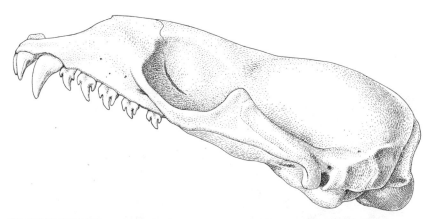

TOOTH PATTERN of the Weddell seal represents one of the animal's adaptations to its environment. At top is a Weddell seal's skull, showing how the upper teeth are shaped for reaming ice. At bottom for comparison are the upper teeth of the nonreaming leopard seal.

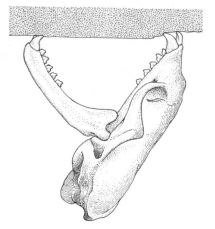

SEAL REOPENS HOLE that has frozen over; the task is done by a reaming action that relies mainly on the upper teeth. The smaller upper incisor teeth engage the ice first.

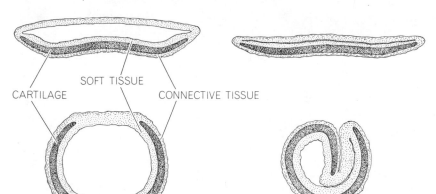

SOFT TISSUE

CARTILAGE   CONNECTIVE TISSUE

COMPARISON OF TRACHEAS of a Weddell seal and a human being indicate one of the seal's physiological adaptations to deep diving. The seal's trachea is normally flat (*top left*), and under pressure it collapses readily without damage to the cartilage (*top right*). A human trachea (*bottom*) is normally ringlike and cannot collapse much without damage.

DOCILE BEHAVIOR of Weddell seal when approached by humans is evident as a female calmly allows Charles M. Drabek of the University of Arizona to draw off milk with a tube.

out of the blood produces the painful embolisms known as caisson disease, or the bends. This sometimes fatal sickness, long recognized as a hazard of tunnel diggers, helmet divers and scuba divers who breathe compressed air, has recently been found to attack breath-holding divers who venture to depths of 60 feet or more in a number of dives made during a short period of time; among divers of the South Sea islands it is called *taravana* [see "The Diving Women of Korea and Japan," by Suk Ki Hong and Hermann Rahn; SCIENTIFIC AMERICAN; Offprint 1072]. Seals are not entirely immune; in one experiment a seal that had submerged to a considerable depth died of embolism when it rose too rapidly. We have found, however, that Weddell seals rise at the rate of 120 meters per minute without any apparent discomfort.

What mechanism protects the seals? Many years ago Scholander suggested that the lungs of marine mammals probably collapse at depth, and that the consequent absence of nitrogen in the lungs prevents absorption of the gas into the bloodstream. On the basis of lung-volume measurements he calculated that the alveoli of the lungs collapse at a depth of about 50 meters in the water. This implies that seals should be safe from attacks of the bends after dives of more than 50 meters but may be vulnerable in shallower dives. We have observed, in fact, that Weddell seals spend most of their time below 50 meters even in their comparatively shallow exploratory dives, and they usually come up slowly from a shallow dive. Moreover, they rarely make prolonged shallow dives serially. Thus their diving behavior, as well as the anatomical structure favoring collapse of the lungs, provides safeguards against the bends.

In the Weddell seal evolution has produced an animal exquisitely suited to its special environment—an environment so inhospitable that no other mammal save technologically armed man has chosen to invade it. Because of its uniqueness we are finding the Weddell seal to be a valuable animal for biological research. We must also observe, however, that for all its hardiness the species has certain disquieting vulnerabilities. One of them is its total lack of any protective fear of man or other predators. It would be an easy prey for commercial hunting. As a source of oil and protein it could provide only an ephemeral supply; on the other hand, as a source of enlightenment about life and evolution the Weddell seal has a value that is inestimable.

Dolphins

by Bernd Würsig
March 1979

*These descendants of land mammals that took to sea
have a large brain, learn quickly and exhibit a rich
vocal repertory. Yet lack of evidence leaves open the
question of how intelligent they are*

Because dolphins have big brains, are quick to learn tricks devised by human trainers and exhibit a rich repertory of vocal signals, they are widely reputed to have a level of intelligence unmatched by any other animal and perhaps even the equal of man's. On the basis of observations of dolphins in several of their natural habitats I believe the effort to position them firmly in the spectrum of animal intelligence is premature. In the present state of marine technology it is simply impossible for human observers to spend more than brief and isolated moments with an animal that lives in the ocean and moves rapidly over great distances. When a great deal more is known about the behavior of dolphins, the question of their intelligence will answer itself. At present the most one can say is that dolphins are gregarious herding animals, comparable in their individual and social behavior to more easily observed herding and flocking mammals on land.

Dolphins evolved at least 50 million years ago from land mammals that may have resembled the even-toed ungulates of today such as cattle, pigs and buffaloes. After taking to the sea dolphins became progressively better adapted to life in the water: their ancestral fur was replaced by a thick coat of blubbery fat, they became sleek and streamlined, they lost all but internal remnants of their hind limbs and grew a powerful tail, their forelimbs were modified into steering paddles and apparently as a further aid in steering and stabilization many species evolved a dorsal fin.

More than 30 species of dolphins can be identified. They belong to the suborder Odontoceti, or toothed whales, of the order Cetacea. (The quite similar porpoise, which is often confused with the dolphin, is also a member of the whale family; it is distinguished from the dolphin mainly by having a less beaklike snout and also by its laterally compressed, spadelike teeth. It should be noted, however, that many American students of marine mammals refer to all small odontocete cetaceans as porpoises, regardless of their physical charac-

teristics.) A few dolphin species live in fresh water, but most species have an ocean habitat. The freshwater species travel in small groups or are nearly solitary, whereas the ocean species (such as the Pacific spotted dolphin) may congregate in aggregations of several thousand. Such numbers are reminiscent of buffalo herds in North America and grazing animals on the Serengeti plains of Africa; one wonders about behavioral and ecological similarities between the dolphins and their distant terrestrial relatives.

In general highly social mammals have complex social signals and a rich behavioral repertoire. Hence they can interact with other members of their group in sophisticated ways. Examples include signals of aggression that are useful in establishing and maintaining dominance hierarchies, signals for courtship, warning sounds or movements at the approach of a potential attacker and many other signals that contribute to the functioning of the group and its individual members.

The effort to accumulate data on the behavior and social systems of dolphins is made difficult by the fact that most of their communication goes on below the waves. It is extremely difficult to approach a group of dolphins in a boat and to stay with them long enough to begin to understand their social system. All one sees is a group of dorsal fins as the animals surface to breathe, and then they are lost to view as they move on underwater. Even in the rare circumstances when the water is calm and clear and the dolphins can be seen for more than a few minutes, the proximity of a boat may disturb them, so that it becomes difficult to separate what is natural in their behavior from what is unnatural, being merely a reaction to the boat.

As a result of these problems and others most of the early observations involved dolphins in captivity. Although captivity must be an awkward situation for animals that are accustomed to a life with few physical boundaries, the captive dolphins nonetheless yielded useful

data simply because observers could watch them for extended periods of time. The best-known work of this kind was done by Margaret C. Tavolga of Marineland of Florida. She observed a group of 12 bottlenose dolphins (*Tursiops truncatus*) in a large tank for a total period of about five years.

Tavolga found that the group had a definite dominance hierarchy. The one adult male, which was the largest animal in the group, was more aggressive and less fearful than any of the females, subadult males or young dolphins. In general the larger animals were dominant over the smaller ones.

Similar data were obtained by Gregory Bateson at the Oceanic Institute of Hawaii. He found a dominance hierarchy in a group consisting of two spotted dolphins (*Stenella attenuata*) and five spinner dolphins (*S. longirostris*). The largest male threatened other dolphins (by lunging at them or showing his teeth) but was never threatened himself. The second-ranking dolphin, also a male, threatened the animals below him, and so on down the line. Bateson's findings showed, as the findings of other workers have shown, that the hierarchy is not as strict as it is among some other mammals. For example, the lowest-ranking male was still able to mate with a female without being challenged by the largest male.

If the dominance hierarchy is not necessarily related to the access of the males to the females (as it is, for example, in the harem of the elephant seal), one wonders what its function is in wild dolphin populations. Kenneth S. Norris and Thomas P. Dohl of the University of California at Santa Cruz have speculated that the function may be to organize the members of the group to deal with a variety of situations. For example, threats and chases by the larger dolphins could cause the smaller females and young animals to be herded into the center of the group, where they would be better protected from such potential predators as sharks and killer whales. Norris believes he has seen such struc-

THREE STAGES OF A LEAP are demonstrated (unintentionally) in the Sea of Cortés (the Gulf of California) off the coast of Mexico by three dolphins of the species *Delphinus delphis*. The repertory of leaps, spins and somersaults executed by dolphins is richly varied.

HERD OF DOLPHINS of the species *Delphinus delphis* was photographed leaping in the Atlantic. Dolphins are social mammals that sometimes congregate in quite large herds, as they are doing here, but more often they are found in subgroups of perhaps 20. Dolphins are also air-breathing animals, so that their leaps serve in part to enable them to breathe. The animals also leap during play and hunting.

turing in spinner dolphins and spotted dolphins.

For such a system to evolve it is helpful and perhaps necessary that the animals possessing it be genetically related. W. D. Hamilton of the Imperial College of Science and Technology and other biologists have argued that closely related animals should tend to protect one another more than they protect distant relatives, since close relatives share more genes. If a mother saves her offspring by herding it inside the group while putting herself in at least some danger, her apparent altruism will be adaptive for the group because a significant proportion of her genes will be preserved.

If dolphins herd group members in this way, it is likely that at least some of the animals in a group are related. Dolphins have also been seen to help an ailing member of the group reach the surface to breathe and to protect a group member from predators or other dangers. These patterns of behavior have often been cited as evidence of human-like altruism and of great intelligence. It appears more likely that they represent an outgrowth of an evolved tendency to help related individuals.

Unfortunately the degrees of consanguinity in a wild dolphin population are not known. Only recently has any idea at all of social structure among dolphins been gained, and the knowledge is complicated by the existence of many different species and many more separate populations within species. From terrestrial mammals it has been learned that the social system represents in part an adaptation to the population's habitat. For example, wolves that live mainly on deer tend to travel in small packs, whereas wolves that hunt moose are found in larger and more highly organized societies. This difference is apparently related to the need for a coordinated effort by several animals to successfully bring down the larger prey.

A few excellent studies that make comparisons possible among dolphin species and populations have been done recently. Norris and Dohl studied Hawaiian spinner dolphins from sea cliffs and from underwater. They found that these highly social mammals traveled in schools averaging 25 members. The structure of the school varied during the day in a predictable manner. In the morning the dolphins moved slowly and in tight groups, with individuals almost touching. They appeared to be resting. Later they became increasingly active, swimming faster, with individuals leaping clear of the water in the spins, somersaults and other displays for which dolphins in oceanariums are famous.

At such times the schools became more spread out, with animals often as much as 20 meters apart. Moreover, groups tended to join, so that 50 or more members might constitute the expanded school, with all the animals moving in the same direction. As night approached, the school moved several kilometers away from the shore, entering deeper water and beginning deep dives in order to feed on fish several hundred meters below the surface. The array consisted of many widely spaced groups within an area with a diameter of several kilometers. Because Norris and Dohl were able to recognize some individuals, they found that a given small group of individuals tended to stay together but often shifted as a unit from one school to another.

Norris and Dohl suggested that dolphins may form close groups while they are resting so that they can employ the combined sensory abilities of all the individuals in the school to scan the environment and to detect potential danger. It is well known that dolphins can scan the water by echo location over much greater distances than would be possible by eyesight. (In echo location a dolphin projects high-frequency sounds in short pulses, much as bats do. The sounds bounce off objects, and the echoes give back information on the distance, size, shape and even texture of the object.) Norris and Dohl's hypothesis about combined sensory abilities, particularly during rest, relies in part on the ability of dolphins to get information from echo location. Presumably each dolphin

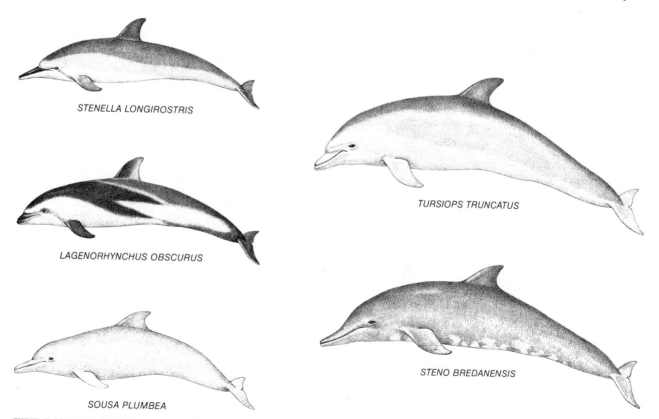

STENELLA LONGIROSTRIS

LAGENORHYNCHUS OBSCURUS

SOUSA PLUMBEA

TURSIOPS TRUNCATUS

STENO BREDANENSIS

FIVE DOLPHIN SPECIES are portrayed to indicate their differences in shape and marking. They are the bottlenose dolphin (*Tursiops truncatus*), the spinner dolphin (*Stenella longirostris*), the South Atlantic dusky dolphin (*Lagenorhynchus obscurus*), the Indo-Pacific humpback dolphin (*Sousa plumbea*) **and the rough-toothed dolphin (*Steno bredanensis*). All of them are drawn to the same scale. About 25 other species of dolphins are known. Porpoises, which are quite similar, have a blunter snout and are ordinarily shorter and fatter.**

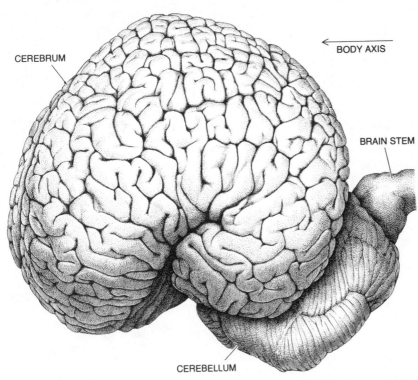

CEREBRUM

BODY AXIS

BRAIN STEM

CEREBELLUM

BRAIN OF DOLPHIN is depicted for the species *Tursiops truncatus*, the bottlenose dolphin. A dolphin brain weighs about 3.5 pounds (1.59 kilograms); a typical human brain weighs about three pounds (1.36 kilograms) but is larger than the dolphin's in proportion to body weight.

also represent a form of communication. A leap is usually followed by a loud slap or splash as the dolphin enters the water. Such sounds travel fairly long distances underwater and may signify the presence of the leaper to others. Indeed, groups of some dolphin species sometimes converge on an active, leaping school from a distance of several kilometers. Nighttime feeding in deep water is also attended by much leaping and loud splashing. At this time the members of a group are quite widely separated, and the assumption once again is that leaping may serve to communicate location and possibly information such as the number of other dolphins nearby and what they are doing.

Graham Saayman and C. K. Tayler of the Port Elizabeth Museum in South Africa studied Indian Ocean bottlenose dolphins (*Tursiops aduncus*), and my wife Melany Würsig and I studied South Atlantic dusky dolphins (*Lagenorhynchus obscurus*). Both species have habitats similar to the habitat of the Hawaiian spinner dolphin. The habitats are coastal-pelagic, meaning that all three populations can often be seen and studied from the shore but that they also move far from the shore, usually to feed. All three populations exhibited similar patterns of behavior and movement. It therefore seems reasonable to say that the habitat of these marine mammals is largely responsible for their way of life.

This assertion can be examined by studying dolphins in a different environment. Three such environments can be found: deep-ocean, coastal and freshwater. The most thoroughly examined populations have been coastal ones. Susan H. Shane of Texas A & M University observed Atlantic bottlenose dolphins off Texas; A. Blair Irvine, Randall S. Wells and Michael Scott of the University of Florida observed them off Florida; Melany Würsig and I observed them off Argentina, and Saayman and Tayler observed Indo-Pacific humpback dolphins (of the genus *Sousa*) off South Africa. Again major similarities were evident. They can be summarized by a description of our study of bottlenose dolphins off Argentina.

The bottlenose dolphin is a coastal species in many parts of its worldwide range. Hence it can be observed readily from the shore. For a period of 21 months my wife and I observed a school of bottlenose dolphins that passed close to the shore (always within a kilometer) in water less than 40 meters deep. We studied them by observations from coastal cliffs and from a small rubber boat, by underwater recordings of their sounds, by photographing their dorsal fins in order to recognize individuals and by tracking their movements with a surveyor's transit on the shore. (With a transit one can determine precisely the

in a closely organized school can hear the echo-location sounds made by other members of the group. Therefore even though any given individual might not make many sounds, much information about the environment would be rapidly and efficiently disseminated to all. It is also probable that a resting group swims

close to shore in order to be in shallow water that is not frequented by large deepwater sharks.

During periods of alertness the spacing of the spinner dolphins increases and the animals do a great deal of leaping. This activity may in part be play, as many people have suggested, but it may

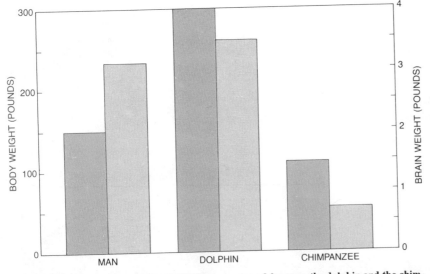

BODY WEIGHT (POUNDS)

BRAIN WEIGHT (POUNDS)

MAN   DOLPHIN   CHIMPANZEE

BRAIN WEIGHT AND BODY WEIGHT are compared for man, the dolphin and the chimpanzee. Man's brain weighs about 2 percent of his body weight, the chimpanzee's less than 1 percent and the dolphin's slightly more than 1 percent. Although rough ideas of comparable intelligence can often be obtained by comparing the ratio of brain weight to body weight, other factors such as the length of the body, the manner in which limbs are used and the complexity of the brain must also be considered. It is not possible to say on the basis of a simple comparison of brain weight to body weight that one kind of organism is more intelligent than another.

animal was the only one that consistently slapped its tail against the surface of the water when a boat approached. Experienced dolphin trainers say that tail slapping can signify anger, aggression or warning. It is tempting to say that the large dolphin was in some sense the leader of the group, but I do not have enough information about the social system and behavior of dolphins to substantiate the hypothesis.

Two other groups of five and six dolphins that we always found together consisted of adults of approximately the same size. It is possible (again I conjecture) that the groups were made up of nonbreeding members of the population, much as bachelor herds of elephants travel together.

The most detailed findings on the age and sex composition of dolphin schools have come from the work of Ir-

vine, Wells and Scott on Florida bottlenose dolphins. They captured 47 dolphins and put tags and other identifying marks on them in order to recognize them later in their natural habitat. As they tagged the animals they were able to determine the sex and size of the individuals.

Once the dolphins had been released Irvine, Wells and Scott found that the home range of the resident herd covered about 85 square kilometers. Females and calves often traveled in groups that included only a few adult males or none. Such males tended to associate more often with calfless females than with mothers and young, and they rarely associated with subadult males. The subadult males were at times found in bachelor groups far from the other dolphins. Several females were seen with their calves for as long as 15 months. Hence a

strong social bond exists between mother and calf, probably continuing long after weaning. No such long-term association between a male and a calf has been observed.

It should be emphasized that the social relations of dolphins are not clearcut or immutable. They are highly variable. Nevertheless, a few major features are apparent. The bonds between mother and calf are strong; the bonds between male and female and male and calf appear to be less so. This comparison suggests that mating is somewhat promiscuous. Subadult males may be excluded from the normal social routine but subadult females are not, which suggests that adult males may copulate more with various females than females do with different males. Such a relation is indicative of a polygynous mating system, which is also common among terrestrial mammals.

Less can be said about the behavior of dolphins than about their social relations, even though a considerable body of accounts of behavior has been built up from observations of dolphins in nature and in captivity. I believe not enough is yet known to support any firm and broad statements. Still, a few major examples of behavior can be cited. Bottlenose dolphins (and other species) appear to engage in courtship and copulation throughout the year, as is often indicated in the wild by belly-to-belly swimming. Yet bottlenose dolphins and some other species have a definite yearly calving peak (sometimes two peaks). Among the dolphins we observed off Argentina all the calves were born in the summer. This finding indicates that a physiological change in the male or the female causes conception to occur in a limited period. Such a change has been documented in seasonal increases in the weight of the testes in the males of several dolphin species.

Yearlong mating also implies that courtship may have more than a sexual connotation. Several investigators have suggested that such interactions may also serve to define and strengthen social hierarchies and bonds. The argument is reinforced by the frequent homosexual activity seen at least among captive dolphins. Future studies may show that "homosocial" might be a better term. A carry-over of sexual signals to dominance hierarchies is seen in many other mammalian groups.

A second behavior found in almost all dolphin species is leaping. I have mentioned that it tends to occur most often when animals are widely separated and so may have a communicative function. Bottlenose dolphins off Argentina leaped far less than dusky dolphins in the same vicinity did, even when both species were hunting fish in essentially the same manner. The bottlenose dolphins, however, moved in one school

MEANS OF IDENTIFICATION of individual dolphins is provided by the pattern of nicks and scars on the trailing edge of the dorsal fin. These photographs represent a sampling of 12 bottlenose dolphins from a group of about 50. All individuals in the group could be identified.

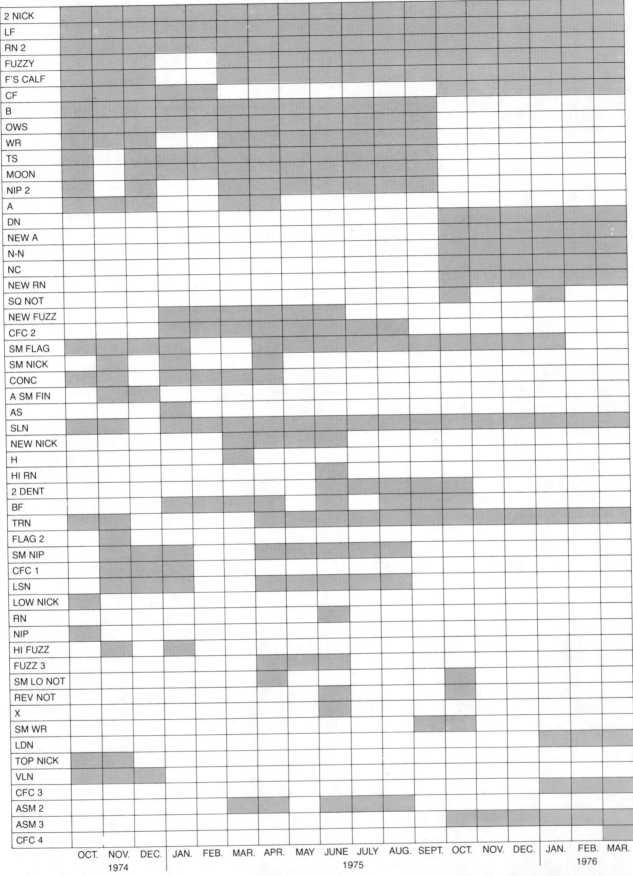

| | OCT. NOV. DEC. 1974 | JAN. FEB. MAR. APR. MAY JUNE JULY AUG. SEPT. OCT. NOV. DEC. 1975 | JAN. FEB. MAR. 1976 |

GROUP COHESIVENESS of dolphins is indicated by this chart of the presence and absence of 53 known bottlenose dolphins off the coast of Argentina over a period of 18 months. The designations at the left are abbreviated names the observers gave the dolphins. A bar opposite a designation indicates that the individual was seen near shore at least once during the corresponding month; a blank space means the animal was not seen during that month. A dolphin shown as having been seen may have been seen more than once in the month.

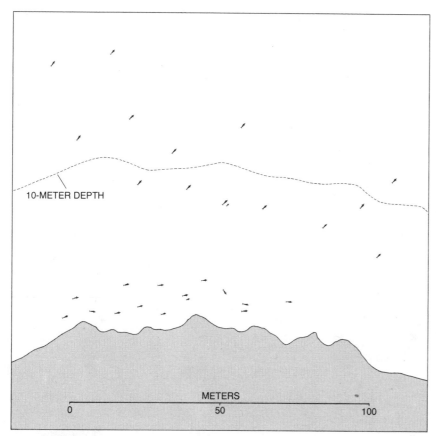

GROUP FORMATIONS of bottlenose dolphins off the coast of Argentina varied according to whether the animals were in shallow water close to shore or in deeper water farther out. In shallow water the dolphins were individually hunting rock-dwelling fishes, whereas in deep water they functioned as a group to find schools of anchovies and to herd them to the surface.

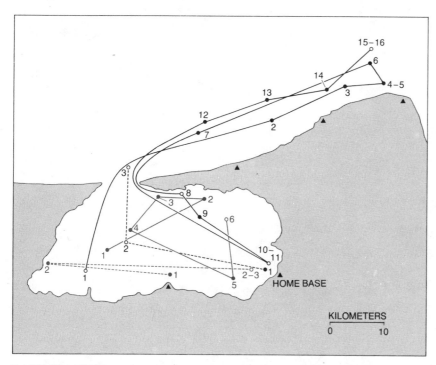

RADIO TRACKING of four dolphins off Argentina (in the Gulf of San José and the adjoining open ocean) produced these patterns of movement. Each dolphin is represented by a different track, and the number beside each circle shows the animal's position that number of days after it was fitted with a radio. Solid circles represent known positions, obtained either by triangulation from the shore points indicated by triangles or by approaching the dolphin in a boat, and open circles represent estimated positions. The animals' daily locations are given as of midday.

and so would have little cause to communicate with other members of the school by leaping. The dusky dolphins moved in as many as 30 small schools in one vicinity, and we often saw schools coalesce when leaping began. Leaps undoubtedly have other functions, such as helping to herd or catch prey, but the function of communication among members of a particular species may be important.

On the basis of the rather modest body of knowledge about dolphins, what can be said of their intelligence? Dolphins are certainly adept at learning complex tasks, as they demonstrate in their tricks in oceanariums, and they remember the tasks for years. They also have been shown to be capable of relatively abstract thinking. For example, Karen Pryor, working at the Oceanic Institute in Hawaii, trained rough-toothed dolphins (*Steno bredanensis*) to perform a new trick for a reward of fish. After several days of training they exhibited ever-different types of leaps and contortions, apparently "realizing" that the forms of behavior they had displayed previously would not be rewarded. Still, various trainers have pointed out that the same thing happens among dogs and other mammals and possibly even among pigeons, which implies that one does not have to invoke superintelligence to explain what the dolphins are doing.

Edward O. Wilson of Harvard University has suggested that the brain of the dolphin may be larger in relation to the size and weight of the body than the brains of most other mammals because of the same reputed imitative abilities that have made dolphins such a favorite with animal trainers and the public. The question then is: Why should an animal be such a superb imitator? R. J. Andrew of the University of Sussex has noted that vocal mimicry may be important to animals that often travel out of sight of one another. Individuals of a widely spaced group could then recognize other members because of an elaborate convergence of signals among the animals of one group or herd. This system, in the form of dialects, has been shown to operate among some primates and birds. It is plausible that the system evolved to an even greater degree among dolphins, which rely heavily on sound.

What about mimicry of movement? Wilson suggests that individual dolphins may imitate the members of the group that are most successful at catching fish and avoiding predation. Furthermore, it is advantageous for animals in social societies that cooperate to hunt food (as has been shown in at least some dolphin species) to know one another's movements well and for individuals to be able to take several roles in the herding of a school of fish. Wilson argues that imitation alone is enough to explain the size of the dolphin's brain and that the social signals of dolphins are probably no more sophisticated than those of most mammals and birds. In my opinion not enough is known about the social signals of dolphins to provide a basis for such a statement. Norris views the imitative powers of dolphins as not being necessarily better than those of many other mammals with brains that are smaller and physically less complex. It seems futile at present to compare the intelligence of dolphins with that of other mammals simply because of the lack of

appropriate information about dolphins and the great differences between their environments and those of terrestrial mammals.

The need for better information about dolphins turns one's mind to better means of obtaining it. One possibility is to try to habituate dolphins to observers to such an extent that they will go about their daily activities as if the observers were not present. George Schaller studied mountain gorillas in this way, and Jane Goodall similarly opened a new era of work on chimpanzees. They moved with the animals and sat patient-

ly until the animals either accepted them or simply ignored them.

How might one follow a group of dolphins in the ocean? Perhaps it is not necessary. Jody Solow of the University of California at Santa Cruz recently learned to make a sound underwater that at times called individuals of a group of nearby Hawaiian spinner dolphins to her. Her achievement opens the possibility that an investigator could eventually recognize all the members of a group, learn their social patterns and interactions and gain a better idea of their natural behavior.

VARIETY OF LEAPS performed by dolphins is suggested by these photographs of three dusky dolphins (*Lagenorhynchus obscurus*) in the open ocean. Dolphins in captivity have been observed to increase the variation in their leaps when they are receiving rewards of food.

BIBLIOGRAPHIES

INTRODUCTORY ESSAYS

II Adaptations: The Body and Movement

Alexander, R. M., FUNCTIONAL DESIGN IN FISHES. Hutchinson University Library, London, 1976.

Alexander, R. M., and Goldspink G., MECHANICS AND ENERGETICS OF ANIMAL LOCOMOTION. Chapman and Hall, London, 1977.

Blight, A. R., "The muscular control of vertebrate swimming movements." *Biol. Rev* 52, 1977, p. 181.

Budker, P., THE LIFE OF SHARKS, Columbia University Press, New York, 1971.

Farner, D. S., and King, J. R., AVIAN BIOLOGY, Vols. I and II. Academic Press, New York, 1972.

Halstead, L. B., VERTEBRATE HARD TISSUES. Wykeham Publications Ltd., London, 1974.

Lieberman, P., THE SPEECH OF PRIMATES. Mouton, Paris, 1972.

Regal, P. J., "The evolutionary origin of feathers," *Quart. Rev. Biol.* 50, 1975, p. 35.

Slijper, E. J., WHALES AND DOLPHINS. University of Michigan Press, Ann Arbor, 1976.

Tuttle, R. H., "Knuckle-walking and the problem of human origins," *Science* 166, 1969, pp. 953–61.

Voitkevich, A. A., THE FEATHERS AND PLUMAGE OF BIRDS. October House, New York, 1966.

Weihs, D., "Hydromechanics of fish schooling," *Nature* 241, 1973, pp. 290–91.

Welty, J. C., THE LIFE OF BIRDS. Philadelphia, Saunders, 1975.

Wu, T. Y-T., Brokaw, C. J., and Brennan, C., SWIMMING AND FLYING IN NATURE, Vols. 1 and 2. Plenum Press, New York, 1974.

III Adaptations: Special Structures

Goss, R. J., PRINCIPLES OF REGENERATION. Academic Press, New York, 1969. (Includes discussion of horns and antlers.)

Lee, R. F., Phleger, C. F., and Horn, M. H., "Composition of oil in fish bones: possible function in neutral buoyancy." *Comp. Biochem. Physiol.*, 50B, 1975, p. 13.

Norris, K. S., and Harvey, G. W., "A theory for the function of the spermaceti organ of the sperm whale." *In* ANIMAL COMMUNICATION AND NAVIGATION. Editor: Galler, S. R. Natl. Aeronaut. Space Admin., Washington, D.C., 1972, p. 397.

Simon, H., THE SPLENDOR OF IRIDESCENCE; STRUCTURAL COLORS IN THE ANIMAL WORLD. Dodd, Mead, New York, 1971.

IV   Adaptations in Reproduction

Lofts, B., and Murton, R. K., "Reproduction in birds." *In* AVIAN BIOLOGY, Vol. III. Editors: Farner, D. S., and King, J. R. Academic Press, New York, 1973.

Marler, P., "Animal communication signals." *Science* 157, 1967, pp. 769–774.

Nottebohm, F., "Ontogeny of bird song." *Science* 167, 1970, pp. 950–56.

Peaker, M., COMPARATIVE ASPECTS OF LACTATION. Academic Press, New York, 1977.

Robertson, R., "Sex changes under the waves." *New Scientist* 31, May 1973. Also, *Science* 177, 1972, p. 1007.

Schaller, G. B., THE SERENGETI LION. University of Chicago Press, Chicago, 1972.

Thorpe, W. H., and North, M. E. W., "Origin and significance of the power of vocal imitation: with special reference to the antiphonal singing of birds." *Nature* 208, 1965, pp. 219–22.

Wickler, W., MIMICRY, McGraw Hill, New York, 1968.

V   Orientation and Navigation

Airapet'yants, E. Sh., and Konstantinov, A. I., ECHOLOCATION IN ANIMALS. Academy of Sciences of the USSR. Israel Program of Scientific Translations, Jerusalem, 1973.

Dethier, V. G., "Other tastes, other worlds." *Science* 201, 1978, p. 224.

Griffin, D. R., LISTENING IN THE DARK. New Haven, Yale University Press, 1958.

Gwinner, E., "Circadian and circannual rhythms in birds." *In* AVIAN BIOLOGY, Vol. V, p. 221. Editors: Farner, D. S., and King, J. A. Academic Press, New York, 1975.

Keeton, W. T., "Avian orientation and navigation." *Ann. Rev. Physiol.* 41, 1979, p. 353.

Menaker, M., Takahashi, J. S., and Eskin, A., "The physiology of circadian pacemakers." *Ann. Rev. Physiol.* 40, 1978, p. 501.

Pfaffman, C., OLFACTION AND TASTE, Vols. I–III. Rockefeller University Press, New York, 1969.

Pfeiffer, W., "The fright reaction of fish." *Biol. Revs.* 37, 1962, pp. 495–511.

Schmidt-Koenig, K., and Keeton, W. T., ANIMAL MIGRATION AND HOMING. Springer-Verlag, New York, 1978.

Walcott, C., Gould, J. L., and Kirschvink, J. L., "Pigeons have magnets." *Science* 205, 1979, pp. 1027–1029.

VI   Vertebrates in Their Habitats

Ali, M. A., VISION IN FISHES. Plenum Press, New York, 1974.

Payne, R. S., and McVay, S., "Songs of humpback whales." *Science* 173, 1971, pp. 585–97.

Poulter, T. C., "Marine Mammals." *In* ANIMAL COMMUNICATION. Editor: Sebeok, T. A., Indiana University Press, 1968, Chapter 18. (Contains information on Weddell seal sounds.)

Rice, C. E., "Human echo perception." *Science* 155, 1967, pp. 656–664.

Simmons, J. A., Fenton, M. B., and O'Farrell, M. J., "Echolocation and pursuit of prey by bats." *Science* 203, 1979, p. 16.

Stonehouse, B. and Gilmore, D., THE BIOLOGY OF MARSUPIALS. University Park Press, Baltimore, 1977.

Varanasi, U., Feldman, H. R. and Malins, D. C., "Molecular basis for formation of lipid sound lens in echolocating cetaceans," *Nature* 255, 1975, p. 340.

ARTICLES

1. Adaptation

ADAPTATION AND NATURAL SELECTION: A CRITIQUE OF SOME CURRENT EVOLUTIONARY THOUGHT. George C. Williams. Princeton University Press, 1966.

EVOLUTION IN CHANGING ENVIRONMENTS: SOME THEORETICAL EXPLORATIONS. Richard Levins. Princeton University Press, 1968.

ADAPTATION AND DIVERSITY: NATURAL HISTORY AND THE MATHEMATICS OF EVOLUTION. E. G. Leigh, Jr. Freeman, Cooper and Company, 1971.

2. How Animals Run

MOTIONS OF THE RUNNING CHEETAH AND HORSE. Milton Hildebrand in *Journal of Mammalogy*, Vol. 40, No. 4, pages 481–495; November, 1959.

QUADRUPEDAL AND BIPEDAL LOCOMOTION OF LIZARDS. Richard C. Snyder in *Copeia*, No. 2, pages 64–70; June, 1952.

SOME LOCOMOTORY ADAPTATIONS IN MAMMALS. J. Maynard Smith and R. J. G. Savage in *Journal of the Linean Society–Zoology*, Vol. 42, No. 288, pages 603–622; February, 1956.

SPEED IN ANIMALS. A. Brazier Howell. University of Chicago Press, 1944.

3. How Snakes Move

LOCOMOTION WITHOUT LIMBS. Carl Gans in *Natural History*, Vol. 75, No. 2, pages 10–17; February, 1966.

LOCOMOTION WITHOUT LIMBS: PART II. Carl Gans in *Natural History*, Vol. 75, No. 3, pages 36–41; March, 1966.

SIDEWINDING AND JUMPING PROGRESSION OF VIPERS. Carl Gans and H. Mendelssohn in *Proceedings of the Second International Symposium on Animals and Plant Toxins*, in press.

4. The Antiquity of Human Walking

THE FOOT AND THE SHOE. J. R. Napier in *Physiotherapy*, Vol. 43, No. 3, pages 65–74; March, 1957.

A HOMINID TOE BONE FROM BED 1, OLDUVAI GORGE, TANZANIA. M. H. Day and J. R. Napier in *Nature*, Vol. 211, No. 5052, pages 929–930; August 27, 1966.

5. The Control of Walking

INSECT WALKING. Donald M. Wilson in *Annual Review of Entomology*, Vol. 11, pages 103–122; 1966.

NERVOUS CONTROL OF WALKING IN THE COCKROACH. K. G. Pearson, C. R. Fourtner and R. K. Wong in *Control of Posture and Locomotion*, edited by R. B. Stein, K. G. Pearson, R. S. Smith and J. B. Redford. Plenum Press, 1973.

LOCOMOTION IN VERTEBRATES: CENTRAL MECHANISMS AND REFLEX INTERACTION. S. Grillner in *Physiological Reviews*, Vol. 55, No. 2. pages 247–304; April, 1975.

6. The Soaring Flight of Vultures

COMPETITION FOR FOOD BETWEEN VULTURES IN EAST AFRICA. Hans Kruuk in *Ardea*, Vol. 55, No. 3–4, pages 171–193; December, 1967.

NEW SOARING PILOT. A. Welch, L. Welch and F. Irving. John Murray [Publishers] Ltd., 1968.

SOARING BEHAVIOUR AND PERFORMANCE OF SOME EAST AFRICAN BIRDS, OBSERVED FROM A MOTOR-GLIDER. C. J. Pennycuick in *Ibis*, Vol. 114, No. 2, pages 178–218; April, 1972.

7. The Evolution of the Hand

MAN THE TOOL-MAKER. Kenneth P. Oakley. British Museum of Natural History, 1950.

THE PREHENSILE MOVEMENTS OF THE HUMAN HAND. J. R. Napier in *The Journal of Bone and Joint Surgery*, Vol. 38-B, No. 4, pages 902–913; November, 1956.

PREHENSILITY AND OPPOSABILITY IN THE HANDS OF PRIMATES. J. R. Napier in *Symposia of the Zoological Society of London*, No. 5, pages 115–132; August, 1961.

8. Paleoneurology and the Evolution of Mind

A STROLL THROUGH THE WORLDS OF ANIMALS AND MEN: A PICTURE BOOK OF INVISIBLE WORLDS. Jakob von Uexküll in *Instinctive Behavior: The Development of a Modern Concept*, edited and translated by Claire H. Schiller. International Universities Press, Inc., 1957.

FACING REALITY: PHILOSOPHICAL ADVENTURES BY A BRAIN SCIENTIST. John C. Eccles. Springer-Verlag New York Inc., 1970.

THE BRAIN IN HOMINID EVOLUTION. Phillip V. Tobias. Columbia University Press, 1971.

OBJECTIVE KNOWLEDGE: AN EVOLUTIONARY APPROACH. Karl R. Popper. Oxford University Press, 1972.

EVOLUTION OF THE BRAIN AND INTELLIGENCE. Harry J. Jerison. Academic Press, 1973.

9. Reflectors in Fishes

RADIANCE DISTRIBUTION AS A FUNCTION OF DEPTH IN AN UNDERWATER ENVIRONMENT. John E. Tyler in *Bulletin of the Scripps Institution of Oceanography*, Vol. 7, No. 5, pages 363–411; 1960.

FUNCTION OF BIOLUMINESCENCE IN MESOPELAGIC ORGANISMS. William D. Clarke in *Nature*, Vol. 198, No. 4887, pages 1244–1246; June 29, 1963.

ON THE ORGANIZATION OF REFLECTING SURFACES IN SOME MARINE ANIMALS. E. J. Denton in *Philosophical Transactions of the Royal Society of London, Series B*, Vol. 258, No. 824, pages 285–313; May 14, 1970.

10. Horns and Antlers

HISTOGENESIS OF BONE IN THE GROWING ANTLER OF THE CERVIDAE. Walter Modell and Charles V. Noback in *The American Journal of Anatomy*, Vol. 49, No. 1, pages 65–95; September 15, 1931.

THE POST-NATAL DEVELOPMENT OF THE HORN TUBULES AND FIBRES (INTERTUBULAR HORN) IN THE HORNS OF SHEEP. A. N. George in *The British Veterinary Journal*, Vol. 112, No. 1, pages 30–34; January, 1956.

STRUCTURE OF RHINOCEROS HORN. M. L. Ryder in *Nature*, Vol. 193, No. 4821, pages 1199–1201; March 24, 1962.

TROPHIC RESPONSES TO TRAUMA IN GROWING ANTLERS. Anthony B. Bubenik and R. Pavlansky in *The Journal of Experimental Zoology*, Vol. 159, No. 3, pages 289–302; August, 1965.

THERMOREGULATORY FUNCTION OF GROWING ANTLERS. Bernard Stonehouse in *Nature*, Vol. 218, No. 5144, pages 870–872; June 1, 1968.

11. The Head of the Sperm Whale

BUOYANCY CONTROL AS A FUNCTION OF THE SPERMACETI ORGAN IN THE SPERM WHALE. M. R. Clarke in *Journal of the Marine Biological Association of the United Kingdom*, Vol. 58, No. 1, pages 27–71; February, 1976.

PHYSICAL PROPERTIES OF SPERMACETI OIL IN THE SPERM WHALE. M. R. Clarke in *Journal of the Marine Biological Association of the United Kingdom*, Vol. 58, No. 1, pages 19–26; February, 1976.

STRUCTURE AND PROPORTIONS OF THE SPERMACETI ORGAN IN THE SPERM WHALE. M. R. Clarke in *Journal of the Marine Biological Association of the United Kingdom*, Vol. 58, No. 1, pages 1–18; February, 1976.

12. The Phalarope

TESTOSTERONE-INDUCED INCUBATION PATCHES OF PHALAROPE BIRDS. J. E. Johns and E. W. Pfeiffer in *Science*, Vol. 140, No. 3572, pages 1225–1226; June 14, 1963.

GONADAL HORMONES IN WILSON'S PHALAROPE (STEGANOPUS TRICOLOR) AND OTHER BIRDS IN RELATION TO PLUMAGE AND SEX BEHAVIOR. E. O. Höhn and S. C. Cheng in *General and Comparative Endocrinology*, Vol. 8, No. 1, pages 1–11; February, 1967.

PROLACTIN AND NESTING BEHAVIOR IN PHALAROPES. Charles S. Nicoll, E. W. Pfeiffer and H. R. Fevold in *General and Comparative Endocrinology*, Vol. 8, No. 1, pages 61–65; February, 1967.

13. Mimicy in Parasitic Birds

CHARACTER DISPLACEMENT. W. L. Brown, Jr., and E. O. Wilson in *Systematic Zoology*, Vol. 5, pages 49–64; 1956.

HOMAGE TO SANTA ROSALIA, OR WHY ARE THERE SO MANY KINDS OF ANIMALS? G. E. Hutchinson in *The American Naturalist*, Vol. 93, No. 870, pages 145–159; May–June, 1959.

EVOLUTION IN CHANGING ENVIRONMENTS. Richard Levins. Princeton University Press, 1968.

ON THE METHODS OF RESOURCE DIVISION IN GRASSLAND BIRD COMMUNITIES. Martin L. Cody in *The American Naturalist*, Vol. 102, No. 924, pages 107–147; March–April, 1968.

14. The Social System of Lions

THE SERENGETI LION: A STUDY OF PREDATOR-PREY RELATIONS. George B. Schaller. University of Chicago Press, 1972.

LION POPULATION REGULATION. B. C. R. Bertram in *East African Wildlife Journal*, Vol. 11, pages 215–225; 1973.

15. Electric Location by Fishes

ECOLOGICAL STUDIES ON GYMNOTIDS. H. W. Lissmann in *Bioelectrogenesis: A Comparative Survey of its Mechanisms with Particular Emphasis on Electric Fishes*. American Elsevier Publishing Co., Inc., 1961.

ON THE FUNCTION AND EVOLUTION OF ELECTRIC ORGANS IN FISH. H. W. Lissmann in *Journal of Experimental Biology*, Vol. 35, No. 1, pages 156–191; March, 1958.

THE MECHANISM OF OBJECT LOCATION IN GYMNARCHUS NILOTICUS AND SIMILAR FISH. H. W. Lissmann and K. E. Machin in *Journal of Experimental Biology*, Vol. 35, No. 2, pages 451–486; June, 1958.

THE MODE OF OPERATION OF THE ELECTRIC RECEPTORS IN GYMNARCHUS NILOTICUS. K. E. Machin and H. W. Lissmann in *Journal of Experimental Biology*, Vol. 37, No. 4, pages 801–811; December, 1960.

16. The Infrared Receptors of Snakes

PROPERTIES OF AN INFRA-RED RECEPTOR. T. H. Bullock and F. P. J. Diecke in *The Journal of Physiology*, Vol. 134, No. 1, pages 47–87; October 29, 1956.

MEN AND SNAKES. Ramona and Desmond Morris. Hutchinson of London, 1965.

RADIANT HEAT RECEPTION IN SNAKES. T. H. Bullock and R. Barrett in *Communication in Behavioral Biology*, Part A, Vol. 1, pages 19–29; January, 1968.

THE PIT ORGANS OF SNAKES. Robert Barrett in *Biology of the Reptilia—Morphology B: Vol II*. Academic Press, 1970.

SNAKE INFRARED RECEPTORS: THERMAL OR PHOTOCHEMICAL MECHANISM? John F. Harris and R. Igor Gamow in *Science*, Vol. 172, No. 3989, pages 1252–1253; June 18, 1971.

17. The Stellar-Orientation System of a Migratory Bird

DIE STERNENORIENTIERUNG NACHTLICH ZIEHENDER GRASMUCKEN. E. G. Franz Sauer in *Zeitschrift für Tierpsychologie*, Vol. 14, pages 29–70; 1957.

BOBOLINK MIGRATORY PATHWAYS AND THEIR EXPERIMENTAL ANALYSIS UNDER NIGHT SKIES. William J. Hamilton III in *The Auk*, Vol. 79, pages 208–233; 1962.

MIGRATORY ORIENTATION IN THE INDIGO BUNTING, PASSERINA CYANEA. PART II: MECHANISM OF CELESTIAL ORIENTATION. Stephen T. Emlen in *The Auk*, Vol. 84, pages 463–489; 1967.

BIRD MIGRATION: INFLUENCE OF PHYSIOLOGICAL STATE UPON CELESTIAL ORIENTATION. Stephen T. Emlen in *Science*, Vol. 165, No. 3894, pages 716–718; August 15, 1969.

CELESTIAL ROTATION: ITS IMPORTANCE IN THE DEVELOPMENT OF MIGRATORY ORIENTATION. Stephen T. Emlen in *Science*, Vol. 170, No. 3963, pages 1198–1201; December 11, 1970.

BIRD MIGRATION. Donald R. Griffin. Dover Publications, Inc., 1974.

18. The Mystery of Pigeon Homing

ORIENTATION BY PIGEONS: IS THE SUN NECESSARY? William T. Keeten in *Science*, Vol. 165, pages 922–928; 1969.

MAGNETS INTERFERE WITH PIGEON HOMING. William T. Keeton in *Proceedings of the National Academy of Sciences of the United States of America*, Vol. 68, No. 1, pages 102–106; January, 1971.

HOMING IN PIGEONS WITH IMPAIRED VISION. K. Schmidt-Koenig and H. J. Schlichte in *Proceedings of the National Academy of Sciences of the United States of America*, Vol. 69, No. 9, pages 2446–2447; September, 1972.

RELEASE-SITE BIAS AS A POSSIBLE GUIDE TO THE "MAP" COMPONENT IN PIGEON HOMING. William T. Keeton in *Journal of Comparative Physiology*, Vol. 86, pages 1–16; 1973.

ORIENTATION OF HOMING PIGEONS ALTERED BY A CHANGE IN THE DIRECTION OF AN APPLIED MAGNETIC FIELD. Charles Walcott and Robert P. Green in *Science*, Vol. 184, No. 4133, pages 180–182; 1974.

THE ORIENTATIONAL AND NAVIGATIONAL BASIS OF HOMING IN BIRDS. William T. Keeton in *Advances in the Study of Behavior*: Vol. 5, edited by D. S. Lehrman, R. Hindle and E. Shaw. Academic Press, 1974.

19. The Nile Crocodile

THE NILE CROCODILE, CROCODILUS NILOTICUS: Notes on the Incubation Period and Growth Rate of Juveniles. A. C. Pooley in *The Lammergeyer*, Vol. 2, No. 1, pages 1–55; February, 1962.

DIVING BRADYCARDIA AND WITHDRAWAL BRADYCARDIA IN CAIMAN CROCODILUS. Abbot S. Gaunt and Carl Gans in *Nature*, Vol. 223, No. 5202, pages 207–208; July 12, 1969.

HOW DOES A BABY CROCODILE GET TO WATER? A. C. Pooley in *African Wildlife*, Vol. 28, No. 4, pages 8–11; Summer, 1974.

20. Kangaroos

LIFE OF MARSUPIALS. H. Tyndale-Biscoe. Edward Arnold, 1973.

BIOGEOGRAPHICAL CONSIDERATIONS OF THE MARSUPIAL-PLACENTAL DICHOTOMY. Jason A. Lillegraven in *Annual Review Ecology and Systematics*, Vol. 5, pages 74–94; 1974.

THE BIOLOGY OF KANGAROOS (MARSUPIALIA-MACROPODIDAE). Eleanor M. Russell in *Mammal Review*, Vol. 4, No. 1 and No. 2, pages 1–59; March/June, 1974.

21. The Archer Fish

THE ARCHER FISH. Hugh M. Smith in *Natural History*, Vol. 38, No. 1, pages 2–11; June, 1936.

HOW THE SHOOTING APPARATUS OF THE ARCHER FISH WAS DISCOVERED. George S. Myers in *The Aquarium Journal*, Vol. 23, No. 10, pages 210–214; October, 1952.

MORPHOLOGISCH UND HISTOLOGISCHE UNTERSUCHUNGEN AM AUGE DES SCHÜTZENFISCHES TOXOTES JACULATRIX (PALLAS 1766) (TOXOTIDAE) NEBST BEMERKUNGEN ZUM SPUCKGEHABEN. K. H. Lüling in *Zeitschrift für Morphologie und Ökologie der Tiere*, Vol. 47, No. 6, pages 529–610; December, 1958.

AUM SCHIESSEN DES SCHÜTZENFISCHES, TOXOTES JACU-
LATRIX. H. Hediger and H. Heusser in *Natur und
Volk*, Vol. 91, No. 7, pages 237–243; July, 1961.

22. The Weddell Seal

ANIMALS IN AQUATIC ENVIRONMENTS: DIVING MAMMALS
AND BIRDS. P. F. Scholander in *Handbook of Physi-
ology, Section 4: Adaptation to the Environment.*
American Physiological Society, 1964.

AN ANALYSIS OF SOME BEHAVIORAL AND PHYSIOLOGICAL
CHARACTERISTICS RELATED TO DIVING IN THE WED-
DELL SEAL. Gerald L. Kooyman in *Antarctic Re-
search Series, Vo. XI: Biology of the Antarctic Seas
III*, edited by George A. Llano and Waldo L.

Schmitt. American Geophysical Union, 1967.

DEEP DIVING. G. L. Kooyman and H. T. Andersen in
The Biology of Marine Mammals, edited by H. T.
Andersen. Academic Press, 1968.

23. Dolphins

MAMMALS OF THE SEA: BIOLOGY AND MEDICINE. Edited
by Sam H. Ridgway. Charles C. Thomas, Publisher,
1972.

OCCURRENCE AND GROUP ORGANIZATION OF ATLANTIC
BOTTLENOSE PORPOISES (TURSIOPS TRUNCATUS) IN
AN ARGENTINE BAY. Bernd Würsig in *The Biological
Bulletin*, Vol. 154, No. 2, pages 348–359; 1978.

INDEX